LONDON MATHEMATICAL SOCIETY LECTURE NOTE SERIES

Managing Editor: Professor J.W.S. Cassels, Department of Pure Mathematics and Mathematical Statistics, University of Cambridge, 16 Mill Lane, Cambridge CB2 1SB, England

The books in the series listed below are available from booksellers, or, in case of difficulty, from Cambridge University Press.

4 Algebraic topology, J.F. ADAMS
17 Differential germs and catastrophes, Th. BROCKER & L. LANDER
27 Skew field constructions, P.M. COHN
34 Representation theory of Lie groups, M.F. ATIYAH *et al*
36 Homological group theory, C.T.C. WALL (ed)
39 Affine sets and affine groups, D.G. NORTHCOTT
40 Introduction to H_p spaces, P.J. KOOSIS
42 Topics in the theory of group presentations, D.L. JOHNSON
43 Graphs, codes and designs, P.J. CAMERON & J.H. VAN LINT
45 Recursion theory: its generalisations and applications, F.R. DRAKE & S.S. WAINER (eds)
46 p-adic analysis: a short course on recent work, N. KOBLITZ
49 Finite geometries and designs, P. CAMERON, J.W.P. HIRSCHFELD & D.R. HUGHES (eds)
50 Commutator calculus and groups of homotopy classes, H.J. BAUES
51 Synthetic differential geometry, A. KOCK
54 Markov processes and related problems of analysis, E.B. DYNKIN
57 Techniques of geometric topology, R.A. FENN
58 Singularities of smooth functions and maps, J.A. MARTINET
59 Applicable differential geometry, M. CRAMPIN & F.A.E. PIRANI
60 Integrable systems, S.P. NOVIKOV *et al*
62 Economics for mathematicians, J.W.S. CASSELS
65 Several complex variables and complex manifolds I, M.J. FIELD
66 Several complex variables and complex manifolds II, M.J. FIELD
68 Complex algebraic surfaces, A. BEAUVILLE
69 Representation theory, I.M. GELFAND *et al*
74 Symmetric designs: an algebraic approach, E.S. LANDER
76 Spectral theory of linear differential operators and comparison algebras, H.O. CORDES
77 Isolated singular points on complete intersections, E.J.N. LOOIJENGA
78 A primer on Riemann surfaces, A.F. BEARDON
79 Probability, statistics and analysis, J.F.C. KINGMAN & G.E.H. REUTER (eds)
80 Introduction to the representation theory of compact and locally compact groups, A. ROBERT
81 Skew fields, P.K. DRAXL
82 Surveys in combinatorics, E.K. LLOYD (ed)
83 Homogeneous structures on Riemannian manifolds, F. TRICERRI & L. VANHECKE
84 Finite group algebras and their modules, P. LANDROCK
85 Solitons, P.G. DRAZIN
86 Topological topics, I.M. JAMES (ed)
87 Surveys in set theory, A.R.D. MATHIAS (ed)
88 FPF ring theory, C. FAITH & S. PAGE
89 An F-space sampler, N.J. KALTON, N.T. PECK & J.W. ROBERTS
90 Polytopes and symmetry, S.A. ROBERTSON
91 Classgroups of group rings, M.J. TAYLOR
92 Representation of rings over skew fields, A.H. SCHOFIELD
93 Aspects of topology, I.M. JAMES & E.H. KRONHEIMER (eds)
94 Representations of general linear groups, G.D. JAMES
95 Low-dimensional topology 1982, R.A. FENN (ed)

96 Diophantine equations over function fields, R.C. MASON
97 Varieties of constructive mathematics, D.S. BRIDGES & F. RICHMAN
98 Localization in Noetherian rings, A.V. JATEGAONKAR
99 Methods of differential geometry in algebraic topology, M. KAROUBI & C. LERUSTE
100 Stopping time techniques for analysts and probabilists, L. EGGHE
101 Groups and geometry, ROGER C. LYNDON
103 Surveys in combinatorics 1985, I. ANDERSON (ed)
104 Elliptic structures on 3-manifolds, C.B. THOMAS
105 A local spectral theory for closed operators, I. ERDELYI & WANG SHENGWANG
106 Syzygies, E.G. EVANS & P. GRIFFITH
107 Compactification of Siegel moduli schemes, C-L. CHAI
108 Some topics in graph theory, H.P. YAP
109 Diophantine Analysis, J. LOXTON & A. VAN DER POORTEN (eds)
110 An introduction to surreal numbers, H. GONSHOR
111 Analytical and geometric aspects of hyperbolic space, D.B.A.EPSTEIN (ed)
112 Low-dimensional topology and Kleinian groups, D.B.A. EPSTEIN (ed)
113 Lectures on the asymptotic theory of ideals, D. REES
114 Lectures on Bochner-Riesz means, K.M. DAVIS & Y-C. CHANG
115 An introduction to independence for analysts, H.G. DALES & W.H. WOODIN
116 Representations of algebras, P.J. WEBB (ed)
117 Homotopy theory, E. REES & J.D.S. JONES (eds)
118 Skew linear groups, M. SHIRVANI & B. WEHRFRITZ
119 Triangulated categories in the representation theory of finite-dimensional algebras, D. HAPPEL
120 Lectures on Fermat varieties, T. SHIODA
121 Proceedings of *Groups - St Andrews 1985*, E. ROBERTSON & C. CAMPBELL (eds)
122 Non-classical continuum mechanics, R.J. KNOPS & A.A. LACEY (eds)
123 Surveys in combinatorics 1987, C. WHITEHEAD (ed)
124 Lie groupoids and Lie algebroids in differential geometry, K. MACKENZIE
125 Commutator theory for congruence modular varieties, R. FREESE & R. MCKENZIE
126 Van der Corput's method for exponential sums, S.W. GRAHAM & G. KOLESNIK
127 New directions in dynamical systems, T.J. BEDFORD & J.W. SWIFT (eds)
128 Descriptive set theory and the structure of sets of uniqueness, A.S. KECHRIS & A. LOUVEAU
129 The subgroup structure of the finite classical groups, P.B. KLEIDMAN & M.W.LIEBECK
130 Model theory and modules, M. PREST
131 Algebraic, extremal & metric combinatorics, M-M. DEZA, P. FRANKL & I.G. ROSENBERG (eds)
132 Whitehead groups of finite groups, ROBERT OLIVER
133 Linear algebraic monoids, MOHAN S. PUTCHA
134 Number thoery and dynamical systems, M. DODSON & J. VICKERS (eds)
135 Operator algebras and applications, 1, D. EVANS & M. TAKESAKI (eds)
136 Operator algebras and applications, 2, D. EVANS & M. TAKESAKI (eds)
137 Analysis at Urbana, I, E. BERKSON, T. PECK, & J. UHL (eds)
138 Analysis at Urbana, II, E. BERKSON, T. PECK, & J. UHL (eds)
139 Advances in homotopy theory, B. STEER & W. SUTHERLAND (eds)
140 Geometric aspects of Banach spaces E.M.PEINADOR and A.R Usan (eds)
141 Surveys in combinatorics 1989, J. SIEMONS (ed)
142 The geometry of jet bundles, D.J. SAUNDERS

London Mathematical Society Lecture Note Series. 138

Analysis at Urbana

Volume II: Analysis in Abstract Spaces

Edited by
E. Berkson, T. Peck & J. Uhl
Department of Mathematics
University of Illinois

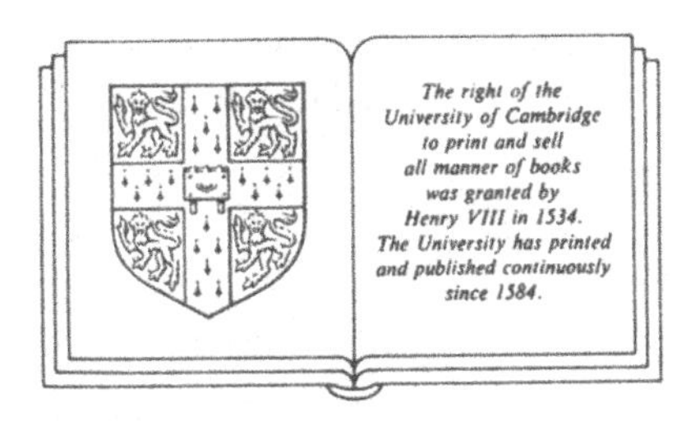

CAMBRIDGE UNIVERSITY PRESS
Cambridge
New York New Rochelle Melbourne Sydney

CAMBRIDGE UNIVERSITY PRESS
Cambridge, New York, Melbourne, Madrid, Cape Town,
Singapore, São Paulo, Delhi, Tokyo, Mexico City

Cambridge University Press
The Edinburgh Building, Cambridge CB2 8RU, UK

Published in the United States of America by Cambridge University Press, New York

www.cambridge.org
Information on this title: www.cambridge.org/9780521364379

First published 1989

A catalogue record for this publication is available from the British Library

Library of Congress Cataloguing in Publication Data

ISBN 978-0-521-36437-9 Paperback

CONTENTS

ACKNOWLEDGEMENTS

The organisers and participants gratefully acknowledge the support of the Special Year in Modern Analysis at the University of Illinois provided by the following agencies;

The Department of Mathematics, University of Illinois at Urbana-Champaign
The National Science Foundation
The Argonne Universities Association Trust Fund
The George A. Miller Endowment Fund (University of Illinois)
The Campus Research Board (University of Illinois at Urbana-Champaign)

PREFACE

The Special Year in Modern Analysis at the University of Illinois was devoted to the synthesis and expansion of modern and classical analysis. The program brought together analysts from around the globe for intensive lectures and discussions, including an International Conference on Modern Analysis, held March 16-19, 1987. The Special Year's success is a tribute to the outstanding merits and professional dedication of the participants. Contributions to these *Proceedings of the Special Year* were solicited from the participants in order to record and disseminate the fruits of their activities. The editors are grateful to the contributors for their response, which accurately reflects the quality and substance of the Special Year. In keeping with the wide scope of topics treated, the contents of these *Proceedings* fell naturally into two interrelated volumes, covering "Analysis in Abstract Spaces" and "Analysis in Function Spaces".

Thanks are due to the National Science Foundation, the Argonne Universities Association Trust Fund, the University of Illinois Campus Research Board, the University of Illinois Miller Endowment Fund, the University of Illinois Department of Mathematics, and J. Bourgain's Chair in Mathematics at the University of Illinois, without whose financial support the Special Year could not have taken place. Special thanks are also due to Professor Bela Bollobas, Consulting Editor at Cambridge University Press, and Mr. David Tranah, Senior Editor in Mathematical Sciences at Cambridge University Press, for the guidance and encouragement which made these *Proceedings* possible.

Earl Berkson
N. Tenney Peck
J. Jerry Uhl

The C_1 Contractions

by Bernard Beauzamy
Institut de Calcul Mathématique
Université de Paris 7

Let E be a Banach space and T a linear continuous operator on it. The operator is said to be a C_1*-contraction* if $\|T\| = 1$ and :

$$T^n x \not\to 0 , \quad n \to +\infty , \qquad \text{for all } x \neq 0.$$

The terminology "C_1 " is a shortening of Nagy-Foiaş teminology "$C_{1.}$ " ; see [7].

Here are three examples :

a) any isometry,

b) on the space $l_2(\mathbb{Z})$, equipped with the canonical basis $(e_n)_{n\in\mathbb{Z}}$, a weighted shift of the following type :

$$Te_n = w_n e_{n+1} ,$$

with $w_n = 1$ for ≥ 0, $w_n = 1/4$ for $n < 0$.

We refer to [2] for a detailed study of this operator.

c) Let K be a connected domain, with regular boundary, contained in the closed unit disk $\overline{D}$, and such that $\partial K \cap C$ is an interval I_1 (C is the unit circle). Let φ be a conformal map from D onto K°, and M_φ the operator of multiplication by φ on the Hardy space H^2. This operator is a C_1-contraction. Indeed, M_φ is of norm 1. Moreover, φ extends to an homeomorphism from $\overline{D}$ onto K ; let $I = \varphi^{-1}(I_1)$. Then, for every function f in H^2, which is not identically 0, we have :

$$\begin{aligned} \|M_\varphi^n f\|_2^2 &= \int_0^{2\pi} |\varphi^n(e^{i\theta})|^2 \, |f(e^{i\theta})|^2 \, \frac{d\theta}{2\pi} \\ &\geq \int_I |f(e^{i\theta})|^2 \frac{d\theta}{2\pi} , \end{aligned}$$

and this last quantity is strictly positive, because no function in H^2 (except 0) can vanish identically on a set of positive measure. We will come back later to this example, with more details.

As it is well-known, the usual bilateral shift : $Te_n = e_{n+1}$, $n \in \mathbb{Z}$, on $l_2(\mathbb{Z})$ has two types of Invariant Subspaces (see for instance Hoffman [6]) :

- type 1 : to every sequence $(a_j)_{j\in\mathbb{Z}}$, we associate the function

$$f(e^{i\theta}) = \sum_{j\in\mathbb{Z}} a_j e^{ij\theta} ,$$

which is in $L_2(\Pi, d\theta/2\pi)$. Let A be a measurable subset of Π, with $0 < P(A) < 1$ (P being the Haar measure on Π). Let :

$$F_A = \{(a_j) \in l_2(\mathbb{Z}) \; ; \; f = 0 \text{ on } A\}.$$

Then obviously, if (a_j) is in F_A, so is (Ta_j).

- type 2 : For $n \in \mathbb{Z}$, let

$$G_n = \{(a_j)_{j\in\mathbb{Z}} \; ; \; a_j = 0 \text{ if } j \leq n\}.$$

Obviously also, $TG_n \subset G_n$.

This second type has a more general description : let m be an inner function (see [6]), then $G = m \cdot H^2$ is the general form of such an Invariant Subspace.

More generally, we will say that a closed subspace $F \subset E$ is *invariant* for T if $TF \subset F$. The subspace F is said to be *non- trivial* if $F \neq 0$ and $F \neq E$. In the sequel, we omit the words "non-trivial" and speak about Invariant Subspaces.

The question whether every C_1-contraction has Invariant Subspaces is still open, even on Hilbert spaces. In what follows, we try to make a formal description of the two types of Invariant Subspaces of the usual shift, that is, to obtain a description which makes sense for a general C_1-contraction. This will be done in the next two paragraphs, under some specific hypotheses.

1. Invariant Subspaces of Functional Type.

Let $\mathcal{A}(\Pi)$ be the vector space of functions with absolutely convergent Fourier series :

$$\mathcal{A}(\Pi) = \{f = \sum_{k\in\mathbb{Z}} a_k e^{ik\theta} \; ; \; \sum_{k\in\mathbb{Z}} |a_k| < \infty\} ,$$

which is an algebra under the norm :

$$\|f\|_{\mathcal{A}} = \sum_{k\in\mathbb{Z}} |a_k|$$

Let T be a C_1-contraction and f a function in $\mathcal{A}(\Pi)$. For any $m \in \mathbb{Z}$, we define :

$$\psi_m(f) = \sum_{k\geq -m} a_k T^{k+m} ,$$

and this series converges since $\|T\| = 1$.

We observe that the operators $(\psi_m(f))_{m\in\mathbb{Z}}$ are uniformly bounded ; indeed :

$$\|\psi_m(f)\| \leq \|f\|_{\mathcal{A}} .$$

Therefore, the set :

$$F_f = \{x \in E \; ; \; \psi_m(f)x \to 0, \; m \to +\infty\},$$

is a closed subspace of E, which is invariant under T (this subspace is even hyper-invariant : that is, invariant also under all operators which commute with T).

Such a subspace is called "of functional type", because it arises from a function in $\mathcal{A}(\Pi)$.

In the sequel, for convenience, we will assume T to be invertible (the condition upon the iterates $\|T^{-n}x_0\|$ may be replaced by a condition upon a chain of approximate backward iterates, when T is not invertible ; see [1]). Then we have :

Theorem 1 (B. B. [1]). – *If there exists a point x_0 such that*

$$\sum_{n\geq 0} \frac{\log \|T^{-n}x_0\|}{1+n^2} < +\infty \tag{1}$$

then T has non-trivial hyperinvariant subspaces of functional type. Moreover, the condition (1) is best possible for this type of invariant subspace : for any sequence $(\rho_n)_{n\geq 0}$, satisfying

$$\rho_n \geq 1 , \qquad \rho_{m+n} \leq \rho_m \cdot \rho_n , \qquad m, n \in \mathbb{N},$$

and

$$\sum_{n\geq 0} \frac{\log \rho_m}{1+m^2} < \infty ,$$

there exists an operator T with $\|T^{-m}\| = \rho_m$, such that, for every $x \neq 0$, there is $C(x) > 0$, with :

$$\|T^{-m}x\| \geq C(x)\rho_m ,$$

and this operator has no Invariant Subspace of functional type.

This result improves two previously known theorems :

- John Wermer (1954) made the assumption that :

$$\sum_{m\geq 0} \frac{\log \|T^{-m}\|}{1+m^2} < \infty ,$$

- Colojoara- Foiaş(1966) made the assumption that both T and ${}^tT$ are C_1 -contractions (this last assumption implies that some point x has a chain of *bounded* inverses, which is of course much stronger than (1)).

We refer the reader to [1] for the proof.

This theorem applies of course to the usual bilateral shift, and, when applied to it, it gives the Invariant Subspaces of the first type, which are called ***spectral*** subspaces. They correspond to the following basic idea : if f and g are two functions on Π, disjointly supported, such that $f(T)$ and $g(T)$ both make sense, the product $f(T) \circ g(T)$ will be zero, and if $f(T) \neq 0$, $g(T) \neq 0$, we have non-trivial Invariant Subspaces : $Ker\, f(T)$ is the required Invariant Subspace. Here, we cannot give a meaning to $f(T) = \sum_{k \in \mathbb{Z}} a_k T^k$, for any function f in $\mathcal{A}(\Pi)$, because this series may diverge, but we replace this "ordinary" functional calculus by an asymptotic one, using $\psi_m(f)$.

So this Theorem provides a large supply of Invariant Subspaces ; however, it does not apply to all C_1-contractions. Condition (1) is required, and we have seen it was best possible. So we now turn to another approach, corresponding to the Invariant Subspaces of type 2.

<u>2. The unitary extension of T.</u>

(Most results in this paragraph are part of a join paper with Michel Rome [5].)

On the space E, we define a norm by the formula :

$$[\![x]\!] = \lim_{n \to +\infty} \|T^n x\|,$$

and denote by $\mathcal{E}$ the completion of E with respect to this norm. From the inequality :

$$[\![\cdot]\!] \leq \|\cdot\|,$$

we deduce that there is a continuous injection from E into $\mathcal{E}$. On the space $\mathcal{E}$, the operator T extends naturally to an isometry, which we denote by $\tilde{T}$. One sees easily that $\tilde{T}$ is surjective when T has dense range (which we may assume, if we are looking for Invariant Subspaces).

The space $\mathcal{E}$ is finitely representable in E. If E is a Hilbert space, so is $\mathcal{E}$, with the scalar product :

$$[x, y] = \lim_{m \to +\infty} \langle T^m x, T^m y \rangle ,$$

where $\langle ., . \rangle$ is the scalar product in E.

From now on, we assume in this paragraph that E is a Hilbert space. For x, y in E, we consider, for $k \in \mathbb{Z}$:

$$\lambda_k(x, y) = \lim_{m \to +\infty} \langle T^{m+k} x, T^m y \rangle ,$$

and the Fourier series

$$\sum_{k\in\mathbb{Z}} \lambda_{-k}(x,y)e^{ik\theta} . \tag{2}$$

Since the coefficients λ_k are bounded, this series defines a pseudo- measure. But this result can be improved :

We say that T is *completely non unitary* (in short c.n.u.) if there is no subspace F such that $TF = F$ and $T|_F$ is unitary. Then we have :

Proposition 2 (B. B. , M. Rome [5]). – *If T is c.n.u., the Fourier series (2) is that of an integrable function.*

Let's denote by $\Lambda_{x,y}$ this function. Since :

$$\lambda_{-k}(x,y) = [\tilde{T}^{-k}x,y] = \int_0^{2\pi} e^{-ik\theta}\Lambda_{x,y}(\theta)\,\frac{d\theta}{2\pi} ,$$

we obtain that, for every f, g in $\mathcal{A}(\Pi)$:

$$[f(\tilde{T})x, g(\tilde{T})y] = \int_0^{2\pi} f(e^{i\theta})\bar{g}(e^{i\theta})\,\Lambda_{x,y}(\theta)\,\frac{d\theta}{2\pi} . \tag{3}$$

Therefore, the functions $\Lambda_{x,y}$ allow us to write scalar products of functions of T, at points x, y. For this reason, these functions will be called *representing functions* (for T) at points x, y.

If $y = x$, we just write Λ_x instead of $\Lambda_{x,x}$, and call it representing function at the point x.

This concept allows us to give an Invariant Subspace theorem, which generalizes those of type 2 :

Theorem 3 ([5]). – *If there is a point x such that :*

$$\int_0^{2\pi} \log \Lambda_x(\theta)\,\frac{d\theta}{2\pi} > -\infty \tag{4}$$

then :

$$dist(x, \overline{\text{span}}\{Tx, T^2x, \ldots\}) > 0. \tag{5}$$

Consequently, $\overline{\text{span}}\{Tx, T^2x, \ldots\}$ is the required non- trivial Invariant Subspace.

Again, this theorem applies to the usual bilateral shift (with $\Lambda_{e_0} = 1$), and gives the fact that $e_0 \notin \overline{\text{span}}\{e_1, e_2, \ldots\}$.

In order to study condition (4), we will list some properties of the functions Λ_x or $\Lambda_{x,y}$. We refer the reader to [5] for the proofs.

Assume that a point x_0 is cyclic for T, that is :

$$E = \overline{\text{span}}\{x_0, Tx_0, T^2x_0, \ldots\}.$$

Let $\mu_0 = \Lambda_{x_0} d\theta/2\pi$ be the corresponding measure. Then the support K_0 of μ_0 is exactly the spectrum of $\tilde{T}$. For a function $f \in L_\infty(K_0, \mu_0)$, the operator $f(\tilde{T})$ is well- defined, since $\tilde{T}$ is normal.

Proposition 4. - *Let x, y be in E, f, g in $L_\infty(K_0, \mu_0)$. Put $x' = f(\tilde{T})x$, $y' = g(\tilde{T})y$. Then :*

$$\Lambda_{x',y'} = f\bar{g}\, \Lambda_{x,y} \,.$$

Corollary. - *All the functions $\Lambda_{x,y}$ vanish a.e. on $C \setminus K_0$.*

This is clear : take $f = g = 1$ on K_0, and 0 on $C \setminus K_0$. Then $f(T) = I$, so $\Lambda_{x,y} = |f|^2 \Lambda_x$, and this function is 0 on $C \setminus K_0$.

We observe also that $\sigma(\tilde{T}) \subset \sigma(T) \cap C$ (and this inclusion can be strict : see [3]). Therefore, if we want condition (4) to hold, we need $\sigma(\tilde{T}) = C$, and a fortiori, $\sigma(T) \supset C$.

Therefore, Theorem 3 applies only to the C_1-contractions such that $\sigma(T) \supset C$ (but even not to all of them).

Thus, so far, we have given two theorems describing formalizations of the Invariant Subspaces of the usual shift : type 1 and type 2 respectively. Each of them requires a specific assumption, so we may ask : does the combination of both cover all possible cases ? We will see that this is not the case. But before that, we mention several other applications of the functions Λ_x : they allow us to obtain a "reverse" functional calculus. For instance :

Proposition 5. -*Let*

$$\sum_{k \in \mathbb{Z}} a_k T^k x = z$$

be a convergent series in E, which means that :

$$\sum_{-M}^{N} a_k T^k x \to z\,, \qquad M, N \to +\infty.$$

Then the series $\sum_{k \in \mathbb{Z}} a_k e^{ik\theta}$ is that of a function ϕ in $L_2(\Lambda_x\, d\theta/2\pi)$, in the sense that :

$$\sum_{-M}^{N} a_k e^{ik\theta} \to \phi\,, \qquad M, N \to +\infty,$$

in this space.

This statement can be wiewed as a reverse functional calculus, in the sense that, starting from a property of $f(T)$, we deduce a property of f, whereas in general one does the converse.

3. Study of an example.

We come back to the example we mentioned at the beginning : ϕ being a conformal map from D onto a regular domain K°, contained in D, such that $\partial K \cap C = I_1$ is an interval. We assume moreover that K does not contain the origin.

Therefore the operator M_ϕ of multiplication by ϕ in the space H^2 is invertible : ϕ is in H^∞ (even in $A(D)$), and is outer.

Moreover, for every function $f \in H^2$, there is a constant $C(f) > 0$ and an $\varepsilon > 0$ such that :

$$\|M_\phi^{-n}\|_2 \geq C(1+\varepsilon)^n , \qquad n \in \mathbb{N},$$

and therefore Theorem 1 does not apply to this operator. Moreover, one can see directly that it has no Invariant Subspaces of functional type.

It follows easily from Runge's theorem that, for every function f in H^2, not identically 0,

$$f \in \overline{\text{span}}\{\phi f, \phi^2 f, \ldots\}.$$

So property (5) does not hold. Indeed, also, since $\sigma(M_\phi) = K$, $\sigma(M_\phi) \cap C$ is not C, and the assumption of Theorem 3 does not hold.

Let $I = \phi^{-1}(I_1)$; this is the set where $|\phi| = 1$. Then one can see easily that $\mathcal{E} = L_2(I)$, and $\tilde{M}_\phi = M_\phi$, multiplication by ϕ on $L_2(I)$: this is a unitary operator.

The Invariant Subspaces for M_ϕ on $L_2(I)$ are easy to describe : they are of the form $L_2(A)$, where A is a measurable subset of I, and they are non-trivial if and only if $P(A) > 0$, $P(I \setminus A) > 0$. Another way of describing them is : the set of functions f which vanish a.e. on $I \setminus A$.

Since no function in H^2 (except 0) can vanish a.e. on such a set, we get :

Theorem 4 ([4]). – *If F is any Invariant Subspace of M_ϕ on $L_2(I)$, then $F \cap H^2$ is $\{0\}$ or H^2.*

So, no Invariant Subspace of M_ϕ comes from an Invariant Subspace of $\tilde{M}_\phi$. Since the extension $\tilde{T}$ has a meaning in Nagy-Foiaş dilation theory (it corresponds to the *-residual part of this extension, see [5] and [7]), we may say that no Invariant Subspace of this operator comes from Nagy-Foiaş dilation theory.

The Invariant Subspaces of M_ϕ are easy to describe : they are of the form $m \cdot H^2$, where m is an inner function. This is so because $M_\phi = \phi(M_{e^{i\theta}})$.

This leads us to the following comments : we now see that Theorems 1 and 3 are insufficient to describe all the C_1-contractions. We also observe that, though their conclusions are invariant if one replaces T by $f(T)$, their assumptions are not. Indeed, the usual shift satisfies the assumptions of both theorems, though its image by ϕ does not. It would be nice to have versions of these theorems which would be invariant under the operation $T \to f(T)$.

This is a short summary of a series of lectures given at the University of Illinois, Urbana-Champaign, during a special year in Modern Analysis, 1986-87. The author wishes to thank the Department of Mathematics for its nice hospitality, and Professor Earl Berkson for having arranged the invitation and the lectures.

References.

[1] BEAUZAMY, Bernard : Sous-espaces invariants de type fonctionnel dans les Espaces de Banach. *Acta math.*, vol. 144, 1-2 (1981), pp. 27-64.

[2] BEAUZAMY, Bernard : A weighted bilateral shift with no cyclic vector. *Journal of Oper. Th.*, 4 (1981), pp. 287-288.

[3] BEAUZAMY, Bernard : Spectre d'une contraction de classe C_1 et de son extension unitaire. *Publications de l'Université Paris VII.* Séminaire d'Analyse fonctionnelle, Universités de Paris VI, Paris VII, 1983-84, pp. 1-8.

[4] BEAUZAMY, Bernard : Propriétés spectrales d'un opérateur de multiplication sur $H^2(\Pi)$. *Publications dé l'Université Paris VII.* Séminaire d'Analyse fonctionnelle, 1981-82, pp. 115-122.

[5] BEAUZAMY, Bernard - ROME, Michel : Extension unitaire et fonctions de représentation d'une contraction de classe C_1. *Arkiv för Mathematik*, vol.23, 1 (1985) pp. 1-17.

[6] HOFFMAN, Kenneth : Banach Spaces of Analytic Functions. *Englewood Cliffs*, N.Y. 1962.

[7] NAGY, Sz. - FOIAS, Ciprian. : Harmonic Analysis of Operators on Hilbert spaces. Akademiai Kiaido, Budapest, 1966.

FACTORIZATION THEOREMS FOR INTEGRABLE FUNCTIONS

by

Hari Bercovici
Department of Mathematics
Indiana University

The research in this paper was supported in part by a grant from the National Science Foundation.

Let $(Z, \mathscr{B}, \mu)$ be a measure space and let $\mathscr{D}$ be a separable, complex Hilbert space. We denote by $L^2(\mu; \mathscr{D})$ the Hilbert space of all (classes of) measurable, square integrable functions $x : Z \longrightarrow \mathscr{D}$. For two functions $x,y \in L^2(\mu; \mathscr{D})$ we can define the function $x \cdot y \in L^1(\mu)$ by setting $(x \cdot y)(\zeta) = <x(\zeta),y(\zeta)>$ for almost every $\zeta \in Z$. ($<\cdot,\cdot>$ denotes the scalar product in any Hilbert space. For instance, for $x,y \in L^2(\mu; \mathscr{D})$ we have $<x,y> = \int_Z <x(\zeta),y(\zeta)>d\mu(\zeta)$.)

In this paper we study the possibility of solving, at least approximately, an equation of the form $x \cdot y = f$, where f is a given function in $L^1(\mu)$, and x,y are required to belong to a given subspace $\mathscr{H} \subset L^2(\mu; \mathscr{D})$. (See Theorem 10 for the precise statement.) Our results here strengthen and put in an abstract framework certain results of Brown [3] and the author [2]. These results were obtained in relation to operator theoretical problems. We will show in a subsequent paper how our results can be used to settle the conjecture made in [1] about the structure of contraction operators.

Throughout this paper $\mathscr{H}$ is a linear subspace of $L^2(\mu; \mathscr{D})$ satisfying the following condition.

1. ASSUMPTION. Given $\sigma \in \mathscr{B}$, $\mu(\sigma) > 0$, a positive number ϵ, and a finite number of vectors $\xi_1, \xi_2, \ldots, \xi_p \in L^2(\mu; \mathscr{D})$, there exists $z \in \mathscr{H}$, $z \neq 0$, such that

(i) z is essentially bounded, i.e., $z \in L^\infty(\mu; \mathscr{D})$;

(ii) $\|\chi_{Z\setminus\sigma} z\| < \epsilon \|\chi_\sigma z\|$; and

(iii) $\langle z, \xi_j \rangle = 0$, $j = 1, 2, \ldots, p$.

We need the following consequence of Assumption 1.

2. PROPOSITION. Let $f \in L^\infty(\mu)$ be a function such that $0 \leq f \leq 1$, let $\xi_1, \xi_2, \ldots, \xi_p \in L^2(\mu; \mathscr{D})$, and let $\epsilon > 0$. If $\|f\|_\infty > 1 - \epsilon$ then there exists $x \in \mathscr{H} \cap L^\infty(\mu; \mathscr{D})$ such that $\langle x, \xi_j \rangle = 0$, $1 \leq j \leq p$, and $\|(1-f)^{1/2} x\| < \frac{\epsilon}{1-\epsilon} \|f^{1/2} x\|^2$.

<u>Proof</u>. Choose $\epsilon' < \epsilon$ such that the set $\sigma = \{\zeta \in Z : f(\zeta) > 1 - \epsilon'\}$ has positive measure, and choose $\delta > 0$ such that

$$\frac{\epsilon' + \delta^2}{1-\epsilon'} < \frac{\epsilon}{1-\epsilon}.$$

By Assumption 1 we can find $x \in \mathscr{H} \cap L^\infty(\mu; \mathscr{D})$, $x \neq 0$, with $\langle x, \xi_j \rangle = 0$, $1 \leq j \leq p$, such that $\|\chi_{Z\setminus\sigma} x\|^2 < \delta \|\chi_\sigma x\|^2$. We will show that this x satisfies the conclusion of our proposition. We have

$$\|f^{1/2} x\|^2 \geq \int_\sigma f(\zeta) \|x(\zeta)\|^2 d\mu(\zeta) \geq (1 - \epsilon') \|\chi_\sigma x\|^2,$$

and hence

$$\|(1-f)^{1/2} x\|^2 = \int_\sigma (1 - f(\zeta)) \|x(\zeta)\|^2 d\mu(\zeta)$$

$$+ \int_{Z\setminus\sigma} (1 - f(\zeta)) \|x(\zeta)\|^2 d\mu(\zeta)$$

$$\leq \epsilon'\|\chi_\sigma x\|^2 + \|\chi_{Z\setminus\sigma}x\|^2 < \epsilon'\|\chi_\sigma x\|^2 + \delta^2\|\chi_\sigma x\|^2$$

$$< \frac{\epsilon'+\delta^2}{1-\epsilon'}\|f^{1/2}x\|^2 \leq \frac{\epsilon}{1-\epsilon}\|f^{1/2}x\|^2 ,$$

as desired.

Let $\sigma \subset Z$ be a set with finite positive measure, let η and δ be two positive numbers, and let $\xi_1,\xi_2,\ldots,\xi_p \in L^2(\mu;\mathscr{D})$.

3. DEFINITION. The set $S(\sigma;\xi_1,\xi_2,\ldots,\xi_p;\eta;\delta)$ consists of those vectors $x \in \mathscr{H}$ such that

(a) $\langle x,\xi_j\rangle = 0$, $1 \leq j \leq p$;

(b) x can be written as $x = g + b$, $g, b \in L^2(\mu;\mathscr{D})$, and

(b_1) $\|g(\zeta)\| \leq \chi_\sigma(\zeta)$ for almost every $\zeta \in Z$;

(b_2) $\|b\| < \eta\|g\|$; and

(b_3) $|\langle g,\xi_j\rangle| < \delta\|x\|$, $1 \leq j \leq p$.

Observe that Assumption 1 implies the existence of nonzero elements in $S(\sigma;\xi_1,\xi_2,\ldots,\xi_p;\eta;\delta)$. If $x = g + b$ belongs to $S(\sigma;\xi_1,\xi_2,\ldots,\xi_p;\eta;\delta)$ then clearly

$$\frac{\|x\|}{1+\eta} \leq \|g\| \leq \frac{\|x\|}{1-\eta}. \tag{4}$$

5. PROPOSITION. If $\eta < \frac{1}{2}$ then

$$\sup\{\|x\| : x \in S(\sigma;\xi_1,\xi_2,\ldots,\xi_p;\eta;\delta)\} > 2^{-3}\eta\mu(\sigma)^{1/2} .$$

Proof. For simiplicity set $S = S(\sigma;\xi_1,\xi_2,\dots,\xi_p;\eta;\delta)$ and $\sup\{\|x\| : x \in S\} = \gamma\mu(\sigma)^{1/2}$. Assume, to get a contradiction, that $\gamma \le 2^{-3}\eta$. Let $x_n \in S$ be a sequence such that $\lim_{n\to\infty} \|x_n\| = \gamma\mu(\sigma)^{1/2}$, and write $x_n = g_n + b_n$, as required by the definition of the set S. Note that

$$\|g_n\| \le \|x_n\|/(1-\eta) \le 2^{-2}\eta\mu(\sigma)^{1/2},$$

and set

$$\omega_n = \left\{\zeta \in \sigma : \|g_n(\zeta)\| < \tfrac{1}{2}\right\}.$$

Then

$$\mu(\sigma\backslash\omega_n) = \int_{\sigma\backslash\omega_n} d\mu(\zeta) \le 4\int_{\sigma\backslash\omega_n} \|g_n(\zeta)\|^2 d\mu(\zeta) \le 4\|g_n\|^2 \le 2^{-2}\eta^2\mu(\sigma),$$

and hence

$$\mu(\omega_n) = \int_Z \chi_{\omega_n}(\zeta)d\mu(\zeta) \ge (1 - 2^{-2}\eta^2)\mu(\sigma).$$

Dropping to a subsequence we may assume that

(i) x_n converge weakly in $\mathscr{H}$ to x;

(ii) $\chi_{\omega_n} g_n$ converge weakly in $L^2(\mu;\mathscr{D})$ to u;

(iii) $\chi_{Z\backslash\omega_n} b_n$ converge weakly in $L^2(\mu;\mathscr{D})$ to v; and

(iv) χ_{ω_n} converge weak* in L^∞ to f, $0 \le f \le X_\sigma$.

Note that

$$\int_Z f(\zeta)d\mu(\zeta) = \lim_{n\to\infty} \int_Z \chi_{\omega_n}(\zeta)d\mu(\zeta) \geq (1-2^{-2}\eta^2)\mu(\sigma),$$

and hence we must have $f(\zeta) > 1 - 2^{-1}\eta^2$ for ζ in a set of positive measure. An application of Proposition 2 yields an element $z \in \mathscr{H} \cap L^\infty(\mu;\mathscr{D})$, $z \neq 0$, such that

(v) $\quad <z,\xi_j> = <z,f\xi_j> = <z,u> = <z,v> = <z,x> = 0, \ 1 \leq j \leq p$, and

(vi) $\quad \|(1-f)^{1/2}z\|^2 < \dfrac{2^{-1}\eta^2}{1-2^{-1}\eta^2}\|f^{1/2}z\|^2 < \eta^2\|f^{1/2}z\|^2$.

Dividing z by a sufficiently large constant we may assume that $\|z(\zeta)\| \leq \frac{1}{2}$ for almost every $\zeta \in Z$. We claim that for n sufficiently large the vector $x_n' = x_n + z$ belongs to S and $\|x_n'\| > \gamma\mu(\sigma)^{1/2}$. The latter inequality is verified because

$$\lim_{n\to\infty} \|x_n'\|^2 = \lim_{n\to\infty}(\|x_n\|^2 + \|z\|^2 + 2\mathrm{Re}<x_n,z>)$$

$$= \gamma^2\mu(\sigma) + \|z\|^2 + 2\mathrm{Re}<x,z>$$

$$= \gamma^2\mu(\sigma) + \|z\|^2 > \gamma^2\mu(\sigma);$$

here we used (i) and (v). To see that $x_n' \in S$ we note first that x_n' satisfies condition (a) of the definition by (v). Furthermore, we can write $x_n' = g_n' + b_n'$, where $g_n' = g_n + \chi_{\omega_n} z$ and $b_n' = b_n + \chi_{Z\backslash\omega_n} z$. Condition (b) is verified by the definition of ω_n, and because $\|z(\xi)\| \leq \frac{1}{2}$ almost everywhere.. To verify (b2) we calculate

$$\|b_n'\|^2 - \eta^2\|g_n'\|^2 = \|b_n\|^2 - \eta^2\|g_n\|^2 + \|\chi_{Z\backslash\omega_n} z\|^2 - \eta^2\|\chi_{\omega_n} z\|^2$$

$$+ 2\mathrm{Re}[<b_n,\chi_{Z\backslash\omega_n} z> - \eta^2<g_n,\chi_{\omega_n} z>]$$

$$\leq \|\chi_{Z\setminus\omega_n} z\|^2 - \eta^2 \|\chi_{\omega_n} z\|^2$$

$$+ 2\mathrm{Re}[<\chi_{Z\setminus\omega_n} b_n, z> - \eta^2 <\chi_{\omega_n} g_n, z>] ,$$

where we used the fact that x_n satisfies (b2). We use now (ii), (iii), (iv), (v) and (vi) to get

$$\limsup_{n\to\infty} (\|b_n'\|^2 - \eta^2 \|g_n'\|^2) \leq$$

$$\leq \|(1-f)^{1/2} z\|^2 - \eta^2 \|f^{1/2} z\|^2$$

$$+ 2\mathrm{Re}[<v,z> - \eta^2 <u,z>]$$

$$= \|(1-f)^{1/2} z\|^2 - \eta^2 \|f^{1/2} z\|^2 < 0 ,$$

so that x_n' satisfies (b2) eventually. Finally,

$$|<x_n', \xi_j>| \leq |<g_n, \xi_j>| + |<\chi_{\omega_n} z, \xi_j>|$$

$$\leq \delta \|x_n\| + |<\chi_{\omega_n} z, \xi_j>|$$

$$\leq \delta\gamma\mu(\sigma)^{1/2} + |<\chi_{\omega_n} z, \xi_j>| ,$$

so that using (iv) and (v) we get

$$\limsup_{n\to\infty} |<g_n', \xi_j>| \le \delta\gamma\mu(\sigma)^{1/2} + |<fz, \xi_j>| = \delta\gamma|\sigma|^{1/2},$$

and hence $|<g_n', \xi_j>| < \delta\|x_n'\|$ for large n, $1 \le j \le p$. Thus $x_n' \in S$ eventually, and this contradiction proves the proposition.

The following result is an improvement of Proposition 5. The proof is essentially the same with that of Proposition 5, except that the norm of the element z chosen during the proof can now be controlled, and this leads to a simplification.

6. PROPOSITION. If $\eta < \frac{1}{2}$ then

$$\sup\{\|x\| : x \in S(\sigma;\xi_1,\xi_2,\ldots,\xi_p;\eta;\delta)\} \ge (1-\eta)\mu(\sigma)^{1/2}.$$

Proof. Denote, as before, $S = S(\sigma;\xi_1,\xi_2,\ldots,\xi_p;\eta;\delta)$, and set $\sup\{\|x\| : x \in S\} = \alpha(1-\eta)\mu(\sigma)^{1/2}$. Suppose, to get a contradiction, that $\alpha < 1$. Let $x = g + b$ be an element of S, and note that $\|g\| \le \alpha\mu(\sigma)^{1/2}$. The set $\omega = \{\zeta \in \sigma : \|g(\zeta)\| < \alpha^{1/2}\}$ has positive measure; in fact

$$|\sigma\backslash\omega| = \int_{\sigma\backslash\omega} d\mu(\zeta) \le \frac{1}{\alpha}\int_{\sigma\backslash\omega} \|g(\zeta)\|^2 d\mu(\zeta) \le \alpha\mu(\sigma),$$

so that $\mu(\omega) \ge (1-\alpha)\mu(\sigma)$. Let now δ' be a positive member, and choose by virtue of Proposition 5 $z = \gamma + \beta \in S' = S(\omega;b,g,\xi_1,\xi_2,\ldots,\xi_p;\eta;\delta')$ such that $\|z\| \ge 2^{-3}\eta\mu(\omega)^{1/2} \ge 2^{-3}\eta(1-\alpha)^{1/2}\mu(\sigma)^{1/2}$. Let us assume for the moment that the vector $x' = x + (1-\alpha^{1/2})z$ has been shown to belong to S. We must have then $\|x'\| \le \alpha(1-\eta)\mu(\sigma)^{1/2}$ or, equivalently since $<x,z> = 0$, $\|x\|^2 + (1-\alpha^{1/2})^2 2^{-6}\eta^2(1-\alpha)\mu(\sigma) \le \alpha^2(1-\eta)^2\mu(\sigma)$, and letting $\|x\|$ approach $\alpha(1-\eta)\mu(\sigma)^{1/2}$ we get $(1-\alpha^{1/2})^2 2^{-6}\eta^2(1-\alpha)\mu(\sigma) \le 0$, a contradiction. Thus, to conclude the proof it suffices to show that $x' \in S$ for sufficiently small δ'. To do this note that condition (a) of Definition 3 is clearly verified. Next we write $x' = g' + b'$, where $g' = g + (1-\alpha^{1/2})\gamma$ and $b' = b + (1-\alpha^{1/2})\beta$. We have

$$\|g'(\zeta)\| \leq \|g(\zeta)\| + (1-\alpha^{1/2})\|\gamma(\zeta)\|$$

$$\leq \|g(\zeta)\| + (1-\alpha^{1/2})\chi_\omega(\zeta) \leq \chi_\sigma(\zeta)$$

by the definition of the set ω; this shows that condition (b1) of Definition 3 holds for x'. To verify (b2) we calculate

$$\|b'\|^2 - \eta^2\|g'\|^2 = \|b\|^2 - \eta^2\|g\|^2 + (1-\alpha^{1/2})^2(\|\beta\|^2 - \eta^2\|\gamma\|^2)$$

$$+ 2(1-\alpha^{1/2})\mathrm{Re}[\langle g,\gamma\rangle - \eta^2\langle b,\beta\rangle]$$

$$\leq \|b\|^2 - \eta^2\|g\|^2 + 2[|\langle g,\gamma\rangle| + |\langle b,\beta\rangle|].$$

Using the fact that $z \in S'$ we get

$$|\langle g,\gamma\rangle| \leq \delta'\|z\| \leq \delta'(1+\eta)\mu(\omega)^{1/2} \leq \delta'(1+\eta)\mu(\sigma)^{1/2},$$

and

$$|\langle b,\beta\rangle| = |\langle b,\gamma\rangle| \leq \delta'\|z\| \leq \delta'(1+\eta)\mu(\sigma)^{1/2},$$

so that

$$\|b'\|^2 - \eta^2\|g'\|^2 \le \|b\|^2 - \eta^2\|g\|^2 + 4\delta'(1+\eta)\mu(\sigma)^{1/2}.$$

Therefore (b2) is verified provided that

$$4\delta'(1+\eta)\mu(\sigma)^{1/2} < \eta^2\|g\|^2 - \|b\|^2. \tag{7}$$

To verify (b3) we use the facts that $x \in S$ and $z \in S'$ to get

$$\begin{aligned} |\langle x',\xi_j\rangle| &\le |\langle x,\xi_j\rangle| + (1-\alpha^{1/2})|\langle z,\xi_j\rangle| \\ &\le |\langle x,\xi_j\rangle| + |\langle z,\xi_j\rangle| \\ &\le \delta\|x\| + \delta'\|z\| \le \delta\|x\| + \delta'(1+\eta)\mu(\sigma)^{1/2}. \end{aligned}$$

Since

$$\begin{aligned} \|x'\| &= (\|x\|^2 + (1-\alpha^{1/2})^2\|z\|)^{1/2} \\ &\ge [\|x\|^2 + (1-\alpha^{1/2})^2 2^{-6}\eta^2(1-\alpha)\mu(\sigma)]^{1/2}, \end{aligned}$$

we see that (b3) is satisfied provided that

$$\delta\|x\| + \delta'(1+\eta)\mu(\sigma)^{1/2} < \delta[\|x\|^2 + (1-\alpha^{1/2})^2 2^{-6}\eta^2(1-\alpha)\mu(\sigma)]^{1/2}. \tag{8}$$

Since both (7) and (8) are satisfied for sufficiently small δ, the proposition follows.

Since the left hand side in Proposition 6 is increasing in η, while the right hand side is decreasing, we have the following immediate consequence.

9. COROLLARY. For all $\eta > 0$ we have

$$\sup\{\|x\| : x \in S(\sigma;\xi_1,\xi_2,\dots,\xi_p;\eta;\delta)\} \geq (1-\eta)\mu(\sigma)^{1/2}.$$

We are now ready to prove the main result of this paper. Recall that Assumption 1 is supposed to hold.

10. THEOREM. Let $f \in L^1(\mu)$, $\epsilon > 0$, and $\xi_1,\xi_2,\dots,\xi_p \in L^2(\mu;\mathcal{D})$. There exist $x,y \in \mathcal{H}$ such that

(i) $\langle x,\xi_j\rangle = \langle y,\xi_j\rangle = 0$, $1 \leq j \leq p$;

(ii) $\|x\| \leq \|f\|_1^{1/2}$, $\|y\| \leq \|f\|_1^{1/2}$;

(iii) $\|f - x\cdot y\|_1 < \epsilon$; and

(iv) if $f \geq 0$ almost everywhere then $x = y$.

Proof. We may, and shall assume without loss of generality that $f \neq 0$. Choose pairwise disjoint sets $\sigma_1,\sigma_2,\dots,\sigma_n$ with positive finite measure, and scalars $\gamma_1,\gamma_2,\dots,\gamma_n$ such that

$$\|f - \sum_{i=1}^{n} \gamma_i\chi_{\sigma_i}\|_1 < \frac{\epsilon}{2}, \tag{11}$$

and

$$\sum_{i=1}^{n} |\gamma_i|\mu(\sigma_i) < \|f\|_1. \tag{12}$$

Fix $\eta,\delta > 0$, and chose inductively, by virture of Corollary 9 vectors $z_i \in S(\sigma_i;\xi_1,\xi_2,\dots,\xi_p,z_1,z_2,\dots,z_{i-1};\eta;\delta)$ such that

$$\|z_i\| \geq (1-\eta)\mu(\sigma_i)^{1/2}\,,\ i = 1,2,\ldots,n\,. \tag{13}$$

Choose for each i a square root α_i of γ_i, and set $x = \sum_{i=1}^{n} \alpha_i z_i$, $y = \sum_{i=1}^{n} \bar{\alpha}_i z_i$. If f is positive then the γ_i can be assumed positive, and hence $x = y$. To conclude the proof we have to show that x and y satisfy conditions (i), (ii), and (iii) provided that η is appropriately small. Condition (i) is immediate. Write $z_i = g_i + b_i$ as required by the fact that $z_i \in S(\sigma_i;\xi_1,\xi_2,\ldots,\xi_p;\eta;\delta)$. Since the z_i are pairwise orthogonal, we have

$$\|x\|^2 = \|y\|^2 = \left[\sum_{i=1}^{n} |\gamma_i|\,\|z_i\|^2\right]^{1/2}$$

$$\leq \left[\sum_{i=1}^{n} |\gamma_i|(\|g_i\| + \|b_i\|)^2\right]^{1/2}$$

$$\leq (1+\eta)\left[\sum_{i=1}^{n} |\gamma_i|\mu(\sigma_i)\right]^{1/2},$$

and by (12) we see that (ii) holds if η is small enough. Finally, note that g_i and g_j have disjoint supports if $i \neq j$, and hence $g_i \cdot g_j = 0$ in that case; thus

$$x \cdot y = \sum_{i=1}^{n} \gamma_i g_i \cdot g_i + \sum_{i,j=1}^{n} \alpha_i \alpha_j (g_i \cdot b_j + b_i \cdot g_j + b_i \cdot b_j)\,.$$

Furthermore,

$$\|\chi_{\sigma_i} - g_i \cdot g_i\|_1 = \int_Z (\chi_{\sigma_i}(\zeta) - \|g_i(\zeta)\|^2) d\mu(\zeta)$$

$$= \mu(\sigma_i) - \|g_i\|^2 \leq \mu(\sigma_i) - \frac{\|x_i\|^2}{(1+\eta)^2}$$

$$\leq \mu(\sigma_i) - \left[\frac{1-\eta}{1+\eta}\right]^2 \mu(\sigma_i),$$

and therefore

$$\left\| \sum_{i=1}^{n} \gamma_i \chi_{\sigma_i} - x \cdot y \right\|_1 \leq$$

$$\leq \sum_{i=1}^{n} \left[1 - \left[\frac{1-\eta}{1+\eta}\right]^2\right] |\gamma_i| \mu(\sigma_i)$$

$$+ \sum_{i,j=1}^{n} |\alpha_i| |\alpha_j| (\|g_i\| \|b_j\| + \|b_i\| \|g_j\| + \|b_i\| \|b_j\|)$$

$$\leq \left[1 - \left[\frac{1-\eta}{1+\eta}\right]^2\right] \|f\|_1$$

$$+ \sum_{i,j=1}^{n} |\alpha_i| |\alpha_j| (2\eta + \eta^2) \mu(\sigma_i)^{1/2} \mu(\sigma_j)^{1/2} .$$

This number can be made $<\frac{\epsilon}{2}$ if η is sufficiently small and (iii) follows now from (11). The theorem is proved.

References

1. C. Apostol, H. Bercovici, C. Foias, and C. Pearcy, Invariant subspaces, dilation theory, and the structure of the predual of a dual algebra, J. Funct. Anal. 63(1985), 369–404.

2. H. Bercovici, A contribution to the theory of operators in the class $\mathbb{A}$, J. Funct. Anal. 78(1988), to appear.

3. S. Brown, Contractions with spectral boundary, Integral Equations Operator Theory, to appear.

SPECTRAL DECOMPOSITIONS AND VECTOR-VALUED TRANSFERENCE

by

Earl Berkson and T.A. Gillespie

1. Introduction. The aim of this paper is to discuss a number of results concerning the "diagonal" representation of operators on Banach spaces. More precisely, we shall consider invertible operators U which can be written as

$$U = \int_{0^-}^{2\pi} e^{i\lambda} dE(\lambda) , \tag{1.1}$$

and strongly continuous one-parameter groups $\{ U_t \}_{t \in \mathbb{R}}$ of the form

$$U_t = \int_{\mathbb{R}} e^{i\lambda t} dE(\lambda) \qquad (t \in \mathbb{R}) . \tag{1.2}$$

In each of (1.1), (1.2) $E(\cdot)$ is a projection-valued function of points in $\mathbb{R}$ acting in the underlying Banach space and possessing certain additional properties (to be specified in §2), and the integrals exist strongly as Riemann-Stieltjes integrals (in the Cauchy principal value sense in (1.2)). Spectral decompositions in terms of such projection-valued functions were first discussed in a Banach space context by Smart and Ringrose [14,16]. They considered operators A on reflexive spaces which can be written

$$A = \int_J \lambda dE(\lambda) \tag{1.3}$$

for some compact interval J in $\mathbb{R}$, calling them *well-bounded* operators. Such operators on reflexive spaces may be characterized in terms of a functional calculus based on the Banach algebra of absolutely continuous functions on J. The general theory of well-bounded operators may also be developed for an arbitrary Banach space X [15], but then the values of $E(\cdot)$ act on the dual space of X, and (1.3) is interpreted in a weak-* sense after formal integration by parts. A more restricted class of well-bounded operators, said to be *of type* (B), was introduced in [2], and for them the values of $E(\cdot)$ do act on the arbitrary Banach space X. Operators expressible in the form (1.1) were first systematically discussed in [3], and are called *trigonometrically well-bounded*. Trigonometrically well-bounded operators can also be characterized in terms of the existence of a functional calculus, this time based on the Banach algebra of absolutely continuous functions on the unit circle $\mathbb{T}$. For a more extended account of the basic theory of well-bounded operators, the reader is referred to [11].

In the present paper, we discuss two main results related to the above circle of ideas. Firstly, a necessary and sufficient condition for a strongly continuous one-parameter group $\{U_t\}_{t \in \mathbb{R}}$ to have a representation of the form (1.2) is obtained in Theorem (2.4). This result is analogous to the aforementioned functional calculus characterization of trigonometrical well-boundedness. Secondly, it was recently shown that, when the underlying Banach space X has the unconditionality property for martingale differences (abbreviated $X \in$ UMD), every invertible power-bounded operator on X is trigonometrically well-bounded, and every uniformly bounded strongly continuous one-parameter group $\{U_t\}_{t \in \mathbb{R}}$ can be written in the form (1.2) ([4, Theorems (4.5) and (5.5)]). In §3 below we show how these results in UMD spaces can be obtained directly from the functional calculus characterization of trigonometrical well-boundedness and its analogue herein for one-parameter groups, without the recourse, needed in [4], to yet another characterization of trigonometrical well-boundedness for power-bounded operators on reflexive spaces ([4, Theorem (3.43)]) or to a generalization of Stone's Theorem for one-parameter groups of trigonometrically well-bounded operators ([1, Theorem (4.20)]). The key notions for our purposes are a vector-valued version [6, Theorem (2.8)] of the Coifman-Weiss transference technique (see Theorem (3.2) below) and Banach space versions (introduced in Theorem (3.4) below) of Stečkin's multiplier theorems [12]. With these tools, the above results on UMD spaces follow easily. In a brief final section, we show how the original Coifman-Weiss transference result [9, Theorem 2.4] can be obtained from the vector-valued version, so that the latter can be viewed as a generalization of the former.

As usual, $\mathbb{R}$, $\mathbb{C}$, $\mathbb{Z}$, $\mathbb{N}$, and $\mathbb{T}$ denote, respectively, the reals, the complexes, the integers, the positive integers, and the circle group. Total variation is abbreviated "var". If $J = [a,b]$ is a compact interval in $\mathbb{R}$, $BV(J)$ denotes the Banach algebra, under pointwise operations, of all functions $f: J \longrightarrow \mathbb{C}$ of bounded variation on J, with norm

$$\| f \|_J = | f(b) | + \operatorname{var}(f,J) ,$$

and $AC(J)$ is the closed subalgebra consisting of the absolutely continuous functions on J. Similarly, $BV(\mathbb{T})$ denotes the Banach algebra of all functions $f: \mathbb{T} \longrightarrow \mathbb{C}$ for which the function $\tilde{f}(t) \equiv f(e^{it})$ is of bounded variation on $[0,2\pi]$, with norm $\| f \|_{\mathbb{T}} \equiv \| \tilde{f} \|_{[0,2\pi]}$, and $AC(\mathbb{T})$ is the closed subalgebra consisting of those functions f such that $\tilde{f}$ is in $AC([0,2\pi])$. The Banach algebra of all functions $f: \mathbb{R} \longrightarrow \mathbb{C}$ of bounded variation on $\mathbb{R}$, with norm given by

$$\| f \|_{\mathbb{R}} = | \lim_{t \longrightarrow +\infty} f(t) | + \operatorname{var}(f,\mathbb{R}) ,$$

is denoted by $BV(\mathbb{R})$, and $AC_0(\mathbb{R})$ is the closed subalgebra consisting of all locally absolutely continuous functions on $\mathbb{R}$ which vanish at $+\infty$ and at $-\infty$, and whose derivative is integrable over

$\mathbb{R}$. The space of infinitely differentiable functions $f: \mathbb{R} \to \mathbb{C}$ having compact support is denoted by $\mathcal{D}$. The Fourier transform of a function f in $L^1(\mathbb{R})$ or $L^1(\mathbb{Z})$ will be written $\hat{f}$ (so that $\hat{f}(\lambda) \equiv \int_{\mathbb{R}} f(t)e^{-i\lambda t}dt$ or $\hat{f}(z) \equiv \sum_{n \in \mathbb{Z}} f(n)z^{-n}$, respectively). Notice that if $f \in \mathcal{D}$, then $\hat{f}$ belongs to the Schwartz class $\mathcal{S}$ of testing functions (see [17, §I.3]). In particular, $\hat{f} \in AC_0(\mathbb{R})$ if $f \in \mathcal{D}$.

Given a complex Banach space X , X^* denotes its dual space, $\mathcal{B}(X)$ the Banach algebra of all bounded linear operators on X , and I the identity operator on X . If $T \in \mathcal{B}(X)$ is invertible, and q is a trigonometric polynomial, that is, $q(e^{it}) = \sum_{n \in \mathbb{Z}} a_n e^{int}$ with all but finitely many of the complex constants a_n vanishing, then $q(T)$ denotes $\sum_{n \in \mathbb{Z}} a_n T^n$. When convenient, the value of $x^* \in X^*$ at $x \in X$ will be written as $\langle x,x^* \rangle$. Unless otherwise stated, integrals of X-valued functions are to be interpreted as Bochner integrals. Given a measure space $(\mathcal{M},\mu)$ and a Banach space X , $L^p(\mathcal{M},X)$ will denote the usual Lebesgue-Bochner space of p-integrable X-valued functions on $(\mathcal{M},\mu)$ for $1 \le p < \infty$. The L^p-norm of any scalar-valued or vector-valued function under discussion will be denoted by $\|\cdot\|_p$ whenever it is clear from the context which particular L^p-space is being considered.

2. <u>Spectral families and well-boundedness</u>. In this section we discuss in more detail the diagonal representations of the form (1.1) and (1.2). Throughout, let X be a fixed complex Banach space. A *spectral family in* X is a projection-valued function $E(\cdot): \mathbb{R} \to \mathcal{B}(X)$ with the following properties:

(i) $\sup\{ \| E(\lambda) \| : \lambda \in \mathbb{R} \} < \infty$;

(ii) $E(\lambda)E(\mu) = E(\mu)E(\lambda) = E(\lambda)$ if $-\infty < \lambda \leq \mu < +\infty$;

(iii) $E(\cdot)$ is right continuous on $\mathbb{R}$ in the strong operator topology of $\mathfrak{B}(X)$;

(iv) at each $\lambda \in \mathbb{R}$, $E(\cdot)$ has a left-hand limit $E(\lambda^-)$ in the strong operator topology of $\mathfrak{B}(X)$;

(v) $E(\lambda) \longrightarrow I$ as $\lambda \longrightarrow +\infty$, and $E(\lambda) \longrightarrow 0$ as $\lambda \longrightarrow -\infty$, both limits being with respect to the strong operator topology of $\mathfrak{B}(X)$.

These properties are not, in fact, independent. For general X , (i) follows from (iii)-(v) by the principle of uniform boundedness, while in the case of reflexive X , (iv) follows from (i) and (ii). If there exist $a,b \in \mathbb{R}$ such that $E(\lambda) = 0$ for $\lambda < a$, and $E(\lambda) = I$ for $\lambda \geq b$, then we say that $E(\cdot)$ is *concentrated* on the interval $[a,b]$.

Given a spectral family $E(\cdot)$ in the Banach space X and a compact interval $J = [a,b]$ in $\mathbb{R}$, the integral

$$\int_J f(\lambda)dE(\lambda)$$

exists in the strong operator topology as a Riemann-Stieltjes integral for each $f \in AC(J)$. Furthermore the mapping $\Phi\colon AC(J) \longrightarrow \mathfrak{B}(X)$ defined by

$$(2.1) \qquad \Phi(f) = f(a)E(a) + \int_J f(\lambda)dE(\lambda) \qquad (f \in AC(J))$$

is an algebra homomorphism satisfying

$$\| \Phi(f) \| \leq \| f \|_J \sup\{ \| E(\lambda) \| : \lambda \in J \} \qquad (f \in AC(J)) .$$

For brevity we write

$$\int_J^{\oplus} f(\lambda)dE(\lambda)$$

for the right-hand side of (2.1). Notice that the homomorphism Φ in (2.1) is identity-preserving if $E(\cdot)$ is concentrated on J . Details of these ideas may be found in [11, Chapter 17] or, in more abbreviated form, in [1, §2].

A bounded linear operator A on X is said to be *well-bounded of type* (B) if there is a spectral family $E(\cdot)$ in X concentrated on a compact interval J such that

$$A = \int_J^{\oplus} \lambda dE(\lambda) .$$

In this case, the spectral family $E(\cdot)$ is uniquely determined by A [11, Theorem 16.3-(i)] and is called the *spectral family of* A. Moreover, A and the range of its spectral family have the same commutants [11, Theorem 16.3-(ii)]. In [3], the class of trigonometrically well-bounded operators on X was introduced. By definition, an operator $U \in \mathfrak{B}(X)$ is *trigonometrically well-bounded* if there is a well-bounded operator A of type (B) on X such that $U = e^{iA}$. In this case, A can be chosen so that its spectrum is contained in $[0,2\pi]$ but 2π does not belong to its point spectrum. These two additional spectral conditions force the type (B) well-bounded operator A to be uniquely determined by U, and, with this normalization, A is called the *argument of* U. If U is trigonometrically well-bounded, then U and its argument have the same commutants [1, Proposition (3.14)-(ii)]. In terms of an integral representation, an operator $U \in \mathfrak{B}(X)$ is trigonometrically well-bounded if and only if there is a spectral family $E(\cdot)$ in X, concentrated on $[0,2\pi]$, such that

$$U = \int_{[0,2\pi]}^{\oplus} e^{i\lambda} dE(\lambda) .$$

The spectral family $E(\cdot)$ in this representation for U can be normalized so that $E((2\pi)^-) = I$, and this additional condition determines $E(\cdot)$ uniquely. We then call $E(\cdot)$ the *spectral decomposition of* U . The spectral decomposition of U is, in fact, the spectral family of the argument of U . These results are discussed in [1, §3] (particularly in [1, Theorem (3.16)]).

An operator is well-bounded of type (B) if and only if it has a weakly compact $AC(J)$-functional calculus for some compact interval J in $\mathbb{R}$ [11, Theorem 17.14 and its proof] (alternatively, see [3, §2, especially pp. 42,43]). There is a characterization of trigonometrical well-boundedness analogous to this in which the algebra $AC(\mathbb{T})$ replaces $AC(J)$. For $n \in \mathbb{Z}$, let $e_n(z) \in AC(\mathbb{T})$ be given by $e_n(z) \equiv z^n$. Thus a trigonometric polynomial is a linear combination of finitely many e_n's .

(2.2) **THEOREM.** ([3, Theorem (2.3) and its proof]). *A bounded linear operator* U *on the Banach space* X *is trigonometrically well-bounded if and only if there exists a norm-continuous homomorphism* Φ *of the Banach algebra* $AC(\mathbb{T})$ *into* $\mathfrak{B}(X)$ *such that:*

(i) $\Phi(e_0) = I$, *and* $\Phi(e_1) = U$;

(ii) for each $x \in X$, *the mapping* $f \longrightarrow \Phi(f)x$ *from* $AC(\mathbb{T})$ *to* X *is weakly compact.*

If this is the case, then

$$\Phi(f) = \int^{\oplus}_{[0,2\pi]} f(e^{i\lambda})dE(\lambda) \qquad (f \in AC(\mathbb{T})) ,$$

where $E(\cdot)$ *is the spectral decomposition of* U , *and* $\| E(\lambda) \| \le 3 \| \Phi \|$ *for all* $\lambda \in \mathbb{R}$.

Apart from its final assertion concerning reflexive spaces, the following corollary is essentially a reformulation of Theorem (2.2) (see [3, Corollary 2.17]). The reflexivity assertion follows immediately from the fact that, when X is reflexive, bounded subsets of $\mathfrak{B}(X)$ are relatively compact in the weak operator topology (that is, have closure in the weak operator topology which is compact in that topology).

(2.3) **COROLLARY.** *A bounded linear operator* U *on a Banach space* X *is trigonometrically well-bounded if and only if it is invertible and:*

(i) there exists a constant K *such that* $\| q(U) \| \leq K \| q \|_{\mathbb{T}}$ *for all trigonometric polynomials* q ;

(ii) the set

$$\{q(U): \ q \ \text{is a trigonometric polynomial and} \ \| q \|_{\mathbb{T}} \leq 1\}$$

is relatively compact in the weak operator topology.

If these conditions are satisfied, then the spectral decomposition $E(\cdot)$ *of* U *satisfies* $\| E(\lambda) \| \leq 3K$ *for all* $\lambda \in \mathbb{R}$. *When* X *is reflexive and* U *is invertible, condition (i) suffices for* U *to be trigonometrically well-bounded.*

Notice that any invertible operator U on X gives rise to the representation of $\mathbb{Z}$ in X defined by $n \longrightarrow U^n$. Also, the inequality in condition (i) in Corollary (2.3) can be rewritten as

$$\Big\| \sum_{n \in \mathbb{Z}} f(n) U^{-n} \Big\| \leq K \big\| \hat{f} \big\|_{\mathbb{T}}$$

for each function $f: \mathbb{Z} \longrightarrow \mathbb{C}$ having finite support. Accordingly, the following result may be viewed as an analogue of Corollary (2.3) for one-parameter groups of operators.

(2.4) **THEOREM.** *A strongly continuous one-parameter group* $\{ U_t \}_{t \in \mathbb{R}}$ *of operators on a Banach space* X *is of the form*

$$(2.5) \qquad U_t = \lim_{a \longrightarrow +\infty} \int_{-a}^{a} e^{i\lambda t} dE(\lambda) \qquad (t \in \mathbb{R}) ,$$

where $E(\cdot)$ *is a spectral family in* X , *and the limit is in the strong operator topology of* $\mathfrak{B}(X)$, *if and only if:*

(i) there exists a constant K *such that*

$$\Big\| \int_{\mathbb{R}} f(t) \, U_{-t} dt \Big\| \leq K \big\| \hat{f} \big\|_{\mathbb{R}} \qquad (f \in \mathfrak{D}) ;$$

(ii) the set

$$\left\{ \int_{\mathbb{R}} f(t)\, U_{-t} dt : \; f \in \mathfrak{D} \; \text{and} \; \left\| \hat{f} \right\|_{\mathbb{R}} \le 1 \right\}$$

is relatively compact in the weak operator topology of $\mathfrak{B}(X)$.

If these conditions are satisfied, then the spectral family $E(\cdot)$ *in* (2.5) *is unique, and* $\| E(\lambda) \| \le 2K$ *for all* $\lambda \in \mathbb{R}$. *When* X *is reflexive, condition (i) suffices for* $\{ U_t \}_{t \in \mathbb{R}}$ *to have a representation of the form* (2.5).

Proof. Suppose first that $\{ U_t \}_{t \in \mathbb{R}}$ has a representation of the form (2.5). Temporarily fix $x \in X_n = \left[E(n) - E(-n) \right] X$ for some $n \in \mathbb{N}$. Then

$$\int_{-a}^{a} e^{i\lambda t} dE(\lambda)x = \int_{-n}^{n} e^{i\lambda t} dE(\lambda)x$$

whenever $a \ge n$. Integrating by parts in (2.5), we thus have

$$U_t x = e^{iat} x - \int_{-a}^{a} i t e^{i\lambda t} E(\lambda) x \, d\lambda$$

for $a \ge n$. Hence, for $x^* \in X^*$, $f \in \mathfrak{D}$, and $a \ge n$,

$$< \left[\int_{\mathbb{R}} f(t)\, U_{-t} dt \right] x, x^* >$$

$$= \hat{f}(a)<x,x^*> + \int_{\mathbb{R}} if(t)t \left\{ \int_{-a}^{a} e^{-i\lambda t} <E(\lambda)x,x^*> d\lambda \right\} dt$$

$$= \hat{f}(a)<x,x^*> - \int_{-a}^{a} \hat{f}'(\lambda) <E(\lambda)x,x^*> d\lambda \; .$$

Letting $a \to +\infty$ and noting that $\bigcup \{ X_n : \; n \in \mathbb{N} \}$ is dense in X , we deduce that for $f \in \mathfrak{D}$,

$$\left\| \int_{\mathbb{R}} f(t)\, U_{-t} dt \right\| \le \left\| \hat{f} \right\|_{\mathbb{R}} \sup \{ \| E(\lambda) \| : \; \lambda \in \mathbb{R} \} .$$

This proves (i) with $K = \sup \{ \| E(\lambda) \| : \; \lambda \in \mathbb{R} \}$.

To prove (ii), note first that, for each $x \in X$, the set $\Omega_x = \{ E(\lambda)x : \; \lambda \in \mathbb{R} \}$ is totally bounded with respect to the norm-induced metric of X . This fact follows from properties (iii)-(v) in the definition of a spectral family (cf. the argument on p. 347 of [11]). Again, letting $a \to +\infty$ in the above computation, we see that

$$(2.6) \qquad \int_{\mathbb{R}} f(t)\, U_{-t}x\,dt = -\int_{\mathbb{R}} \hat{f}'(\lambda)E(\lambda)x\, d\lambda$$

for $x \in \bigcup \{ X_n : n \in \mathbb{N} \}$ and $f \in \mathcal{D}$. The integrals here exist as Bochner integrals and are norm-continuous as functions of x. Since $\bigcup \{ X_n : n \in \mathbb{N} \}$ is dense in X, it follows that (2.6) is valid for all $x \in X$ and all $f \in \mathcal{D}$. (2.6) now shows that, for each $x \in X$, the set $\left\{ \int_{\mathbb{R}} f(t)\, U_{-t}x\, dt : f \in \mathcal{D},\ \| \hat{f} \|_{\mathbb{R}} \le 1 \right\}$ is contained in the closed absolutely convex hull of Ω_x. This gives (ii) with, in fact, relative compactness in the strong operator topology.

The uniqueness statement in the theorem also follows from (2.6). To see this, suppose that $\tilde{E}(\cdot)$ is another spectral family in X such that (2.5) holds with with $E(\cdot)$ replaced by $\tilde{E}(\cdot)$. Then (2.6) will also be valid for $\tilde{E}(\cdot)$ in place of $E(\cdot)$, and, since the left-hand side of this latter equation involves neither $E(\cdot)$ nor $\tilde{E}(\cdot)$ explicitly, we deduce that, for $x \in X$, $x^* \in X^*$, and $f \in \mathcal{D}$,

$$\int_{\mathbb{R}} \hat{f}'(\lambda)\langle E(\lambda)x, x^*\rangle\, d\lambda = \int_{\mathbb{R}} \hat{f}'(\lambda)\langle \tilde{E}(\lambda)x, x^*\rangle\, d\lambda .$$

It follows by standard considerations [17, I.(3.4), (3.8), (3.9)] that in this last equation we can replace $\hat{f}'$ by φ', where φ is an arbitrary element of the Schwartz class $\mathcal{S}$. Further standard reasoning now shows that if $x \in X$, and $x^* \in X^*$, then $\left[\langle E(\cdot)x, x^*\rangle - \langle \tilde{E}(\cdot)x, x^*\rangle \right]$ is equal a.e. in $\mathbb{R}$ to a constant function. From properties (iii) and (v) in the definition of spectral family, we infer that $E(\lambda) = \tilde{E}(\lambda)$ for all $\lambda \in \mathbb{R}$.

To prove the "if" part, suppose that conditions (i) and (ii) in the statement of the theorem hold, and let $\hat{\mathfrak{D}} = \left\{ \hat{f}: \ f \in \mathfrak{D} \right\}$. Then $\hat{\mathfrak{D}}$ is a dense subalgebra of $AC_0(\mathbb{R})$, and the mapping $\hat{f} \longrightarrow \int_{\mathbb{R}} f(t)\, U_{-t} dt$ is a well-defined algebra homomorphism of $\hat{\mathfrak{D}}$ into $\mathfrak{B}(X)$ which, by (i), is norm-continuous. Let $\Psi: AC_0(\mathbb{R}) \longrightarrow \mathfrak{B}(X)$ be the continuous extension of this mapping. Then $\| \Psi \| \le K$ by (i), and (ii) implies that the set

$$\mathcal{F}_{\mathbb{R}} \equiv \left\{ \Psi(g): \ g \in AC_0(\mathbb{R}), \ \| g \|_{\mathbb{R}} \le 1 \right\}$$

is relatively compact in the weak operator topology of $\mathfrak{B}(X)$.

Now let ϕ be any strictly increasing, infinitely differentiable function mapping $\mathbb{R}$ onto the open interval $(0, 2\pi)$ and such that $\phi' > 0$ on $\mathbb{R}$. Given $f \in AC_0(\mathbb{T}) \equiv \{ G \in AC(\mathbb{T}): \ G(1) = 0 \}$, define $\theta_f: \mathbb{R} \longrightarrow \mathbb{C}$ by setting

$$\theta_f(t) = f(e^{i\phi(t)}) \qquad (t \in \mathbb{R}).$$

The mapping $\theta: f \longrightarrow \theta_f$ is an isometric algebra isomorphism of $AC_0(\mathbb{T})$ onto $AC_0(\mathbb{R})$. Let $\Phi_0 = \Psi \circ \theta$, so that Φ_0 is an algebra homomorphism of $AC_0(\mathbb{T})$ into $\mathfrak{B}(X)$ with $\left\| \Phi_0 \right\| \le K$. Extend Φ_0 to an algebra homomorphism $\Phi: AC(\mathbb{T}) \longrightarrow \mathfrak{B}(X)$ by defining

$$\Phi(\alpha + f) = \alpha I + \Phi_0(f) \qquad (\alpha \in \mathbb{C}, \ f \in AC_0(\mathbb{T})).$$

Then Φ is identity-preserving with $\| \Phi \| \le \max \{K, 1\}$. Also, the set

$$\mathcal{F}_{\mathbb{T}} \equiv \{ \Phi(f): \ f \in AC(\mathbb{T}), \ \| f \|_{\mathbb{T}} \le 1 \}$$

equals

$$\left\{ \alpha I + \Psi(g): \ \alpha \in \mathbb{C}, \ g \in AC_0(\mathbb{R}), \ | \alpha | + \| g \|_{\mathbb{R}} \le 1 \right\}.$$

Hence $\mathcal{F}_{\mathbb{T}}$ is relatively compact in the weak operator topology, since $\mathcal{F}_{\mathbb{R}}$ has this property. By Theorem (2.2) there exists a unique spectral family $F(\cdot)$ in X, concentrated on $[0,2\pi]$ and with $F((2\pi)^-) = I$, such that

$$\Phi(f) = \int_{[0,2\pi]}^{\oplus} f(e^{i\lambda})\, dF(\lambda) \qquad (f \in AC(\mathbb{T})) .$$

Upon multiplying this equation by $F(0)$, we see that, in particular, $\Phi(f)F(0) = 0$ for all $f \in AC_0(\mathbb{T})$, and hence $\Psi(g)F(0) = 0$ for all $g \in AC_0(\mathbb{R})$. Thus

$$\int_{\mathbb{R}} f(t)\, U_{-t}x\, dt = 0 \qquad (f \in \mathcal{D},\ x \in F(0)X) ,$$

from which we deduce, using the strong continuity of the group $\{ U_t \}_{t \in \mathbb{R}}$, that $F(0) = 0$. Define

$$E(\lambda) = F(\phi(\lambda)) \qquad (\lambda \in \mathbb{R}) .$$

Then $E(\cdot)$ is a spectral family in X.

Given $f \in \mathcal{D}$,

$$\begin{aligned}\int_{\mathbb{R}} f(t)\, U_{-t}dt = \Psi(\hat{f}) &= \Phi_0(\theta^{-1}(\hat{f})) \\ &= \int_0^{2\pi} \left[\theta^{-1}(\hat{f})\right](e^{i\lambda})\, dF(\lambda) \\ &= \lim_{\delta \to 0^+} \int_{\delta}^{2\pi-\delta} \left[\theta^{-1}(\hat{f})\right](e^{i\lambda})\, dF(\lambda) \\ &= \lim_{b \to +\infty} \int_{-b}^{b} \hat{f}(\lambda)dE(\lambda) ,\end{aligned}$$

where the limits are in the strong operator topology. Let $y \in \left\{ E(a) - E(-a) \right\}X$ for some $a > 0$, let $x^* \in X^*$, and let $f \in \mathcal{D}$. Since

$$\int_{-b}^{b} \hat{f}(\lambda)dE(\lambda)y = \int_{-a}^{a} \hat{f}(\lambda)dE(\lambda)y$$

for all $b \geq a$,

$$\int_{\mathbb{R}} f(t)\, U_{-t}y\, dt = \int_{-a}^{a} \hat{f}(\lambda)dE(\lambda)y . \tag{2.7}$$

Evaluation of the right-hand side of (2.7) at x^* gives

$$(2.8)\qquad \left\langle \left[\int_{\mathbb{R}} f(t)\, U_{-t}dt \right] y, x^* \right\rangle = \hat{f}(a)\langle y, x^*\rangle - \int_{-a}^{a} \hat{f}'(\lambda)\langle E(\lambda)y, x^*\rangle d\lambda .$$

Substituting

$$\hat{f}(a) = \int_{\mathbb{R}} e^{-iat} f(t)dt , \quad \text{and} \quad \hat{f}'(\lambda) = -i\int_{\mathbb{R}} te^{-i\lambda t} f(t)dt ,$$

we deduce that

$$(2.9)\qquad \int_{-\infty}^{\infty} f(t)\, U_{-t}y\, dt = \int_{-\infty}^{\infty} f(t) \left\{ \int_{-a}^{a} e^{-i\lambda t}\, dE(\lambda)y \right\} dt .$$

The expression in braces on the right-hand side of (2.9) is continuous on $\mathbb{R}$ as a function of t (by [11, Theorem 17.5]) , and the function $t \longrightarrow U_{-t}x$ is continuous. Since (2.9) is valid for all $f \in \mathfrak{D}$, it follows that

$$U_{-t}y = \int_{-a}^{a} e^{-i\lambda t}\, dE(\lambda)y$$

for all $t \in \mathbb{R}$. Hence

$$U_t\left\{ E(a) - E(-a) \right\}x = \int_{-a}^{a} e^{i\lambda t}\, dE(\lambda)x$$

for all $t \in \mathbb{R}$, all $x \in X$, and all $a \geq 0$. Letting $a \longrightarrow \infty$, we obtain (2.5).

To obtain the bound on $\| E(\cdot) \|$, let $\lambda \in (0,2\pi)$, and, for $n > \max\left\{ \lambda^{-1} , (2\pi - \lambda)^{-1} \right\}$, let $g_n \in AC([0,2\pi])$ be such that $g_n = 1$ on $[n^{-1},\lambda]$, $g_n = 0$ on $[\lambda + n^{-1},2\pi]$, $g_n(0) = 0$, and g_n is linear on $[0,n^{-1}]$ and on $[\lambda,\lambda + n^{-1}]$. Then $\| g_n \|_{[0,2\pi]} = 2$ for all n , and $\{ g_n \}$ converges pointwise on $[0,2\pi]$ to the characteristic function χ of the interval $(0,\lambda]$. Define $G_n \in AC_0(\mathbb{T})$ by putting $G_n(e^{it}) = g_n(t)$ for $0 \leq t \leq 2\pi$. By [11, Theorem 17.5] and the fact that $F(0) = 0$,

$$\Phi_0(G_n) = \int_0^{2\pi} g_n(\mu)dF(\mu) \longrightarrow \int_0^{2\pi} \chi(\mu)dF(\mu) = F(\lambda)$$

in the strong operator topology as $n \longrightarrow +\infty$. Hence $\| F(\lambda) \| \leq 2K$, which gives the required bound for $\| E(\cdot) \|$. The final result for reflexive spaces is clear, and so the proof of the theorem is complete.

(2.10) **COROLLARY.** *Suppose that the strongly continuous one-parameter group* $\{ U_t \}_{t \in \mathbb{R}}$ *of operators on a Banach space* X *satisfies* (2.5) *for some spectral family* $E(\cdot)$. *Then:* (i) *an operator* $S \in \mathfrak{B}(X)$ *commutes with* U_t *for all* $t \in \mathbb{R}$ *if and only if* S *commutes with* $E(\lambda)$ *for all* $\lambda \in \mathbb{R}$; (ii) *a closed subspace* M *of* X *is invariant under* $\{ U_t \}_{t \in \mathbb{R}}$ *if and only if* M *is invariant under* $E(\cdot)$.

Proof. (i) The "if" assertion is evident from (2.5). Conversely, suppose S commutes with each U_t . Then, in the notation of the proof of Theorem (2.4), we obviously have that S commutes with $\Phi(f)$ for all $f \in AC(\mathbb{T})$. In particular, S commutes with $\Phi(e_1)$, where, as previously, $e_1(z) \equiv z$. The spectral family $F(\cdot)$ defined in the proof of Theorem (2.4) is obviously the spectral decomposition of the trigonometrically well-bounded operator $\Phi(e_1)$. Hence S commutes with each value of $F(\cdot)$, and consequently with each value of the unique spectral family $E(\cdot)$ in (2.5). The equivalence asserted in (ii) is established similarly.

(2.11) **PROPOSITION.** *Let* $\{ U_t \}_{t \in \mathbb{R}}$ *be a strongly continuous one-parameter group of operators on a Banach space* X *such that* (2.5) *holds for some spectral family* $E(\cdot)$. *Then the group* $\{ U_t \}_{t \in \mathbb{R}}$ *is continuous in the uniform operator topology of* $\mathfrak{B}(X)$ *if and only if* $E(\cdot)$ *is concentrated on a compact interval.*

Proof. The "if" assertion is evident. Conversely, suppose that $\{ U_t \}_{t \in \mathbb{R}}$ is continuous with respect to the uniform operator topology. Then there is $A \in \mathfrak{B}(X)$ such that $U_t = e^{itA}$ for all $t \in \mathbb{R}$. Let $\lambda_0 \in \mathbb{R}$ with $|\lambda_0| > \| A \|$. We claim that for some

$\epsilon > 0$, $E(\lambda_0 + \epsilon) = E(\lambda_0 - \epsilon)$. Suppose to the contrary that for each $\epsilon > 0$ there is $x_\epsilon \in X$ such that $x_\epsilon = \{ E(\lambda_0 + \epsilon) - E(\lambda_0 - \epsilon) \}x_\epsilon$, and $\| x_\epsilon \| = 1$. For $\epsilon > 0$, $t > 0$, we have

$$t^{-1}\left[U_t x_\epsilon - x_\epsilon \right] = \int_{\lambda_0 - \epsilon}^{\lambda_0 + \epsilon} t^{-1}\left[e^{i\lambda t} - 1 \right] dE(\lambda)x_\epsilon .$$

Upon letting $t \to 0^+$, we obtain

$$Ax_\epsilon = \int_{\lambda_0 - \epsilon}^{\lambda_0 + \epsilon} \lambda \, dE(\lambda)x_\epsilon .$$

Hence

$$(\lambda_0 - A)x_\epsilon = \int_{\lambda_0 - \epsilon}^{\lambda_0 + \epsilon} (\lambda_0 - \lambda) \, dE(\lambda)x_\epsilon ,$$

and consequently,

$$\left\| (\lambda_0 - A)x_\epsilon \right\| \le K \left\| \lambda_0 - \lambda \right\|_{[\lambda_0 - \epsilon, \lambda_0 + \epsilon]} = 3K\epsilon ,$$

where $K = \sup \{ \| E(\lambda) \| : \lambda \in \mathbb{R} \}$. Hence $(\lambda_0 - A)x_\epsilon \to 0$ as $\epsilon \to 0^+$, which implies that λ_0 belongs to the spectrum of A . This contradicts the fact that $|\lambda_0| > \| A \|$, and establishes our claim. Hence, if $|\lambda_0| > \| A \|$, then λ_0 is the mid-point of an open interval on which $E(\cdot)$ is constant. A simple connectedness argument applied to each of the intervals $(-\infty , -\| A \|)$, $(\| A \|, +\infty)$, in conjunction with (v) in the definition of spectral family, now shows that $E(\cdot)$ is concentrated on the interval $[-\| A \| , \| A \|]$.

(2.12) **COROLLARY.** *Let* $\{ U_t \}_{t \in \mathbb{R}}$ *be a one-parameter group of operators on a Banach space* X *which is continuous with respect to the uniform operator topology of* $\mathfrak{B}(X)$. *Let* iA *be the infinitesimal generator of* $\{ U_t \}_{t \in \mathbb{R}}$. *Then* $\{ U_t \}_{t \in \mathbb{R}}$ *has a representation of the form* (2.5) *for some spectral family* $E(\cdot)$ *if and only if* A *is well-bounded of type* (B) . *If this is the case, then* $E(\cdot)$ *is unique and is the spectral family of* A .

Proof. If A is well-bounded of type (B), and $E(\cdot)$ denotes its spectral family, then there is a compact interval J such that $A = \int_J^{\oplus} \lambda dE(\lambda)$ and $E(\cdot)$ is concentrated on J. Hence for all $t \in \mathbb{R}$, $U_t = e^{itA} = \int_J^{\oplus} e^{it\lambda} dE(\lambda)$. In view of the uniqueness conclusion in Theorem (2.4), it remains only to show the "only if" assertion of this corollary. But in this case, it follows from Proposition (2.11) that $E(\cdot)$ is concentrated on a compact interval J_0. It follows readily that $i\int_{J_0}^{\oplus} \lambda dE(\lambda)$ is the infinitesimal generator of $\{ U_t \}_{t \in \mathbb{R}}$.

Remarks. The constructions reproduced in [11, Proposition 18.5] provide a well-bounded operator T of type (B) on the Hilbert space ℓ^2 such that $\sup \{ \| U_t \| : t \in \mathbb{R} \} = +\infty$, where $U_t = e^{itT}$ for all $t \in \mathbb{R}$. This gives an example of a group $\{ U_t \}_{t \in \mathbb{R}}$ satisfying (2.5) for some spectral family $E(\cdot)$ such that $\{ U_t \}_{t \in \mathbb{R}}$ is not uniformly bounded.

We conclude this section by using Theorem (2.4) to give a simplified proof (and strengthened version) of the Generalized Stone's Theorem for one-parameter groups of trigonometrically well-bounded operators ([1, Theorem (4.20)]). Before doing so it will be convenient to record the following obvious consequence of Corollary (2.3).

(2.13) **LEMMA.** *Let U be a trigonometrically well-bounded operator on a Banach space X, let $E(\cdot)$ be the spectral decomposition of U, and let $K \equiv \sup \{ \| E(\lambda) \| : \lambda \in \mathbb{R} \}$. If $\alpha \in \mathbb{T}$, then αU is trigonometrically well-bounded, and the spectral decomposition $E_\alpha(\cdot)$ of αU satisfies* $\sup \left\{ \left\| E_\alpha(\lambda) \right\| : \lambda \in \mathbb{R} \right\} \le 6K$.

(2.14) GENERALIZED STONE'S THEOREM ([1, Theorem (4.20)]). *Let* $\{ U_t \}_{t \in \mathbb{R}}$ *be a strongly continuous one-parameter group of trigonometrically well-bounded operators on a Banach space* X *such that for some* $\delta > 0$,

$$K_\delta \equiv \sup \left\{ \left\| E_t(\lambda) \right\| : \ \lambda \in \mathbb{R}, \quad 0 < t \le \delta \right\} < +\infty ,$$

where $E_t(\cdot)$ *is the spectral decomposition of* U_t, *for each* $t \in \mathbb{R}$. *Then there is a unique spectral family* $E(\cdot)$ *in* X *such that* (2.5) *holds. Moreover,*

$$\sup \{ \| E(\lambda) \| : \ \lambda \in \mathbb{R} \} \le 12K_\delta .$$

Proof. Let A_1 be the argument of U_1, and define $W_t = U_t \exp(-itA_1)$ for $t \in \mathbb{R}$, where we have used "exp" to stand for "exponential of". Thus, $\{ W_t \}_{t \in \mathbb{R}}$ is a strongly continuous one-parameter group on X such that $W_1 = I$. For $n \in \mathbb{Z}$, let

$$P_n x = \int_0^1 e^{-2\pi int} W_t x \, dt , \qquad \text{for } x \in X .$$

Then it is well-known and easy to see that: each P_n is a bounded idempotent operator on X; $W_t P_n = e^{2\pi int} P_n$ for $t \in \mathbb{R}$, $n \in \mathbb{Z}$; and $P_n P_m = 0$ if $m \ne n$. Moreover, an obvious separation argument in conjunction with the injectivity of the Fourier transform (in this instance, for periodic functions) shows that X_∞, the span of $\{ P_n X : \ n \in \mathbb{Z} \}$ is dense in X. Temporarily fix $N \in \mathbb{N}$, and let X_N denote the closed subspace $\bigoplus_{n=-N}^{N} (P_n X) = \left[\sum_{n=-N}^{N} P_n \right] X$. For any operator $T \in \mathfrak{B}(X)$ which has X_N as an invariant subspace, the restriction $T \mid X_N$ of T to X_N will be written $T^{(N)}$. In particular, since

$$\{ A_1 \} \cup \{ U_t : t \in \mathbb{R} \} \cup \{ W_t : t \in \mathbb{R} \} \cup \{ P_n : n \in \mathbb{Z} \}$$

is a commutative family, X_N is invariant under the groups $\{ U_t \}_{t \in \mathbb{R}}$ and $\{ W_t \}_{t \in \mathbb{R}}$, and

$$U_t^{(N)} = \sum_{n=-N}^{N} \left\{ \exp \left[it \left[A_1^{(N)} + 2\pi n \right] \right] \right\} P_n^{(N)} , \quad \text{for } t \in \mathbb{R} .$$

Hence $\{ U_t^{(N)} \}_{t \in \mathbb{R}}$ is continuous in the uniform operator topology of $\mathfrak{B}(X_N)$, with infinitesimal generator $i \bigoplus_{n=-N}^{N} (A_1 + 2\pi n) \Big| (P_n X)$. Since $\bigoplus_{n=-N}^{N} (A_1 + 2\pi n) \Big| (P_n X)$ is obviously well-bounded of type (B) on X_N, we can apply Corollary (2.12) to infer the existence of a spectral family F_N concentrated on a compact interval $J_N = [a_N, b_N]$ such that

$$(2.15) \qquad U_t^{(N)} = \int_{J_N}^{\oplus} e^{it\lambda} \, dF_N(\lambda) , \quad \text{for all } t \in \mathbb{R} .$$

By expanding the interval J_N toward the right, we can assume without loss of generality that $F_N(b_N^-) = I$, and that $2\pi(b_N - a_N)^{-1} < \delta$. Define $\varphi_N : \mathbb{R} \to \mathbb{R}$ by setting $\varphi_N(\lambda) = 2\pi (b_N - a_N)^{-1} (\lambda - a_N)$ for all $\lambda \in \mathbb{R}$. If we define $G_N(\cdot)$ on $\mathbb{R}$ by putting $G_N(\varphi_N(\lambda)) = F_N(\lambda)$ for all $\lambda \in \mathbb{R}$, then it is easy to see that $G_N(\cdot)$ is a spectral family concentrated on $[0, 2\pi]$, and $G_N((2\pi)^-) = I$. Putting $\alpha_N = \exp \left\{ -i2\pi a_N / (b_N - a_N) \right\}$, we get, with the aid of (2.15),

$$(2.16) \qquad \alpha_N U^{(N)}_{2\pi / (b_N - a_N)} = \int_{J_N}^{\oplus} \left\{ \exp(i\varphi_N(\lambda)) \right\} dF_N(\lambda) = \int_{[0,2\pi]}^{\oplus} e^{i\mu} dG_N(\mu) .$$

Since X_N is invariant under the spectral decomposition $E_{2\pi / (b_N - a_N)}$ of $U_{2\pi / (b_N - a_N)}$, it follows that $E^{(N)}_{2\pi / (b_N - a_N)}$ is the spectral decomposition of $U^{(N)}_{2\pi / (b_N - a_N)}$. In view of this fact, application of Lemma (2.13) to (2.16) together with the definition of $G_N(\cdot)$ gives us:

(2.17) $$\| F_N(\lambda) \| \le 6K_\delta , \quad \text{for all } \lambda \in \mathbb{R} .$$

By virtue of (2.15) and (2.17), we can apply the proof of Theorem (2.4)-(i),(ii) to the group $\{ U^{(N)}_t \}_{t \in \mathbb{R}}$ to get:

(2.18) $$\left\| \int_{\mathbb{R}} f(t)\, U_{-t}x \, dt \right\| \le 6K_\delta \left\| \hat{f} \right\|_{\mathbb{R}} \| x \| ,$$

$$\text{for } f \in \mathfrak{D} , \quad x \in X_N ,$$

and,

(2.19) $$\left\{ \int_{\mathbb{R}} f(t)\, U^{(N)}_{-t} \, dt : \ f \in \mathfrak{D} \text{ and } \left\| \hat{f} \right\|_{\mathbb{R}} \le 1 \right\}$$ is relatively compact in the strong operator topology of $\mathfrak{B}(X_N)$.

Since X_∞ is dense in X , and $N \in \mathbb{N}$ is arbitrary, we infer from (2.18) that

(2.20) $$\left\| \int_{\mathbb{R}} f(t)\, U_{-t} dt \right\| \le 6K_\delta \left\| \hat{f} \right\|_{\mathbb{R}} , \quad \text{for } f \in \mathfrak{D} .$$

Moreover, the density of X_∞ in X and (2.20) allow us to deduce from (2.19) that for $x \in X$,

$$\left\{ \int_{\mathbb{R}} f(t)\, U_{-t}x \, dt : \ f \in \mathfrak{D} , \ \left\| \hat{f} \right\|_{\mathbb{R}} \le 1 \right\}$$

is totally bounded in X . We have thus shown that the conditions (2.4)-(i),(ii) hold for the given group $\{ U_t \}_{t \in \mathbb{R}}$ on X , and hence there is a unique spectral family $E(\cdot)$ in X such that (2.5) holds. Moreover, we also infer from (2.20) and Theorem (2.4) that

$$\sup \{ \| E(\lambda) \| : \ \lambda \in \mathbb{R} \} \le 12K_\delta .$$

3. <u>Vector-valued transference and norm estimates</u>. To use the criteria of Corollary (2.3) and Theorem (2.4) to obtain spectral representations of the form (1.1) or (1.2), we need a method for estimating appropriate

operator norms. Our first such tool is a vector-valued extension of the General Transference Result of Coifman and Weiss [9, Theorem (2.4)]. Its proof is a mild adaptation of the original result and was given in [6]. To state it precisely, let G be a locally compact abelian group with Haar measure m, and let X be a Banach space. Given $k \in L^1(G)$ and $1 \le p < \infty$, let $N_{p,X}(k)$ denote the norm of convolution by k as an operator on the Lebesgue-Bochner space $L^p(G,X)$. Suppose that $u \longrightarrow R_u$ is a strongly continuous representation of G in X such that

$$\| R_u \| \le c \qquad (u \in G),$$

where c is a real constant. This representation gives rise to a representation in X, $k \longrightarrow T_k$, of the convolution algebra $L^1(G)$ by means of the formula

$$(3.1) \qquad T_k x = \int_G k(u)\, R_{-u} x \, dm(u) \qquad (k \in L^1(G),\ x \in X).$$

It is clear that $T_k \in \mathfrak{B}(X)$ and that $\| T_k \| \le c \| k \|_1$.

(3.2) **THEOREM** ([6, Theorem (2.8)]). *With the above notation,*

$$\| T_k \| \le c^2 N_{p,X}(k)$$

for $k \in L^1(G)$ *and* $1 \le p < \infty$.

<u>**Remark**</u>. The original Coifman-Weiss result applies to locally compact <u>**amenable**</u> groups, but for simplicity we have confined attention here to the smaller class of locally compact abelian groups, since this class suffices for the applications below.

To apply the above vector-valued transference result (3.2) in the settings of (2.3) and (2.4), we take $G = \mathbb{Z}$ and $\mathbb{R}$, respectively, and seek estimates for appropriate convolution norms $N_{p,X}(k)$. In the scalar case $X = \mathbb{C}$, such estimates are given by results due to Stečkin

[12, Theorems 6.4.4 and 6.2.5] when $1 < p < \infty$ and $\hat{k}$ is of bounded variation. Furthermore, these estimates are of the form $N_{p,\mathbb{C}}(k) \le C_p \| \hat{k} \|$, with $\| \hat{k} \|$ denoting the relevant variation norm of $\hat{k}$ and C_p a constant depending only on p.

The crucial fact behind Stečkin's results is the boundedness of the Hilbert transform on $L^p(\mathbb{R})$, or its discrete analogue on $L^p(\mathbb{Z})$, for $1 < p < \infty$. The Banach spaces X such that the vector-valued Hilbert transform is bounded on $L^p(\mathbb{R},X)$ for $1 < p < \infty$ have recently been characterized as those with the unconditionality property for martingale differences (written more briefly as $X \in$ UMD). More precisely $X \in$ UMD by definition if and only if, for $1 < p < \infty$, X-valued martingale difference sequences are unconditional in $L^p([0,1],X)$ with a uniform unconditionality constant that depends only on X and p. To state the connection with the Hilbert transform, let h_ϵ be the truncated Hilbert kernel on $\mathbb{R}$ defined for $\epsilon > 0$ by

$$h_\epsilon(t) = (\pi t)^{-1} \quad (|t| \ge \epsilon), \quad h_\epsilon(t) = 0 \text{ otherwise,}$$

and let d_N, for $N \in \mathbb{N}$, be its discrete analogue on $\mathbb{Z}$ defined by

$$d_N(n) = n^{-1} \quad (0 < |n| \le N), \quad d_N(n) = 0 \text{ otherwise.}$$

For a Banach space X and a strongly measurable function $f: \mathbb{R} \longrightarrow X$, let $H_{X,\epsilon} f = h_\epsilon * f$ provided this convolution exists almost everywhere on $\mathbb{R}$, and let

$$(H_X f)(t) = \lim_{\epsilon \longrightarrow 0^+} (h_\epsilon * f)(t)$$

whenever the limit exists (in the norm of X) almost everywhere on $\mathbb{R}$. Similarly, for $f: \mathbb{Z} \longrightarrow X$, let

$$(D_X f)(n) = \lim_{N \longrightarrow +\infty} (d_N * f)(n)$$

whenever the limit exists for all $n \in \mathbb{Z}$. The following theorem gives the required characterization of the UMD property in terms of the boundedness of the Hilbert transform. The equivalence of (i) and (ii), which is at the heart of the result, is due to Bourgain and Burkholder ([7, Lemma 2], [8, Lemma 3]), while the equivalence of (ii) and (iii) is given by [4, Theorem (2.8)].

(3.3) **THEOREM.** *Let* X *be a Banach space, and let* $1 < p < \infty$. *The following statements are equivalent:*

(i) $X \in$ UMD;

(ii) H_X *is a bounded linear operator* $L^p(\mathbb{R},X) \longrightarrow L^p(\mathbb{R},X)$;

(iii) D_X *is a bounded linear operator* $L^p(\mathbb{Z},X) \longrightarrow L^p(\mathbb{Z},X)$.

Remarks. (a) Notice that if (ii) or (iii) holds for a single value of p in the range $1 < p < \infty$, then it holds for all such p , since (i) is a condition which does not depend on a particular value of p . (b) If $X \in$ UMD and $1 < p < \infty$, then each truncated convolution operator $H_{X,\epsilon}$ is a bounded linear mapping of $L^p(\mathbb{R},X)$ into itself, with operator norm not exceeding that of H_X on $L^p(\mathbb{R},X)$. This is proved exactly as in the scalar case (see [13, pp. 240-241]). (c) The class UMD contains the L^p-spaces associated with an arbitrary measure space for $1 < p < \infty$, as well as their non-commutative counterparts, including the von Neumann-Schatten p-classes for p in the same range. Also, the class UMD is closed under the formation of dual spaces, quotient spaces, and subspaces, and each UMD space is reflexive (in fact, super-reflexive). For more detailed background information and further references, see [4,5,8].

We can now obtain vector-valued versions of Stečkin's theorems.

(3.4) **THEOREM.** *Let* $X \in$ UMD *and let* $1 < p < \infty$.

(i) *If* $k \in L^1(\mathbb{Z})$ *and* $\hat{k} \in BV(\mathbb{T})$, *then*

$$N_{p,X}(k) \leq (1 + \alpha_p) \| \hat{k} \|_{\mathbb{T}} ,$$

where $(2\pi\alpha_p)$ *is the norm of* D_X *as an operator on* $L^p(\mathbb{Z},X)$.

(ii) *If* $k \in L^1(\mathbb{R})$ *and* $\hat{k} \in BV(\mathbb{R})$, *then*

$$N_{p,X}(k) \leq \beta_p \| \hat{k} \|_{\mathbb{R}} ,$$

where β_p *is the norm of* H_X *as an operator on* $L^p(\mathbb{R},X)$.

Proof. (i) Let q be the index conjugate to p , and fix two finitely supported functions $f: \mathbb{Z} \to X$ and $g: \mathbb{Z} \to X^*$. For $0 \leq t \leq 2\pi$, let $\phi(t) = \hat{k}(e^{it})$ and define $f_t: \mathbb{Z} \to X$, $g_t: \mathbb{Z} \to X^*$ by

$$f_t(n) = e^{-int}f(n) , \quad g_t(n) = e^{int}g(n) \qquad (n \in \mathbb{Z}) .$$

After the substitution $k(n) = (2\pi)^{-1}\int_0^{2\pi} \phi(t)e^{int}\, dt$, a straightforward calculation involving one integration by parts gives

$$\sum_{m,n \in \mathbb{Z}} k(n-m)\langle f(m),g(n)\rangle = k(0) \sum_{n \in \mathbb{Z}} \langle f(n),g(n)\rangle +$$
$$i(2\pi)^{-1}\int_0^{2\pi} \sum_{n \in \mathbb{Z}} \langle \left[D_X f_t\right](n),g_t(n)\rangle d\phi(t) .$$

It follows that

$$\left| \sum_{m,n \in \mathbb{Z}} k(n-m)\langle f(m),g(n)\rangle \right| \leq \left[\| \hat{k} \|_\infty + \alpha_p \| \hat{k} \|_{\mathbb{T}} \right] \| f \|_p \| g \|_q .$$

Since $L^q(\mathbb{Z},X^*)$ is the dual space of $L^p(\mathbb{Z},X)$, (i) now follows. (The details of this proof are a mild variant of the proof of the scalar result given in [11, pp. 377-378].)

The proof of (ii) is similar to that of (i), with a little extra care being needed because of the singularity of the kernel $(\pi t)^{-1}$ at $t = 0$. Firstly, assume that $\hat{k}$ is integrable and regard k as a continuous function, *viz.* the inverse Fourier transform of $\hat{k}$. Then

$$k(t) = (2\pi)^{-1}\int_{\mathbb{R}} e^{itu}\, \hat{k}(u)du = i(2\pi)^{-1}\int_{\mathbb{R}} t^{-1}e^{itu}\, d\hat{k}(u) \tag{3.5}$$

for $t \neq 0$. Let $f \in L^p(\mathbb{R},X)$ and $g \in L^q(\mathbb{R},X^*)$ be continuous and have compact support, and, for $u \in \mathbb{R}$, set

$$f_u(t) = e^{-itu}f(t) , \quad g_u(t) = e^{itu}g(t) \qquad (t \in \mathbb{R}) .$$

Fix $\epsilon > 0$. By (3.5) and an application of Fubini's Theorem,

$$\iint_{|s-t| \geq \epsilon} k(t - s)\langle f(s), g(t)\rangle \, dsdt$$

$$= i(2\pi)^{-1}\int_{\mathbb{R}} \left\{ \iint_{|s-t| \geq \epsilon} (t - s)^{-1}\langle f_u(s), g_u(t)\rangle \, dsdt \right\} d\hat{k}(u)$$

$$= i2^{-1}\int_{\mathbb{R}} \left\{ \int_{\mathbb{R}} \langle \left[H_{X,\epsilon} f_u\right](t), g_u(t)\rangle \, dt \right\} d\hat{k}(u) .$$

As remarked after Theorem (3.3), the norm of $H_{X,\epsilon}$ on $L^p(\mathbb{R},X)$ is dominated by β_p . Hence

$$\left| \iint_{|s-t| \geq \epsilon} k(t - s)\langle f(s), g(t)\rangle \, dsdt \right| \leq 2^{-1}\beta_p \| \hat{k} \|_{\mathbb{R}} \| f \|_p \| g \|_q .$$

The integrand on the left-hand side of this inequality is integrable over $\mathbb{R} \times \mathbb{R}$; we may thus let $\epsilon \longrightarrow 0^+$ and deduce that

$$(3.6) \qquad \left| \int_{\mathbb{R}} \langle (k * f)(t), g(t)\rangle \, dt \right| \leq 2^{-1}\beta_p \| \hat{k} \|_{\mathbb{R}} \| f \|_p \| g \|_q .$$

By continuity, it follows that (3.6) holds for all $f \in L^p(\mathbb{R},X)$ and all $g \in L^q(\mathbb{R},X^*)$. Under the natural duality, $L^q(\mathbb{R},Y^*)$ is a norm-determining subspace of the dual space of $L^p(\mathbb{R},Y)$ for an arbitrary Banach space Y (adapt the reasoning of [10, p. 98]), and so (3.6) implies that

$$(3.7) \qquad N_{p,X}(k) \leq 2^{-1}\beta_p \| \hat{k} \|_{\mathbb{R}} .$$

(It can easily be deduced from [10, Theorem 1 on p. 98] that $L^q(\mathbb{R},X^*)$ in fact equals the dual space of $L^p(\mathbb{R},X)$ since X is reflexive. However, this is less elementary than the corresponding result for $L^p(\mathbb{Z},X)$ used in (i), which is actually valid if X is replaced by an arbitrary Banach space.)

Now suppose only that $\hat{k} \in BV(\mathbb{R})$, and let $k_n = v_n * k$ for $n \in \mathbb{N}$, where $v_n(t) = (\sqrt{n / \pi})e^{-nt^2}$. Then $\hat{k}_n \in L^1(\mathbb{R})$, and, since $\| \hat{v}_n \|_{\mathbb{R}} = 2$, $\| \hat{k}_n \|_{\mathbb{R}} \le 2 \| \hat{k} \|_{\mathbb{R}}$ for all n. Also, $N_{p,X}(k_n) \longrightarrow N_{p,X}(k)$ since $\| k_n - k \|_1 \longrightarrow 0$. Thus, applying (3.7) to k_n and letting $n \longrightarrow \infty$, we obtain the required inequality $N_{p,X}(k) \le \beta_p \| \hat{k} \|_{\mathbb{R}}$. This completes the proof of (ii).

We can now obtain the main results concerning spectral decomposability on UMD spaces originally proved in [4] by a different route.

(3.8) **THEOREM.** *Let X be a UMD space.*

(i) *Suppose that U is an invertible operator on X such that $\| U^n \| \le c < +\infty$ ($n \in \mathbb{Z}$) for some constant c. Then U is trigonometrically well-bounded, that is*

$$U = \int^{\oplus}_{[0,2\pi]} e^{i\lambda} dE(\lambda) ,$$

where $E(\cdot)$ is the spectral decomposition of U. Furthermore, there is a constant γ_X, which depends on X but not on U, such that

$$\| E(\lambda) \| \le c^2 \gamma_X \qquad (\lambda \in \mathbb{R}) .$$

(ii) *Suppose that $\{ U_t \}_{t \in \mathbb{R}}$ is a strongly continuous one-parameter group of operators on X such that $\| U_t \| \le c < +\infty$ ($t \in \mathbb{R}$) for some constant c. Then*

$$U_t = \lim_{a \longrightarrow +\infty} \int_{-a}^{a} e^{i\lambda t} dE(\lambda) \qquad (t \in \mathbb{R})$$

for some unique spectral family $E(\cdot)$ in X, where the limit exists in the strong operator topology of $\mathfrak{B}(X)$. Furthermore, there is a constant γ'_X, which depends on X but not on $\{ U_t \}_{t \in \mathbb{R}}$, such that

$$\| E(\lambda) \| \le c^2 \gamma'_X \qquad (\lambda \in \mathbb{R}) .$$

Proof. For (i), consider the representation R of $G = \mathbb{Z}$ in X defined by $R_n = U^n \quad (n \in \mathbb{Z})$, and let $q(e^{it}) = \sum_{n \in \mathbb{Z}} a_n e^{int}$ be a trigonometric polynomial (the sequence $\{a_n\}_{n=-\infty}^{\infty}$ is finitely supported). Defining $k: \mathbb{Z} \to \mathbb{C}$ by $k(n) = a_{-n}$, we have $\hat{k} = q$ and, in the notation of (3.1), $q(U) = T_k$. Fix p in the range $1 < p < \infty$. Then, by Theorems (3.2) and (3.4)-(i)

$$\| q(U) \| \leq c^2 N_{p,X}(k) \leq c^2(1 + \alpha_p) \| \hat{k} \|_{\mathbb{T}} = c^2(1 + \alpha_p) \| q \|_{\mathbb{T}},$$

where $(2\pi\alpha_p)$ is the norm of D_X on $L^p(\mathbb{Z},X)$. Since UMD spaces are reflexive, (i) now follows from Corollary (2.3), with $\gamma_X = 3(1 + \alpha_p)$.

Similarly, (ii) follows by considering the representation R of $\mathbb{R}$ in X given by $R_t = U_t$ and applying Theorems (3.2), (3.4)-(ii), and (2.4). In this case, the constant γ'_X can be taken to be $2\beta_p$ for any p in the range $1 < p < \infty$, where β_p is the norm of H_X on $L^p(\mathbb{R},X)$. Alternatively, (ii) follows from (3.8)-(i) and Theorem (2.14).

4. Scalar-valued transference revisited. In this final section, we show how the original Coifman-Weiss transference result for spaces of scalar-valued functions may be obtained from the vector-valued version discussed in the previous section. As before, we restrict attention to the abelian group case for simplicity, rather than consider the more general result for amenable groups.

The setting is as follows. Let G be a locally compact abelian group with Haar measure m, let $(\mathcal{M},\mu)$ be an arbitrary measure space, and let X be a closed subspace of $L^p(\mathcal{M},\mu)$, where $1 \leq p < \infty$. Suppose further that $u \to R_u$ is a strongly continuous representation of G in X satisfying

$$\| R_u \| \leq c < +\infty \qquad (u \in G),$$

for some constant c. For $k \in L^1(G)$, define the transferred operator T_k on X as in (3.1) by the formula

$$T_k x = \int_G k(u) R_{-u} x \, dm(u) \qquad (x \in X).$$

Let $N_p(k)$ denote the norm of the convolution operator $f \longrightarrow k * f$ on $L^p(G)$.

(4.1) **THEOREM** (Coifman-Weiss General Transference Result [9, Theorem 2.4]). *With the above hypotheses and notation,*

$$\| T_k \| \le c^2 N_p(k)$$

for $k \in L^1(G)$.

__Remarks__. Coifman and Weiss state their result for $X = L^p(\mathcal{M},\mu)$, but their proof applies when X is only assumed to be a subspace of $L^p(\mathcal{M},\mu)$. They also assume that μ is σ-finite in order to apply Fubini's Theorem conveniently, but it is possible to circumvent such technical restrictions (see [6, §2]).

Theorem (4.1) is an immediate corollary of Theorem (3.2), once the following lemma has been established.

(4.2) **LEMMA**. *Given* $k \in L^1(G)$ *and* X *a closed subspace of* $L^p(\mathcal{M},\mu)$, *where* μ *is an arbitrary measure and* $1 \le p < \infty$, *we have*

$$N_{p,X}(k) \le N_p(k).$$

__Proof__. At the outset, we can assume without loss of generality that the set of points where k is non-zero has the form $\bigcup_{m=1}^{\infty} \tau_m$, where each τ_m is a compact subset of G. Let $f\colon G \longrightarrow X$ be a simple function in $L^p(G,X)$ of the form

$$f = \sum_{j=1}^{n} x_{\sigma_j} \phi_j ,$$

where $\sigma_1, \sigma_2, \cdots, \sigma_n$ are compact subsets of G, and

$\phi_1, \phi_2, \cdots, \phi_n$ belong to X. Put

$$F(u,\omega) = \sum_{j=1}^{n} \chi_{\sigma_j}(u)\phi_j(\omega), \qquad \text{for } u \in G, \ \omega \in \mathcal{M}.$$

Thus F is jointly measurable in (u,ω) on $G \times \mathcal{M}$, where $G \times \mathcal{M}$ is viewed as the product of the measurable spaces of m and μ. Abbreviating $dm(u)$ to du, we have:

$$\begin{aligned}\| k * f \|^p_{L^p(G,X)} &= \int_G \left\| \int_G k(u-v)f(v)dv \right\|^p_{L^p(\mu)} du \\ &= \int_G \left\{ \int_{\mathcal{M}} \left| \int_G k(u-v)F(v,\omega)dv \right|^p d\mu(\omega) \right\} du .\end{aligned}$$

Let $\mathcal{M}_0$ be a σ-finite subset of $\mathcal{M}$ such that each ϕ_j vanishes off $\mathcal{M}_0$, and let $K \equiv \bigcup_{m=1}^{\infty} \bigcup_{j=1}^{n} (\tau_m + \sigma_j) \subseteq G$. Thus, K is σ-compact, and for each $v \in \bigcup_{j=1}^{n} \sigma_j$, $k(u-v) = 0$ for all $u \in G\backslash K$. We have:

$$\begin{aligned}\| k * f \|^p_{L^p(G,X)} &= \int_K \left\{ \int_{\mathcal{M}_0} \left| \int_G k(u-v)F(v,\omega)dv \right|^p d\mu(\omega) \right\} du \\ &= \int_{\mathcal{M}_0} \left\{ \int_K \left| \int_G k(u-v)F(v,\omega)dv \right|^p du \right\} d\mu(\omega) \\ &\le \int_{\mathcal{M}_0} \left\{ (N_p(k))^p \int_G |F(v,\omega)|^p dv \right\} d\mu(\omega) \\ &= \int_{\mathcal{M}_0} \left\{ (N_p(k))^p \int_{\bigcup_{j=1}^{n} \sigma_j} |F(v,\omega)|^p dv \right\} d\mu(\omega) \\ &= (N_p(k))^p \int_{\bigcup_{j=1}^{n} \sigma_j} \left\{ \int_{\mathcal{M}_0} |F(v,\omega)|^p d\mu(\omega) \right\} dv \\ &= N_p(k)^p \int_{\bigcup_{j=1}^{n} \sigma_j} \| f(v) \|^p_{L^p(\mu)} dv \\ &= N_p(k)^p \| f \|^p_{L^p(G,X)} .\end{aligned}$$

The desired result now follows by standard approximation arguments.

Acknowledgements

The work of the first author was supported by a grant from the National Science Foundation. The second author is grateful to the organizers for the financial support which enabled him to visit the University of Illinois during the Special Year in Modern Analysis (1986-87).

REFERENCES

1. H. Benzinger, E. Berkson, and T.A. Gillespie, *Spectral families of projections, semigroups, and differential operators*, Trans. Amer. Math. Soc., 275(1983), 431-475.

2. E. Berkson and H.R. Dowson, *On uniquely decomposable well-bounded operators*, Proc. London Math. Soc., (3), 22(1971), 339-358.

3. E. Berkson and T.A. Gillespie, AC *functions on the circle and spectral families*, J. Operator Theory, 13(1985), 33-47.

4. E. Berkson, T.A. Gillespie, and P.S. Muhly, *Abstract spectral decompositions guaranteed by the Hilbert transform*, Proc. London Math. Soc. (3), 53(1986), 489-517.

5. E. Berkson, T.A. Gillespie, and P.S. Muhly, *A generalization of Macaev's Theorem to non-commutative* L^p*-spaces*, Integral Equations and Operator Theory, 10(1987), 164-186.

6. E. Berkson, T.A. Gillespie, and P.S. Muhly, *Generalized analyticity in* UMD *spaces*, Arkiv för Matematik, to appear.

7. J. Bourgain, *Some remarks on Banach spaces in which martingale difference sequences are unconditional*, Arkiv för Matematik, 21(1983), 163-168.

8. D.L. Burkholder, *A geometric condition that implies the existence of certain singular integrals of Banach-space-valued functions*, Proc. Conference on Harmonic Analysis in Honor of Antoni Zygmund (Chicago, 1981), ed. by W. Beckner *et al.*, Wadsworth Publishers, Belmont, California, 1983, pp. 270-286.

9. R.R. Coifman and G. Weiss, *Transference methods in analysis*, Regional Conference Series in Math., No. 31, Amer. Math. Soc., Providence, 1977.

10. J. Diestel and J.J. Uhl, *Vector measures*, Math. Surveys, No. 15, Amer. Math. Soc., Providence, 1977.

11. H.R. Dowson, *Spectral theory of linear operators*, London Math. Soc. Monographs, No. 12, Academic Press, London, 1978.

12. R.E. Edwards and G.I. Gaudry, *Littlewood-Paley and multiplier theory*, Ergebnisse der Mathematik und ihrer Grenzgebiete 90, Springer-Verlag, Berlin, 1977.

13. M. Riesz, *Sur les fonctions conjuguées*, Math. Zeitschrift, 27(1928), 218-244.

14. J.R. Ringrose, *On well-bounded operators*, J. Australian Math. Soc., 1(1960), 334-343.

15. J.R. Ringrose, *On well-bounded operators II*, Proc. London Math. Soc. (3), 13(1963), 613-638.

16. D.R. Smart, *Conditionally convergent spectral expansions*, J. Australian Math. Soc., 1(1960), 319-333.

17. E.M. Stein and G. Weiss, *Introduction to Fourier analysis on Euclidean spaces*, Princeton Math. Series, No. 32, Princeton Univ. Press, Princeton, N.J., 1971.

DEPARTMENT OF MATHEMATICS
UNIVERSITY OF ILLINOIS
1409 W. GREEN ST.
URBANA, ILLINOIS 61801
U.S.A.

DEPARTMENT OF MATHEMATICS
UNIVERSITY OF EDINBURGH
JAMES CLERK MAXWELL BUILDING
THE KING'S BUILDINGS
EDINBURGH EH9 3JZ
SCOTLAND

VECTOR-VALUED HARDY SPACES FROM OPERATOR THEORY

Oscar Blasco[1)]

ABSTRACT. The classical Hardy space in the disc for functions taking values in a Banach space, $H^p{}_B(D)$, $1 < p < \infty$, is identified with special classes of operators from $L^{p'}$ into B.

1.- INTRODUCTION AND PREVIOUS RESULTS.

It is a very well known result in Harmonic Analysis that the boundary values of holomorphic functions in Hardy spaces $H^p(D)$, $1 < p < \infty$, are given by the functions in $L^p(\mathbf{T})$ whose negative Fourier coefficients vanish (see [9],[8]).

We shall try to find the corresponding result when the functions are allowed to take values in a general Banach space. To do that we shall use not only tools from Harmonic Analysis but also we shall need some special classes of operators.

The theorems we are going to prove can be deduced from those proved in [1] where the author solved the analoguous problem for harmonic functions. A different approach, by using vector-valued measures, can also be found in [2]. Here we shall present a proof which does not assume the scalar-valued result (see [1]) and which does not use the argument of w*-compactness of the unit ball in a dual space (see [2]).

Throughout this paper $(B, \|\ \|)$ denotes a complex Banach space and p, p' are real numbers with $1 < p < \infty$ and $1/p + 1/p' = 1$. D stands for the unit disc $\{ |z| < 1 \}$ and $(\mathbf{T}, \mathcal{B}(\mathbf{T}), m)$ is the Lebesgue measure space with $m(\mathbf{T}) = 1$.

We are concerned with the following spaces:

$$H^\infty{}_B(D) = \{ F : D \to B \quad \text{bounded and holomorphic} \}$$
$$|F|_\infty = \sup \{ \|F(z)\| : z \in D \} = \sup \{ \|F_r\|_\infty : 0<r<1 \}$$

where $F_r(t) = F(re^{it})$.

[1)] 1980 Mathematical Subject Classification (1985 Revision): 46E40, 47B38
Partially supported by C.A.I.T.Y.C. n. PB85-0338

For $1 < p < \infty$ the natural extension is

$$H^p_B(D) = \{ F: D \to B \text{ holomorphic} : |F|_p = \sup\{ \|F_r\|_p : 0<r<1\} < \infty \}$$

where

$$\|F_r\|_p = \left((2\pi)^{-1} \int_0^{2\pi} \|F_r(t)\|^p \, dt \right)^{1/p} .$$

Let me first state several properties which still hold in the vector-valued setting for any Banach space.

PROPOSITION 1.- *Let $1 < p < \infty$ and F be a function in $H^p_B(D)$. Then $g(t) = \lim \|F_r(t)\|$ exists a.e. and belongs to $L^p(\mathbf{T})$. Furthermore*

(1.1) $|F|_p = \|g\|_p$

Proof.- It is sufficient to realize that $h(z) = \|F(z)\|$ is subharmonic and that $\sup \; (2\pi)^{-1} \int_0^{2\pi} h_r(t)\, dt \le |F|_p$ for any value of p, $1 \le p \le \infty$, and then apply Littlewood's theorem (see [10]) . #

In the vector valued setting the classical proof (see [9]) also works out and we can state the following result

PROPOSITION 2.- *Let $1 < p < \infty$. If f belongs to $L^p_B(\mathbf{T})$ and $\hat{f}(n) = 0$ for $n < 0$, then $F(re^{i\theta}) = (2\pi)^{-1} \int_0^{2\pi} P_r(\theta - t)\, f(t)\, dt$ belongs to $H^p_B(D)$ and*

(1.2) $\|f\|_p = |F|_p$.

The first difference, when we are dealing with B-valued functions, appears looking at the following example.

REMARK - There exists a Banach space B and a function F in $H^p_B(D)$ which is not the Poisson integral of any function in $L^p(\mathbf{T})$.

Take $B = c_0(\mathbf{Z})$ and $F(z) = (z^n)_{n\in\mathbf{Z}}$.

It is quite obvious that F is holomorphic since

$F(z) = \sum_{n=0}^{\infty} e_n z^n$ where e_n is the canonic basis in c_0. F is clearly bounded, but if $F_r = P_r * f$ for some f in $L^1_B(\mathbf{T})$ then $F_r(\theta)$ would converge to $f(\theta)$ a.e. and in our case $F_r(\theta)$ goes to $(e^{in\theta})_{n\in\mathbf{Z}}$ as r goes to 1, which does not belong to c_0.

From these two facts we have that the Poisson integral embeds the space $\{ f \in L^p_B(\mathbf{T}) : \hat{f}(n) = 0 \text{ for } n < 0 \}$ into $H^p_B(D)$ but it is not surjective unless the Banach space has some property. In [3] Bukhvalov and Danilevich considered the following definition.

DEFINITION 1.- [3]. A complex Banach space B is said to have the *analytic Radon-Nikodym property* (ARNP) if every B-valued bounded holomorphic function from the disc has boundary limits almost everywhere, in other words if the Poisson integral is an isometry between $\{ f \in L^\infty_B(\mathbf{T}) : \hat{f}(n) = 0 \text{ for } n < 0 \}$ and $H^\infty_B(D)$.

There they showed that is equivalent to consider the same property for any value of p, $1 < p < \infty$ and also that RNP implies ARNP, proving that $L^1(\mathbf{T})$ which fails to have RNP does have this other property.

Our objective here will be to look for a substitute for $\{ f \in L^p_B(\mathbf{T}) : \hat{f}(n) = 0 \text{ for } n < 0 \}$ which is valid independently of the Banach space considered, and here Operator Theory plays a role.

2.- SOME SPACES OF OPERATORS.

One of the key points of this papers consists of looking at the functions in L^p_B as operators from $L^{p'}$ into B in the following natural way:

Let f belong to $L^p_B(\mathbf{T})$ for some p, $1 < p \le \infty$, and define

$$(2.1) \quad T_f(\phi) = (2\pi)^{-1} \int_0^{2\pi} f(t)\, \phi(t)\, dt \quad \text{for all } \phi \text{ in } L^{p'}.$$

For $p = 1$ we do the same but with $C(\mathbf{T})$, continuous functions, instead of $L^p(\mathbf{T})$.

Clearly from Holder's inequality, $\|T_f\| \le \|f\|_p$. Let us mention that in the case $p = \infty$ we have equality

$$\|T_f\| = \|f\|_\infty \text{ for } f \text{ in } L^\infty_B(\mathbf{T}).$$

Let me now exhibit the relationship existing between operators and holomorphic functions.

Suppose we have a linear map T from the space of trigonometric polynomials into B. Let us denote by $\mathbf{P}$ the space of polynomials $\{ \sum_{-N}^{M} e^{ik\theta}, N,M \text{ in } \mathbf{N} \}$. Then given T from $\mathbf{P}$ into B it makes sense to define the Fourier coefficient of T as follows

$$(2.2) \quad T(n) = T(e^{-int})$$

Hence anytime we have an operator T from $\mathbf{P}$ into B with $\|T(n)\| \leq C$ for all n we can define the following holomorphic function

$$(2.3) \quad F(z) = \sum_{n=0}^{\infty} T(n)\, z^n$$

On the other hand once we are given a B-valued holomorphic function $F(z) = \sum_{n=0}^{\infty} a_n z^n$ we can define an linear map T from $\mathbf{P}$ into B:

$$(2.4) \quad T\Big(\sum_{-N}^{M} \lambda_k e^{ik\theta} \Big) = \sum_{k=0}^{n} \lambda_{-k} a_k$$

In order to get functions F in $H^p_B(D)$ from (2.3) we have to require T to belong to special class of operators.

DEFINITION 2 [11] . Let X, Y be Banach spaces and T a bounded operator from X into Y. T is called *absolutely summing*(a.s) if T maps unconditionally convergent series in X into absolutely convergent series in Y. In other words, if there exists a constant $C > 0$ such that for every $x_1, x_2, \ldots, x_n$ in X we have

$$(2.5) \quad \sum_{i=1}^{n} \| Tx_i \|_Y \leq C \sup \Big\{ \sum_{i=1}^{n} |\langle \xi, x_i \rangle| : \|\xi\|_{X^*} = 1 \Big\}$$

We shall denote by $\Pi^1(X,Y)$ the space of a.s. operators and the norm in it is given by the infimum of the constants satistying (2.5).

The only case we shall consider here is $X = C(\mathbf{T})$ and $Y = B$. For this special case the Pietsch factorization Theorem [10] says:

An operator T from $C(\mathbf{T})$ into B is absolutely summing if and only if there exist a constant $C > 0$ and a probalility measure μ on $\mathbf{T}$ such that

$$(2.6) \quad \| T(\varphi) \| < C \int |\varphi(t)| \, d\mu(t) \quad \text{for all } \varphi \text{ in } C(\mathbf{T}).$$

Furthermore $\| T \|_{\Pi^1} = \inf \{ C \text{ verifying } (2.6) \}$

DEFINITION 3.- ([6],[4]). Let $1 < p < \infty$. Given an operator T from $L^{p'}$ into B we define

$$(2.7) \quad \| T \|_p = \sup \{ \sum_{i=1}^{n} | \lambda_i | \, \| T(\chi_{E_i}) \| : \| \sum_{i=1}^{n} \lambda_i \chi_{E_i} \|_{p'} \leq 1 \}$$

We shall denote by $\Lambda(L^{p'},B)$ the space of operators with $\| T \|_p < \infty$. This space is a Banach space under the norm given by $\| \ \|_p$ and it was studied by Dinculeanu who connected it with vector valued measures of bounded p-variation ([6], [7])

Let us establish here two properties of this space we shall use later on (see [7], [1]).

PROPOSITION 3.- *Let $1 < p < \infty$.*
$L^p_B(\mathbf{T})$ is isometrically embedded in $\Lambda(L^{p'},B)$.

PROPOSITION 4.- *Let $1 < p < \infty$ and let T be an operator from $L^{p'}$ into B. T belongs to $\Lambda(L^{p'},B)$ f and only if there exists a positive function g in $L^p(\mathbf{T})$ such that*

$$(2.8) \quad \| T(\varphi) \| < \int | \varphi(t) | \, g(t) \, dt \quad \text{for all } \varphi \text{ in } L^{p'}.$$

Moreover g can be chosen with $\| g \|_p = \| T \|_p$.

3.- THE MAIN THEOREM

Now we shall use the general procedure mentioned in formulas (2.3) and (2.4) which relates operators and holomorphic functions to get the extension of the classical result about boundary values in the vector-valued setting. The case $p = \infty$ is essentially contained in [5] and it can be rephrased as follows

THEOREM 1. *Denoting by* $\mathcal{L}(L^1,B)$ *the space of all bounded operators*

$$H^\infty{}_B(D) = \{ T \in \mathcal{L}(L^1,B) : T(n) = 0 \text{ for } n < 0 \}.$$

THEOREM 2.

(a) $H^1{}_B(D) = \{ T \in \Pi^1(C(\mathbf{T}),B) : T(n) = 0 \text{ for } n < 0 \}$

(b) $H^p{}_B(D) = \{ T \in \Lambda(L^{p'},B) : T(n) = 0 \text{ for } n < 0 \} \; (1 < p < \infty)$

Proof. Both proofs can be done simultaneously. Take a function F in $H^p{}_B(D)$ for some p, $1 < p < \infty$ Assume F is given by

$F(z) = \sum_{n=0}^{\infty} a_n z^n$ with a_n in B. By using (2.4) we define a linear map T on $\mathbf{P}$ given by

$$T(\phi) = \sum_{n=0}^{v} \lambda_{-k} a_k \quad \text{for} \quad \phi(t) = \sum_{n=-\mu}^{v} \lambda_{-k} e^{ikt}.$$

Since $a_k = \lim \int e^{-ikt} F(re^{it})\, dt$ then we can express

$$(3.1) \quad T(\phi) = \lim \int \phi(t)\, F_r(t)\, dt$$

for all trigonometric polynomial ϕ.

Now use Proposition 1 to get a function g in L^p, $1 < p < \infty$, given by $g(\theta) = \lim \| F_r(\theta) \|$, and then we have

$$\| T(\phi) \| \le \lim \int | \phi(t) | \, \| F_r(t) \| \, dt = \int | \phi(t) | \, \| g(t) \| \, dt .$$

This allows us to extend our operator to $\mathfrak{L}(L^{p'},B)$ or $\mathfrak{L}(C(\mathbf{T}),B)$. And now we just have to use either Proposition 4 (for the case $1 < p < \infty$) or Pietsch's factorization theorem (2.6) with $d\mu(t) = g(t)\,dt/\|g\|_1$ (case $p = 1$) to verify that T belongs to $\Lambda(L^{p'},B)$ and $\Pi^1(C(\mathbf{T}),B)$ respectively. And we also get that $\|T\|_p \le |F|_p$ for $1 < p < \infty$ or $\|T\|_{\Pi^1} \le |F|_1$.

Conversely, let us assume T belong to $\Pi^1(C(\mathbf{T}),B)$ (resp. to $\Lambda(L^{p'},B)$) and $T(n) = 0$ for $n < 0$. Obviously then $\|T(n)\| \le \|T\|$ for all n, what allows us to use (2.3) and consider the following holomorphic function

$$F(z) = \sum_{n=0}^{\infty} T(n)\, z^n$$

The main point here is that the Poisson kernel has the following expansion convergent in $C(\mathbf{T})$ (resp. in $L^{p'}$):

$$P_z(t) = P_r(\theta-t) = \sum_{n=-\infty}^{\infty} r^{|k|}\, e^{ik(\theta-t)} \qquad \text{for} \quad z = re^{i\theta}$$

From this and the continuity of the operator we can see that $F(z) = T(P_z)$. Now from (2.6) (resp. (2.8)) we have a probability measure μ (resp. a function g in L^p) such that

$$\|F(re^{i\theta})\| = \|T(P_r(\theta-.)\| \le C.\,(P_r * \mu(\theta))$$

$$\text{(resp. } \|F(re^{i\theta})\| = \|T(P_r(\theta-.)\| \le P_r * g(\theta) \text{)}$$

Finally using Fubbini's Theorem (resp. Minkowski's inequality) we get

$$\|F_r\|_1 \le \|T\|_{\Pi^1} |\mu|_1 \|P_r\|_1 = \|T\|_{\Pi^1}$$

(resp. $\| F_r \|_p < \| g \|_p \| P_r \|_1 = \| g \|_p = \| T \|_p$

what completes the proof. #

REMARK.- The reader could think that the cases $p=1$, $1 < p < \infty$ and $p=\infty$ are of different character. Actually all them can be expressed in terms of some special class of operators which unifies Theorem 1 and parts (a) and (b) of Theorem 2 (see [1]).

To finish let me reformulate the result of Bukhvalov and Danilevich [3] in terms of operators.

COROLLARY . *The following statements are equivalent*
(a) B has the analytic Radon-Nikodym property
(b) Every absolutely summing opeartor T from $C(\mathbf{T})$ into B with $T(n) = 0$ for $n < 0$ is representable.
(c) For any p , $1 < p < \infty$ every T in $A(L^{p'},B)$ with $T(n) = 0$ for $n < 0$ is representable.
(d) Every operator from L^1 into B with $T(n) = 0$ for $n < 0$ is representable.

AKNOWLEDGEMENT: I am very grateful to P. Dowling for his valuable comments. I wish to thank also the Department of Mathematics of the University of Illinois at Urbana, where I was visiting during the special year in Modern Analysis (1986-87), for providing me with such an excellent atmosphere for my research.

REFERENCES.

[1] BLASCO, O., "Boundary values of vector-valued harmonic functions considered as operators". Studia Math.86 (1987), 19-33.

[2] BLASCO, O., "Boundary values of functions in vector-valued Hardy spaces and geometry on Banach spaces". Journal of Funct. Analysis (to appear)

[3] BUKHVALOV, A.V. and DANILEVICH, A.A. "Boundary properties of analytic and harmonic functions with values in Banach space" Mat. Zametki 31 (1982), 302-214. English Translation: Mat. Notes 31 (1982) 104-110.

[4] DIESTEL ,J. and UHL, J.J. , Vector Measures. Amer. Math. Soc. MathematicalSurveys 15 (1977).

[5] DOWLING, P.M., "Representable operators and the Analytic Radon-Nikodym property". Proc. Royal Irish Math. Soc. 85,(1985),143-150.

[6] DINCULEANU, N. , "Linear operators on L^p-spaces. Vector and operator valued measures and applications". Proc. Sympos. Snowbird Resorl, Alta. Utah. Academic Press, New York (1972) 109-124.

[7] DINCULEANU, N., Vector Measures. Pergamon Press, New York 1967.

[8] DUREN, P. L. , Theory of H^p spaces. Academic Press, New York , 1970.

[9] KATZNELSON, Y. , An Introduction to Harmonic Analysis, John Wiley and Sons, Inc 1968.

[10] LITTLEWOOD, J. E. "On functions subharmonic in the circle, II" Proc. London Math. Soc.. (2) Vol. 28 (1968), 383-394.

[11] PIETSCH, A., Nuclear locally convex spaces. Springer-Verlag, Berlin 1972.

DEPARTAMENTO DE MATEMATICAS
UNIVERSIDAD DE ZARAGOZA
ZARAGOZA,50009. SPAIN

Current address:

Department of Mathematics
University of Illinois
Urbana, IL ,61801

RESTRICTED INVERTIBILITY OF MATRICES AND APPLICATIONS

J. Bourgain and L. Tzafriri

0. INTRODUCTION.

The present paper stems from a series of lectures delivered by both authors at the University of Illinois under the auspices of the Special Year in Analysis 1986/87.

The topic considered is that of restricted invertibility of linear operators acting on finite dimensional euclidean spaces, ℓ_p^n-spaces; $1 < p < \infty$, or on other classes of Banach spaces with a nice basis. The typical situation investigated in the paper is that of a $n \times n$ matrix T acting on an ℓ_p^n-space; $1 < p < \infty$, which is not trivial in the sense that it has either a "large" diagonal or "large" rows. In order to simplify the notation we usually assume that T has 1's on the diagonal or that it maps the unit vectors of ℓ_p^n into norm one vectors. The purpose is to prove the existence of a submatrix whose rank is proportional to that of the original matrix and which is well invertible in the sense that the norm of the inverse depends only on the norm of the given matrix (and not on the dimension of the underlying space). The restricted invertibility of a matrix with 1's on the diagonal is achieved in general by selecting a "large" submatrix whose off-diagonal part has norm < 1. However, there are cases, in spaces other than the ℓ_p^n-spaces, when the restricted invertibility and the supression of the off-diagonal part have considerably different behaviors.

A key factor in the proofs is the simple but remarkable fact that an operator bounded on an ℓ_p^n-space (or on a space with a 1-symmetric basis) is also bounded on ℓ_2^n when restricted to a suitable subspace of dimension $> n/2$.

The results obtained in the euclidean case, as well as in the ℓ_p^n-case; $1 < p < \infty$, are quite complete except for some numerical estimates which could be perhaps improved. For matrices with 1's on the diagonal, which act on ℓ_p^n-spaces; $1 < p \neq 2$, one can prove even an unrestricted invertibility theorem provided their norm is close to 1. On the other hand, the invertibility results obtained for operators acting on more general finite dimensional spaces (e.g. with a 1-symmetric or 1-unconditional basis) seem not to be the best possible and more research is needed in order to clarify the situation.

The interest in the results of the type described above lies, of course, in their applications. The theorems proved in the euclidean case yield a nice application to harmonic analysis and are also connected to a problem of Kadison and Singer [19] on C^*-algebras. In addition, they allow to improve some estimates obtained by B. S. Kashin [20] on a local variant of Kolmogorov's problem whether, for every orthonormal system $\{u_n\}_{n=1}^\infty$ in $L_2(0,1)$, there exists a permutation π of the integers which makes the system into a convergence system (i.e., for any choice of coefficients $\{a_n\}_{n=1}^\infty$ with $\sum_{n=1}^\infty |a_n|^2 < \infty$, the series $\sum_{n=1}^\infty a_n u_{\pi(n)}$ converges a.e. on [0,1]). The invertibility theorems proved in the ℓ_p^n-case have even more interesting applications. For instance, they yield positive solutions to two questions raised in [18] on the structure of complemented subspaces of an L_p-space which have extremal euclidean distance. Furthermore, the invertibility theorem for matrices with 1's on the diagonal whose norm is close to 1 yields a new simple proof of a very nice result of G. Schechtman [33] on the complementation of subspaces of L_p ; $p \neq 2$, that are almost isometric to ℓ_p^n.

Most of the proofs combine a probabilistic selection with a combinatorial argument after which a factorization result is used in order to conclude the proof.

Many of the theorems presented below have already appeared in print in [7] so, in some sense, the present article is a survey. For sake of completeness, we chose to repeat here also those results which are already published there. However, their proofs are not given in the present paper unless a better proof is available now. This is the case with many results for which sharper estimates or different proofs have meanwhile been found. The article also contains, of course, many new results.

In order to make the paper easier to read, we present in the first sections only the results and their applications together with proofs of a simple nature which, in our opinion, do not disturb the continuous reading. The more elaborated proofs are presented in the last section of the paper.

The results presented in the paper apply to both real and complex spaces. The proofs are usually the same with few exceptions.

1. OPERATORS ON ℓ_1^n AND ℓ_∞^n-SPACES.

We begin with this case which, besides being easy, presents a complete picture and can be used as a model for further research on operators acting on other finite dimensional spaces. What makes the matrices acting on ℓ_1^n and ℓ_∞^n-spaces to be the easiest to handle is the fact that the problems considered are always of a positive nature (since the norm of a matrix operating on such a space remains the same if we replace its entries by their absolute values).

The invertibility results discussed in this section do not represent new contributions but are, in fact, reformulations of some theorems of a combinatorial nature proved in other contexts.

The main result on the restricted invertibility of operators on the spaces ℓ_1^n and ℓ_∞^n is a consequence of the following supression theorem.

THEOREM 1.1. *Let S be a linear operator on the space ℓ_1^n whose matrix relative to the unit vector basis has 0's on the diagonal. Then, for each $\varepsilon > 0$, there exists a subset σ of $\{1,2,...,n\}$ of cardinality*

$$|\sigma| > \varepsilon n/2\|S\|_1$$

such that

$$\|R_\sigma \, S \, R_\sigma\|_1 < \varepsilon,$$

where R_σ denotes the restriction operator defined by $R_\sigma e_i = e_i$, for $i \in \sigma$, and $R_\sigma e_i = 0$, otherwise. The theorem remains valid if ℓ_1^n is replaced by its dual ℓ_∞^n.

COROLLARY 1.2. *Let T be a linear operator on the space ℓ_1^n whose matrix relative to the unit vector basis has 1's on the diagonal. Then, for each $0 < \varepsilon < 1$, one can select a subset σ of $\{1,2,...,n\}$ of cardinality*

$$|\sigma| > \varepsilon n/2\|T\|_1,$$

such that the operator $R_\sigma \, T \, R_\sigma$ is invertible (on $\ell_1^{|\sigma|}$) and

$$\|(R_\sigma \, T \, R_\sigma)^{-1}\|_1 < (1-\varepsilon)^{-1}.$$

Moreover, the result remains valid when ℓ_1^n is replaced by ℓ_∞^n.

Theorem 1.1. is obviously a reformulation of the following combinatorial fact.

THEOREM 1.1'. *Let* $(a_{i,j})_{i,j=1}^n$ *be a* $n \times n$ *matrix satisfying the conditions:*

(i) $a_{i,j} \geq 0$, *for all* $i \leq i,j \leq n$, *and* $a_{i,i} = 0$; $1 \leq i \leq n$,

(ii) $\sum_{j=1}^n a_{i,j} \leq 1$, *for all* $1 \leq i \leq n$.

Then, for each $\varepsilon > 0$, *there exists a subset* σ *of the integer* $\{1,2,\dots,n\}$ *so that* $|\sigma| \geq \varepsilon n/2$ *and*

$$\sum_{j \in \sigma} a_{i,j} \leq \varepsilon, \quad i \in \sigma.$$

This result, with slightly worse estimates for the cardinality of σ, was originally proved by G. Schechtman (see [18]). One can easily deduce from Theorem 1.1 that, for any $\varepsilon > 0$ and any $n \times n$ matrix S with 0's on the diagonal and $\|S\|_1 \leq 1$, there exists a partition $\{\sigma_h\}_{h=1}^k$ of the integers $\{1,2,\dots,n\}$ such that $\|R_{\sigma_h} S R_{\sigma_h}\|_1 < \varepsilon$; $1 \leq h \leq k$, where k is of order of magnitude $\log n$. It turns out that this splitting result can be considerably improved as to allow the choice of a partition $\{\sigma_h\}_{h=1}^k$ with k independent of n.

THEOREM 1.3. *Let* S *be a linear operator on the space* ℓ_1^n *whose matrix relative to the unit vector basis has* 0's *on the diagonal. Then, for each integer* k, *there exists a partition* $\{\sigma_h\}_{h=1}^k$ *of* $\{1,2,\dots,n\}$ *into* k *mutually disjoint subsets such that*

$$\Big\| \sum_{h=1}^k R_{\sigma_h} S R_{\sigma_h} \Big\|_1 \leq \frac{2}{k} \|S\|_1.$$

The result remains valid in ℓ_∞^n *as well.*

COROLLARY 1.4. *Let* T *to be a linear operator on the space* ℓ_1^n *whose matrix relative to the unit vector basis has* 1's *on the diagonal. Then, for each* $0 < \varepsilon < 1$, *one can find a partition* $\{\sigma_h\}_{h=1}^k$ *of* $\{1,2,\dots,n\}$ *into mutually disjoint subsets with* $k = [2\|T\|_1/\varepsilon] + 1$ *such that the operators* $\{R_{\sigma_h} T R_{\sigma_h}\}_{h=1}^k$ *are invertible and*

$$\|(R_{\sigma_h} T R_{\sigma_h})^{-1}\|_1 < (1-\varepsilon)^{-1},$$

for all $1 \leq h \leq k$. *Moreover, the result is valid when* ℓ_1^n *is replaced by* ℓ_∞^n.

Theorem 1.3 can be also reformulated in a non-operator language.

THEOREM 1.3'. *Let* $(a_{i,j})_{i,j=1}^n$ *be a matrix satisfying the conditions*

(i) $a_{i,j} \geq 0$, *for all* $1 \leq i, j \leq n$, *and* $a_{i,i} = 0$; $1 \leq i \leq n$,

(ii) $\sum_{j=1}^n a_{i,j} \leq 1$, *for all* $1 \leq i \leq n$.

<u>Then, for each integer</u> k, <u>there exists a partition</u> $\{\sigma_h\}_{h=1}^k$ <u>of</u> $\{1,2,\ldots,n\}$ <u>into</u> k <u>mutually disjoint subsets such that</u>

$$\sum_{j\in\sigma_h} a_{i,j} < \frac{2}{k}, \quad i \in \sigma_h,$$

<u>for all</u> $1 \leq h \leq k$.

Theorem 1.3' was originally proved in [5] (see also [4]). However, the estimate given there for $\sum_{j\in\sigma_h} a_{i,j}$ is less sharp. We present here an extremal argument due to K. Ball [2] which proves Theorem 1.3' exactly as stated above. This argument is closely related to the proof of a similar statement for matrices with positive entries acting on Hilbert spaces, a topic which will be discussed in the next section (cf. [3]).

Proof. We first notice that there is no loss of generality in assuming that

1° $\quad a_{i,j} > 0$, for all $1 \leq i \neq j \leq n$,

2° $\quad \sum_{j=1}^n a_{i,j} = 1$, for all $1 \leq i \leq n$.

These conditions make the matrix $A = (a_{i,j})_{i,j=1}^n$ into a transition matrix, by right multiplication, of a discrete time Markov chain and its steady state vector (i.e. left eigenvector corresponding to the eigenvalue 1) $\gamma = (\gamma_1,\gamma_2,\ldots,\gamma_n)$ satisfies $\gamma_i > 0$, for all $1 > i \leq n$. This simple fact can also be deduced directly. Indeed, by condition 2°, the number 1 is and eigenvalue of A corresponding to the right eigenvector $(1,1,\ldots,1)$. Therefore, there exists also a left eigenvector $\gamma = (\gamma_1,\gamma_2,\ldots,\gamma_n)$ with $\sum_{i=1}^n |\gamma_i| \neq 0$. Since

$$\sum_{i=1}^n \gamma_i a_{i,j} = \gamma_j; \quad 1 \leq j \leq n,$$

and

$$\sum_{j=1}^n |\gamma_j| \leq \sum_{j=1}^n \sum_{i=1}^n |\gamma_i| a_{i,j} = \sum_{i=1}^n |\gamma_i|$$

it follows that $\{\gamma_i\}_{i=1}^n$ must all have the same sign so we may as well assume that $\gamma_i > 0$, for all $1 \leq i \leq n$, and $\sum_{i=1}^n \gamma_i = 1$.

Now, fix an integer k and, for any partition $\Delta = \{\delta_h\}_{h=1}^k$ of the integers $\{1,2,\ldots,n\}$ into exactly k mutually disjoint subsets, define

$$f(\Delta) = \sum_{h=1}^{k} \sum_{i,j\in\delta_h} \gamma_i a_{i,j}$$

and choose a partition $\Sigma = \{\sigma_h\}_{h=1}^{k}$ for which f attains its minimum.

We claim that the partition Σ satisfies the assertion of Theorem 1.3'. If not then at least for one of the subsets $\{\sigma_h\}_{h=1}^{k}$, say σ_1, there exists an integer $\ell \in \sigma_1$, so that

$$\theta = \sum_{j\in\sigma_1} a_{\ell,j} > 2/k.$$

For each $2 \leq s \leq k$, let $\Sigma^s = \{\sigma_h^s\}_{h=1}^{k}$ be the partition of $\{1,2,\ldots,n\}$ obtained by putting:

$$\sigma_1^s = \sigma_1 \sim \{\ell\},$$

$$\sigma_s^s = \sigma_s \cup \{\ell\},$$

$$\sigma_h^s = \sigma_h \quad \text{if} \quad h \neq 1 \quad \text{and} \quad h \neq s,$$

and let $2 \leq t \leq k$ be chosen so that the expression

$$f(\Sigma) - f(\Sigma^t) = \gamma_\ell \sum_{j\in\sigma_1} a_{\ell,j} + \sum_{i\in\sigma_1} \gamma_i a_{i,\ell} - \gamma_\ell \sum_{j\in\sigma_t} a_{\ell,j} - \sum_{i\in\sigma_t} \gamma_i a_{i,\ell}$$

is maximal (i.e. $f(\Sigma) - f(\Sigma^t) \geq f(\Sigma) - f(\Sigma^s)$, for all $2 \leq s \leq k$). Then, by averaging, we obtain that

$$f(\Sigma) - f(\Sigma^t) \geq f(\Sigma) - \frac{1}{k-1} \sum_{s=2}^{k} f(\Sigma^s)$$

$$= \gamma_\ell \sum_{j\in\sigma_1} a_{\ell,j} + \sum_{i\in\sigma_1} \gamma_i a_{i,\ell} - \frac{1}{k-1} [\gamma_\ell \sum_{j\notin\sigma_1} a_{\ell,j} + \sum_{i\notin\sigma_1} \gamma_i a_{i,\ell}]$$

$$\geq \gamma_\ell \theta - \frac{1}{k-1} [\gamma_\ell(1-\theta) + \gamma_\ell] = \gamma_\ell(\theta k-2)/(k-1) > 0.$$

This contradicts the minimality of $f(\Sigma)$, thus completing the proof. □

A linear operator T acting on ℓ_1^n with 1's on the diagonal clearly satisfies $\|T\|_1 \geq 1$. In the case when the norm of T does not exceed the number 1 by a large amount one can prove a (trivial) unrestricted invertibility result.

PROPOSITION 1.5. *Let S be a linear operator on the space ℓ_1^n whose matrix $(a_{i,j})_{i,j=1}^n$ relative to the unit vector basis has 0's on the diagonal and let $T = I + S$. If, for some $0 < \varepsilon < 1$, $\|T\|_1 < 1 + \varepsilon$ then $\|S\|_1 < \varepsilon$ and, thus, T is invertible and*

$$\|T^{-1}\|_1 \le (1-\varepsilon)^{-1}$$

The result is valid in ℓ_∞^n as well.

Proof. Let $\{e_i\}_{i=1}^n$ denote the unit vector basis of the space ℓ_1^n. Then

$$\|S\|_1 = \max_{1\le i\le n} \sum_{\substack{j=1\\ j\ne i}}^n |a_{i,j}| = \max_{1\le i\le n} (\|Te_i\|_1 - 1) \le \varepsilon$$

which completes the argument. []

We conclude this section by discussing some applications of Corollaries 1.2 and 1.4 to the geometry of Banach spaces.

A famous result of F. John (see e.g. [13]) asserts that the euclidean distance

$$d_X = d(X, \ell_2^n)$$

of any n-dimensional Banach space X is $\le n^{1/2}$. The spaces ℓ_1^n and ℓ_∞^n are examples of spaces whose euclidean distance is maximal. In connection with this fact, it was asked whether ℓ_1^n is the only n-dimensional subspace of L_1 whose euclidean distance is maximal. T. Figiel and W. B. Johnson [11] answered this question in positive. In the corresponding isomorphic case, one cannot really expect to prove that a n-dimensional subspace X of L_1, whose euclidean distance satisfies

$$d_X \ge cn^{1/2},$$

for some $c > 0$ independent of n, is well isomorphic to ℓ_1^n. However, by using Corollary 1.2, W. B. Johnson and G. Schechtman [18] prove that such an X contains a subspace Y of dimension k proportional to n which is well isomorphic to ℓ_1^k and well complemented in L_1. Another fact, which follows from a result of L. Dor [10] and Corollary 1.2, asserts that if $\{f_i\}_{i=1}^n$ is a sequence of functions in L_1 such that

$$c \sum_{c=1}^n |a_i| \le \|\sum_{i=1}^n a_i f_i\|_1 \le \sum_{i=1}^n |a_i|,$$

for all $\{a_i\}_{i=1}^n$ and some $c > 0$, then there exists a subset σ of $\{1,2,\ldots,n\}$ with $|\sigma|$ proportional to n so that $[f_i]_{i\in\sigma}$ is well complemented in L_1. By using Corollary 4, a much stronger assertion was proved in [4]. Namely, it was shown there that the integers $\{1,2,\ldots,n\}$ can be partitioned into a union of mutually disjoint subsets $\{\sigma_h\}_{h=1}^k$, with k depending only on c, such that $[f_i]_{i\in\sigma_h}$; $1 \le h \le \ell$, are all well complemented. Let us add that $[f_i]_{i=1}^n$ itself need not be well complemented (cf. [6]). However, for c close to 1, L. Dor proved in [10] that $[f_i]_{i=1}^n$ is the range of a projection of norm $M = M(c)$ with $M(c) \to 1$, as $c \to 1$. It is the attempt to extend these theorems to the case $p > 1$ that motivated on some extent the present investigation.

2. OPERATORS ON EUCLIDEAN SPACES.

The case of matrices acting on the euclidean space is certainly the most important. The invertibility theorems discussed in this section, besides being interesting in themselves and having applications, are also used in order to prove similar invertibility results for matrices acting on ℓ_p^n-spaces; $1 < p \neq 2$, or on more general finite dimensional spaces with a symmetric basis.

We present first a restricted invertibility theorem for matrices with "large" rows. We maintain the notation $\{e_i\}_{i=1}^n$, for the unit vector basis, in this case, of ℓ_2^n and R_σ, for the natural projection from ℓ_2^n onto $[e_i]_{i\in\sigma}$.

THEOREM 2.1. There exists a constant $c > 0$ such that, whenever T is a linear operator on a real or complex ℓ_2^n for which $\|Te_i\|_2 = 1$; $1 \leq i \leq n$, one can find a subset σ of $\{1,2,\dots,n\}$ so that

$$|\sigma| \geq cn/\|T\|_2^2$$

and

$$\|\sum_{i\in\sigma} a_i\, Te_i\|_2 \geq c(\sum_{i\in\sigma} |a_i|^2)^{1/2},$$

for all $\{a_i\}_{i\in\sigma}$.

A simple model for the understanding of Theorem 2.1 is provided by the nilpotent operators $S_n: \ell_2^n \to \ell_2^n$; $n = 1,2,\dots$, defined by $S_n e_i = e_{i+1}$, for $1 \leq i < n$, and $S_n e_n = 0$. These operators are not invertible even in an algebraic sense. On the other hand, with $\eta_n = \{1,2,\dots,n-1\}$ we easily see that S_n restricted to $[e_i]_{i\in\eta_n}$ is already an isometry.

For matrices with "large" rows, one cannot find in general a square submatrix which is "well" invertible. This fact is already false for the nilpotent operator S_n considered above. However, this becomes possible when we deal with operators having a "large" diagonal. In this case, we can take as a model the operator $T_n = I_n + S_n$, which is algebraically invertible but whose inverse T_n^{-1} has norm $\geq \sqrt{n/2}$. If σ_n denotes the subset of $\{1,2,\dots,n\}$ consisting e.g. of all the odd numbers between 1 and n then $|\sigma_n| \geq (n-1)/2$ and $R_{\sigma_n} T_n R_{\sigma_n}$ is again an isometry (actually, the identity on $[e_i]_{i\in\sigma_n}$).

As in the case of operators acting on ℓ_1^n and ℓ_∞^n-spaces, the restricted invertibility is obtained as a consequence of a supression result.

THEOREM 2.2. For every $\varepsilon > 0$, there exist an integer $n(\varepsilon)$ and a positive number $\delta_2(\varepsilon)$ such that, whenever $n \geq n(\varepsilon)$ and S is a linear operator on ℓ_2^n whose matrix relative to the unit vector basis has 0's on the diagonal, then one can find a subset σ of $\{1,2,\ldots,n\}$ so that

$$|\sigma| \geq \delta_2(\varepsilon)n$$

and

$$\|R_\sigma S R_\sigma\|_2 \leq \varepsilon\|S\|_2.$$

COROLLARY 2.3. For every $M < \infty$ and $0 < \varepsilon < 1$, there is a constant $d = d(M,\varepsilon) > 0$ such that, whenever $n \geq 1/d$ and T is a linear operator on ℓ_2^n of norm $\|T\|_2 \leq M$ whose matrix relative to the unit vector basis has 1's on the diagonal, then there exists a subset σ of $\{1,2,\ldots,n\}$ of cardinality $|\sigma| \geq dn$ so that $R_\sigma T R_\sigma$ is invertible and

$$\|(R_\sigma T R_\sigma)^{-1}\|_2 < (1-\varepsilon)^{-1}.$$

Both Theorems 2.1 and 2.2 are proved in detail in [7] Section 1 and we do not reproduce their proofs here.

The proof of Theorem 2.1 consists of three parts: the first is a probabilistic selection of a subset σ_1 of $\{1,2,\ldots,n\}$ of cardinality $|\sigma_1| \geq c_1 n/\|T\|_2^2$, for some $c_1 > 0$, so that each vector Te_i; $i \in \sigma_1$, is "far" from $[Te_j]_{j\in\sigma_1\sim\{i\}}$. Then one uses a well-known combinatorial result of Sauer [32] (see also [34] and [36]) in order to select a subset σ_2 of σ_1 of cardinality $|\sigma_2| \geq |\sigma_1|/2$ so that

$$\|\sum_{i\in\sigma_2} a_i Te_i\|_2 \geq c_2 \sum_{i\in\sigma_2} |a_i|/\sqrt{|\sigma_2|},$$

for all $\{a_i\}_{i\in\sigma_2}$ and some constant $c_2 > 0$. This inequality puts in evidence the role of the space ℓ_1^n in the proof of Theorem 2.1 - a result which deals only with Hilbert spaces - and can be interpreted as meaning that the map $Te_i \to e_i$; $i \in \sigma_2$, considered as an operator from ℓ_2^n into $\ell_1^{|\sigma_2|}$, is bounded by $\sqrt{|\sigma_2|}/c_2$. The adjoint map takes therefore $\ell_\infty^{|\sigma_2|}$ into ℓ_2^n and the proof is concluded by using a factorization argument based on a theorem of Grothendieck (see e.g. [22] 2.b.7) asserting that any linear operator from L_∞ into L_2 is 2-absolutely summing and on a result of Pietsch [29].

Theorem 2.2 is proved in [7] in a different manner by using a probabilistic selection based on a decoupling principle and then, again, a factorization argument similar to the one above. The estimate given by the proof for the function δ_2 is in the form $\delta_2(\varepsilon) = e^{-C^2/\varepsilon^2}$, for some constant $C < \infty$. However, it follows easily from the definition of δ_2 that it satisfies the condition

$$\delta_2(\varepsilon\varepsilon') \geq \delta_2(\varepsilon)\delta_2(\varepsilon')$$

and, thus, that

$$\delta_2(\varepsilon) \geq \varepsilon^k,$$

for some k and all $0 < \varepsilon < 1/2$. This implies that the dependence of the constant $d(M,\varepsilon)$, appearing in the statement of Corollary 2.3, on ε is of polynomial type. A more elaborated argument presented in Section 8 of [7] yields that

$$\delta_2(\varepsilon) \geq c\varepsilon^8,$$

for some $c > 0$ and all $0 < \varepsilon < 1$. We have not checked the details but it turns out that with additional work one can show that, for any $\rho > 2$, there exists a constant c_ρ so that $\delta_2(\varepsilon) \geq c_\rho\varepsilon^\rho$; $0 < \varepsilon < 1$. As far as we know, it is still an open question if this inequality holds also with $\rho = 2$.

Results of a nature somewhat similar to that of Theorem 2.2 were obtained before by B. S. Kashin [20]. He proved e.g. the existence of a constant $B < \infty$ so that, for any $n \leq m$ and any operator $A: \ell_2^n \to \ell_2^m$, of norm ≤ 1, there exists a subset σ of $\{1,2,\ldots,m\}$ of cardinality equal to n for which

$$\|R_\sigma A\|_2 \leq B/\sqrt{\log(m/n)}.$$

Kashin's result does not yield Theorem 2.2 unless one applies in advance the decoupling principle already used in its proof. Kashin used the afore mentioned result in order to prove some estimates connected with a finite dimensional version of Kolmogorov's problem whether, for any orthogonal system in L_2, there exists a permutation of the integers which makes the system into a convergence one. The method of proof for Theorem 2.2 yields actually a simpler proof for a stronger version of Kashin's result. In order to state the theorem, we shall use the notation A^+ for the upper triangular projection of a $n \times n$ matrix A. Furthermore, if Λ stands for the symmetric group on the integers $\{1,2,\ldots,n\}$ endowed with the normalized invariant measure λ and if, for $\pi \in \Lambda$ and $A = (a_{i,j})_{i,j=1}^n$, A_π denotes the matrix $(a_{i,\pi(j)})_{i,j=1}^n$ then we have the following result.

THEOREM 2.4. <u>For every</u> $1 \le q < 2$, <u>there exists a constant</u> $C_q < \infty$ <u>such that</u>, <u>whenever</u> A <u>is a linear operator on</u> ℓ_2^n, <u>then</u>

$$\int_\Lambda \|(A_\pi)^+\|_{2\to q}\, d\lambda(\pi) \le C_q n^{1/q-1/2}\|A\|_2.$$

The proof is given in [7] Section 8 and it will not be reproduced here.

It is quite interesting to compare Theorem 2.1 with Theorem 2.2. Suppose that T is a linear operator satisfying the assumption of Theorem 2.1 i.e. $\|Te_i\|_2 = 1$, for all $1 \le i \le n$. Then the matrix associated to T^*T has 1's on the diagonal and, by Corollary 2.3, there exists a subset σ of $\{1,2,\ldots,n\}$ of cardinality proportional to n so that $(R_\sigma T^*TR_\sigma)^{-1}$ has norm bounded e.g. by 2. This implies that

$$2\|T^*\|_2 \|\sum_{i\in\sigma} a_i Te_i\|_2 \ge 2\|\sum_{i\in\sigma} a_i R_\sigma T^*Te_i\|_2 \ge (\sum_{i\in\sigma}|a_i|^2)^{1/2},$$

for all $\{a_i\}_{i\in\sigma}$, which means that Theorem 2.2 implies Theorem 2.1 though not with the sharp numerical estimates given in its statement.

It turns out that, conversely, also Theorem 2.1 implies Theorem 2.2. This fact, which is far less trivial than the observation made above, was proved by K. Ball [2]. We present now his argument.

Let S be a linear operator on ℓ_2^n and suppose that its matrix $(a_{i,j})_{i,j=1}^n$, relative to the unit vector basis of ℓ_2^n, has 0's on the diagonal i.e. $a_{i,i} = 0$, for all $1 \le i \le n$. Since

$$\sum_{j=1}^n |a_{i,j}|^2 = \|Se_i\|_2^2 \le \|S\|_2^2; \quad 1 \le i \le n,$$

it follows from Theorem 1.1' that, for any $\varepsilon > 0$, there exists a subset η of $\{1,2,\ldots,n\}$ such that

$$|\eta| \ge \varepsilon^2 n/8\|S\|_2^2$$

and

$$\|R_\eta Se_i\|_2^2 = \sum_{j\in\eta}|a_{i,j}|^2 \le \varepsilon^2/4; \quad i \in \eta.$$

Put

$$x_i = 2R_\eta Se_i/\varepsilon$$

and observe that

a) $\|x_i\|_2 \le 1; \quad i \in \eta,$

b) $\|\sum_{i\in\eta} c_i x_i\|_2 \le M(\sum_{i\in\eta}|c_i|^2)^{1/2},$

for any choice of $\{c_i\}_{i\in\eta}$, where $M = 2\|S\|_2/\varepsilon$.

The main part of the proof consists of an inductive argument which is meant to reduce M to 2 by passing from η to a smaller set. To this end, define a matrix $(b_{i,j})_{i,j\in\eta}$, by setting

$$b_{i,j} = \frac{M^2}{M^2-1}\delta_{i,j} - \frac{\langle x_i x_j\rangle}{M^2-1}\ ; \quad i,j \in \eta$$

and notice that

$$\sum_{i,j\in\eta} b_{i,j}c_i\bar{c}_j = \frac{M^2}{M^2-1}\sum_{i\in\eta}|c_i|^2 - \frac{\|\sum_{i\in\eta} c_i x_i\|_2^2}{M^2-1} \geq 0,$$

for all $\{c_i\}_{i\in\eta}$, i.e. $(b_{i,j})_{i,j\in\eta}$ is positive semidefinite. Hence, there exist vectors $\{z_i\}_{i\in\eta}$ is an arbitrary Hilbert space so that

$$\langle z_i, z_j\rangle = b_{i,j}; \quad i,j \in \eta,$$

and one can easily verify that

$$\|\sum_{i\in\eta} c_i z_i\|_2^2 = \Big(M^2\sum_{i\in\eta}|c_i|^2 - \|\sum_{i\in\eta} c_i x_i\|_2^2\Big)/(M^2-1),$$

for any choice of $\{c_i\}_{i\in\eta}$. This implies that

(i) $\|z_i\|_2 \geq 1;\quad i\in\eta,$

(ii) $\|\sum_{i\in\eta} c_i z_i\|_2 \leq 2\Big(\sum_{i\in\eta}|c_i|^2\Big)^{1/2},$

for all $\{c_i\}_{i\in\eta}$. Consider now the operator $T\colon \ell_2^{|\eta|} \to \ell_2^{|\eta|}$, defined by $Te_i = z_i/\|z_i\|_2;\ i\in\eta$, which, by (i) and (ii), has norm ≤ 2, and apply Theorem 2.1. It follows that there exists a subset η_1 of η of cardinality $|\eta_1| \geq c|\eta|/4$ so that

$$\|\sum_{i\in\eta_1} c_i z_i\|_2 \geq c\Big(\sum_{i\in\eta_1}|c_i|^2\Big)^{1/2},$$

for any choice of $\{c_i\}_{i\in\eta_1}$. This yields that

(b_1) $\|\sum_{i\in\eta_1} c_i x_i\|_2 \leq M_1\Big(\sum_{i\in\eta_1}|c_i|^2\Big)^{1/2},$

for all $\{c_i\}_{i\in\eta_1}$, where

$$M_1 = [1 + (M^2-1)(1-c)]^{1/2}.$$

By repeating the argument, we construct a subset η_2 of η_1 such that $|\eta_2| \geq c|\eta_1|/4$ and

(b_2) $\|\sum_{i\in\eta_1} c_i x_i\|_2 \leq M_2\Big(\sum_{i\in\eta_2}|c_i|^2\Big)^{1/2},$

for all $\{c_i\}_{i\in\eta_2}$, where

$$M_2 = [1 + (M_1^2-1)(1-c^2)]^{1/2} = [1 + (M^2-1)(1-c^2)^2]^{1/2}.$$

Continuing so for m times, where m is chosen to satisfy the condition

$$(M^2-1)(1-c^2)^m < 3,$$

we find eventually a subset η_m so that

$$|\eta_m| > c^m|\eta|/4^m > c^m\varepsilon^2 n/8.4^m\|S\|_2^2$$

and

$$\|\sum_{i\in\eta_m} c_i x_i\|_2 < 2(\sum_{i\in\eta_m} |c_i|^2)^{1/2},$$

for all $\{c_i\}_{i\in\eta_m}$. It follows that we also have

$$\|\sum_{i\in\eta_m} c_i R_{\eta_m} Se_i\|_2 < \varepsilon(\sum_{i\in\eta_m} |c_i|^2)^{1/2},$$

i.e. $\|R_{\eta_m} SR_{\eta_m}\|_2 < \varepsilon$. This completes the proof that Theorem 2.1 implies Theorem 2.2. □

Theorem 2.1 has some nice applications to harmonic analysis and hilbertian systems. In order to state the results, we recall that, for a subset Λ of positive integers, the notions of upper density, $\overline{\text{dens}}\,\Lambda$, and of lower density, $\underline{\text{dens}}\,\Lambda$, are defined as the $\overline{\lim}_{n\to\infty}$, respectively $\underline{\lim}_{n\to\infty}$, of the sequence

$$\frac{|\Lambda \cap \{1,2,\ldots,n\}|}{n}\ ; \quad n = 1,2,\ldots\ .$$

In the case when $\overline{\text{dens}}\,\Lambda = \underline{\text{dens}}\,\Lambda$, their common value $\text{dens}\,\Lambda$ is simply called the density of Λ. We can state now the result.

THEOREM 2.5. _Every subset_ B _of the circle_ $\mathbb{T}$ _of positive measure is a set of isomorphism in_ L_2, _for a family of characters of positive density, in the following sense: there exists a constant_ $c > 0$ _so that, for any_ $B \subset \mathbb{T}$, _one can find a subset_ Λ _of the integers with_ $\text{dens}\,\Lambda > c\nu(B)$ (ν _denotes the normalized Lebesgue measure on_ $\mathbb{T}$), _for which_

$$\|f\|_2 > \|f\chi_B\|_2 > c\sqrt{\nu(B)}\ \|f\|_2,$$

whenever $f \in L_2^\Lambda(\mathbb{T},\nu)$ _i.e._ f _is a function in_ $L_2(\mathbb{T},\nu)$ _whose Fourier coefficients are supported by_ Λ.

In order to prove Theorem 2.5, one considers the operator T on $L_2(\mathbb{T},\nu)$, defined by,

$$Tf = f\chi_B/\sqrt{\nu(B)}; \quad f \in L_2(\mathbb{T},\nu),$$

which satisfies the conditions of Theorem 2.1 whenever it is restricted to the linear span of a set of characters of the form $\{e^{inx}\}_{n=1}^{N-1}$; $N = 1,2,\ldots$. By applying Theorem 2.1, one constructs, for each N, a subset $\sigma_N \subset \{0,1,\ldots,N-1\}$ of cardinality $|\sigma_N| > c\nu(B)N$ so that the assertion of the theorem holds for $f \in [e^{inx}]_{n\in\sigma_N}$; $N = 1,2,\ldots$. The construction of Λ is achieved by using a result of I. Z. Ruzsa [31] which yields a set of integers having the property that each of its finite subsets is contained in a translate of one of the sets σ_N; $N = 1,2,\ldots$. For a simple proof of Ruzsa's result which is based on ergodic arguments, see Y. Peres [28]. The proof of Theorem 2.5 with all the details is given in [7] Section 2.

There are some interesting aspects concerning Theorem 2.5. The first quite suprising fact is that Theorem 2.5 is false in L_p; $p > 2$. This is an immediate consequence of the following result, also proved in [7] Section 2.

THEOREM 2.6. For $p > 2$, a subset B of the circle $\mathbb{T}$ is a set of isomorphism in L_p, for some family of characters of positive density (i.e. the map $f \to f\chi_B$ is an isomorphism in $L_p(\mathbb{T},\nu)$ when it is restricted to functions f belonging to a subspace of $L_p(\mathbb{T},\nu)$ of the form $[e^{inx}]_{n\in\Lambda}$ with Λ being a subset of the integers having positive density), if and only if the union of finitely many translates of B covers the whole circle $\mathbb{T}$, up to a set of measure zero.

It is easily seen that e.g. any Cantor set C on the circle, which has positive measure, provides an example of a set of isomorphism in L_2 but not in L_p; $p > 2$.

Another interesting fact is related to the observation that if Λ is a subset of integers of positive density then, in some sense, $L_2^\Lambda(\mathbb{T},\nu)$ "almost" coincides with $L_2(\mathbb{T},\nu)$. In order to explain the meaning of this statement, fix a subset B of $\mathbb{T}$ of measure $0 < \nu(B) < 1$ and let Λ be the corresponding set of integers of positive density given by Theorem 2.5. By the famous result of E. Szemeredi [35], Λ contains arbitrarily long arithmetic progressions i.e., for each N, there is a sequence of the form $\{\lambda \pm jd\}_{j=-N}^{N}$ which is entirely contained in Λ (where λ and d are fixed integers). Then, by the assertion of Theorem 2.5, any function of the form $f_N(x) = \sum_{j=-N}^{N} a_j e^{ijdx}$ will satisfy the inequality

$$\|f_N\chi_B\|_2 > \sqrt{\nu(B)}\, \|f_N\|_2.$$

On the other hand, since $\nu(B) < 1$ there are plenty of functions of the form $g_N(x) = \sum_{j=-N}^{N} a_j e^{ijx}$, for which $\|g_N \chi_B\|_2 / \|g_N\|_2$ tends to zero, as $N \to \infty$. The difference between these two situations reflects the fact that the parameter d appearing in the arithmetic progression $\{\lambda \pm jd\}_{j=-N}^{N}$ actually depends on N and $d = d(N)$ satisfies the condition $\sup_N d(N) = \infty$.

Theorem 2.5 is an assertion concerning the system of functions $\psi_n(x) = \chi_B e^{inx}$; $n = 0, \pm 1, \pm 2, \ldots$ in $L_2(\mathbb{T}, \nu)$ which clearly satisfies the condition

$$\| \sum_{n=-\infty}^{+\infty} a_n \psi_n \|_2 \leq \left(\sum_{n=-\infty}^{+\infty} |a_n|^2 \right)^{1/2},$$

for all $\{a_n\}_{n=-\infty}^{+\infty}$. It turns out that Theorem 2.5 remains valid with almost the same statement when this system is replaced by any other Hilbertian system. A normalized system of vectors $\{x_n\}_{n=1}^{\infty}$ in a Banach space X is called Hilbertian if there exists a constant $M < \infty$ so that

$$\| \sum_{n=1}^{\infty} a_n x_n \| \leq M \left(\sum_{n=1}^{\infty} |a_n|^2 \right)^{1/2},$$

for any choice of scalars $\{a_n\}_{n=1}^{\infty}$.

THEOREM 2.7. _There exists a constant $D < \infty$ such that, whenever $\{x_n\}_{n=1}^{\infty}$ is a Hilbertian system with constant M in a Hilbert space, then one can find a subset Λ of integers with $\overline{\text{dens}}\, \Lambda \geq 1/DM^2$ such that $\{x_n\}_{n \in \Lambda}$ is DM-equivalent to an orthonormal system._

The proof given in [7] Section 2 consists of a direct construction of Λ by using a sort of gliding hump argument which yields only a set of positive upper density. We do not know whether Λ in Theorem 2.7 can be chosen as to actually have positive density.

We conclude this section with a discussion concerning the problem of splitting matrices acting on euclidean spaces, in analogy with the splitting result given in Corollary 1.4 for the ℓ_1^n and the ℓ_∞^n-case.

The following conjecture of Kadison and Singer [19] is still open.

CONJECTURE 2.8. _For every $\varepsilon > 0$, there exists an integer $k = k(\varepsilon)$ such that, whenever S is a linear operator on ℓ_2^n whose matrix relative to the unit vector basis has 0's on the diagonal, then there exists a partition $\{\sigma_h\}_{h=1}^{k}$ of the integers $\{1,2,\ldots,n\}$ into k mutually disjoint subsets such that_

$$\| \sum_{h=1}^{k} R_{\sigma_h} S R_{\sigma_h} \|_2 \leq \varepsilon \|S\|_2.$$

There are some indications that Conjecture 2.8 is valid. The first such indication is the fact that Theorem 2.2, which is clearly a consequence of their conjecture, is true. Furthermore, Conjecture 2.8 has been verified by K. Berman, H. Halpern, V. Kaftal and G. Weiss [3] and also by K. Gregson [15] for some classes of operators among which the most notable is that of matrices with non-negative entries. This case behaves like that of matrices acting on ℓ_1^n and ℓ_∞^n, and both cases are proved by similar extremal arguments. On the negative side, we should mention the fact noticed in [3] that while, for matrices with non-negative entries, $k(\varepsilon)$ behaves like ε^{-1}, in general, if there is such a $k(\varepsilon)$ at all then it must satisfy $k(\varepsilon) \geq \varepsilon^{-2}$, for all $\varepsilon > 0$.

Since Conjecture 2.8, if true, yields a splitting of the integers $\{1,2,\dots,n\}$ into $k = k(\varepsilon)$ parts, where k is independent of n, we easily conclude that its assertion implies a similar splitting result for operators acting on infinite dimensional Hilbert spaces. More precisely, if $\varepsilon > 0$ is given and S is a bounded linear operator on ℓ_2, whose matrix relative to the unit vector basis of ℓ_2 has 0's on the diagonal, then there exists a splitting $\{\sigma_h\}_{h=1}^k$ of the integers such that $\|R_{\sigma_h} S R_{\sigma_h}\| \leq \varepsilon$, for all $1 \leq h \leq k$.

Conjecture 2.8 is actually equivalent to a long standing open problem of Kadison and Singer [19] on the extension property of pure states on the algebra $\mathbf{D}$ of all the diagonal operators on ℓ_2. More precisely, the problem raised in [19] is whether any pure state on $\mathbf{D}$ has a unique extension to a (pure) state on $B(\ell_2)$.

In order to elaborate on their question, we consider first the so-called relative Dixmier property, in short RDP. A bounded linear operator T on ℓ_2 is said to have the Dixmier property relative to $\mathbf{D}$ if the norm closed convex hull

$$K(T) = \overline{\mathrm{conv}}\{UTU^*;\ U \in \mathbf{D},\ U \ \text{unitary}\}$$

has a non-empty intersection with $\mathbf{D}$. If this is the case then

$$K(T) \cap \mathbf{D} = \{E(T)\},$$

where $E(T)$ denotes the diagonal of T (since all the operators of the form UTU^*, with U as above, have the same diagonal $E(T)$).

It is easily verified that Conjecture 2.8 is equivalent to the assertion that every $T \in B(\ell_2)$ has RDP. One direction follows from the fact that any operator of the form $\sum_{h=1}^k R_{\sigma_h} T R_{\sigma_h}$, with $\{\sigma_h\}_{h=1}^k$ being a partition of the integers, can be expressed as a convex combination of the form $\sum_{i=1}^m \alpha_i U_i T U_i^*$, with $\alpha_i \geq 0$, $1 \leq i \leq m$, $\sum_{i=1}^m \alpha_i = 1$ and $\{U_i\}_{i=1}^m$ being unitary diagonal operators. Indeed, if σ is a subset of the integers and $U = R_\sigma - R_{\sigma^c}$ then U is unitary and diagonal, and

$$(T + UTU^*)/2 = R_\sigma T R_\sigma + R_{\sigma^c} T R_{\sigma^c}.$$

This argument can then be continued by induction. The converse, i.e. the fact that RDP implies Conjecture 2.8, is proved easily along the some lines.

The connection between the problem raised by Kadison and Singer and RDP was already pointed out in their paper [19] and, later, in a more explicit way in [1]. Namely, it was shown that every pure state on **D** has a unique extension to a state on $B(\ell_2)$ if and only if each $T \in B(\ell_2)$ has RDP. Thus, the extension property of pure states on **D** is also equivalent to a positive solution to Conjecture 2.8. The fact that RDP implies the uniqueness of the extensions can be proved in the following manner. Recall first that a state ϕ on a C^*-algebra **A** is a linear functional on **A** satisfying the condition $\phi(a^*a) \geq 0$, for all $a \in$ **A**, which is normalized on the identity e of **A**. Any state has norm equal to one and a pure state is just an extreme point in the set of the states on **A**. A state on **D**, which is a commutative C^*-algebra, is a probability measure and a pure state is a point evaluation on the spectrum of **D**. In particular, any pure state on **D** is multiplicative. This is, of course, not the case with pure states on $B(\ell_2)$.

We shall prove now that, whenever $D \in$ **D**, $T \in B(\ell_2)$ and ϕ is an extension of a pure state on **D** to a state on $B(\ell_2)$, then

$$\phi(DT) = \phi(TD) = \phi(D)\phi(T).$$

This would imply that

$$\phi(UTU^*) = \phi(T),$$

for any diagonal unitary U. Hence, if an operator T has RDP then $\phi(T) = \phi(E(T))$ i.e. the unique extension property holds.

In order to prove the above identity, we shall use the Gelfand-Neimark-Segal representation formula for states which, in our case, asserts the existence of a linear multiplicative *-preserving map π form $B(\ell_2)$ into the algebra $B(H)$ of all the bounded linear operators on a certain abstract Hilbert space H and of a vector $\xi \in H$ with $\|\xi\| = 1$ so that

$$\phi(S) = \langle \pi(S)\xi, \xi \rangle; \; S \in B(\ell_2).$$

Thus, if $D \in$ **D** then

$$\|\pi(D)\xi\|^2 = \langle \pi(D^*)\pi(D)\xi, \xi \rangle = \langle \pi(D^*D)\xi, \xi \rangle = \phi(D^*D)$$

which, by the multiplicativity of ϕ on **D**, implies that

$$\|\pi(D)\xi\|^2 = |\phi(D)|^2.$$

It follows that

$$|\langle\pi(D)\xi,\xi\rangle| = \|\pi(D)\xi\| \cdot \|\xi\|$$

i.e. that we have equality in the Cauchy-Schwarz inequality, which yields that

$$\pi(D)\xi = \phi(D)\xi.$$

Hence, by the fact that π is multiplicative, we finally get that

$$\phi(DT) = \langle\pi(D)\pi(T)\xi,\xi\rangle = \langle\pi(T)\xi, \overline{\phi(D)}\xi\rangle = \phi(D)\phi(T),$$

and, similarly, for $\phi(TD)$.

3. OPERATORS ON ℓ_p^n-SPACES; $1 < p \neq 2$.

The main restricted invertibility results for matrices acting on ℓ_1^n, ℓ_∞^n and ℓ_2^n-spaces can be extended to ℓ_p^n-spaces for any value of $p > 1$. However, the proofs in this case are considerably more complicated. On the other hand, in terms of applications to the geometry of Banach spaces, the case $p > 1$ is certainly the most useful.

The most important restricted invertibility result stated in this section is, as in the euclidean case, a consequence of a supression theorem.

THEOREM 3.1. For every $\varepsilon > 0$ and $1 < p < \infty$, there exist an integer $n(\varepsilon,p)$ and a positive real $\delta_p(\varepsilon)$ such that, whenever $n \geq n(\varepsilon,p)$ and S is a linear operator on ℓ_p^n whose matrix relative to the unit vector basis has 0's on the diagonal, then one can select a subset σ of $\{1,2,\ldots,n\}$ so that

$$|\sigma| \geq \delta_p(\varepsilon)n$$

and

$$\|R_\sigma S R_\sigma\|_p \leq \varepsilon\|S\|_p$$

COROLLARY 3.2. For every $0 < \varepsilon < 1$, $1 < p < \infty$ and $M < \infty$, there exists a constant $c = c(\varepsilon,p,M) > 0$ such that, whenever $n \geq 1/c$ and T is a linear operator on the space ℓ_p^n of norm $\|T\|_p \leq M$ whose matrix relative to the unit vector basis has 1's on the diagonal, then one can find a subset σ of $\{1,2,\ldots,n\}$ of cardinality $|\sigma| \geq cn$ so that $R_\sigma T R_\sigma$ is invertible and

$$\|(R_\sigma T R_\sigma)^{-1}\|_p \leq (1 - \varepsilon)^{-1}.$$

Theorem 3.1 was proved in [7] Section 3 but, meanwhile, we have modified the proof in such a way that it gives, on one hand, better numerical estimates for $\delta_p(\varepsilon)$ and, on the other hand, it can be interpolated as to yield a similar supression theorem for some class of spaces with a symmetric basis. The new proof of Theorem 3.1, for $1 < p \neq 2$, is based on the following result which, quite suprisingly, does not involve the norm of the operator on ℓ_p^n.

THEOREM 3.3. _For every_ $\varepsilon > 0$ _and_ $1 < p < 2$, _there exist an integer_ $n(\varepsilon,p)$ _and reals_ $\rho(\varepsilon,p) > 0$ _and_ $C(p) < \infty$ _such that, whenever_ $n \geq n(\varepsilon,p)$ _and_ S _is a linear operator on_ ℓ_2^n _of norm_ $\|S\|_2 \leq \rho(\varepsilon,p)$, _then one can find a subset_ η _of_ $\{1,2,\ldots,n\}$ _so that_

$$|\eta| = [\varepsilon n]$$

and

$$\|R_\eta Sx\|_p \leq C(p)\varepsilon^{1/p}\left(\|x\|_p + \|Sx\|_p\right),$$

for all $x \in \ell_p^n$.

The proof of Theorem 3.3 uses, essentially speaking, the same ideas as the original proof of Theorem 3.1 but it is organized in a better and more efficient way. To facilitate the reading of their article in a continuous manner, we present the proof of Theorem 3.3, separately, in Section 5.

In order to be able to apply Theorem 3.3, one first has to prove that an operator bounded on ℓ_p^n; $1 \leq p \leq \infty$, is also bounded on ℓ_2^n when is restricted to a suitable subspace. This fact is a consequence of a theorem of W.B. Johnson and L. Jones [17] and can be stated, as follows.

PROPOSITION 3.4. _For every_ $1 \leq p \leq \infty$ _and every linear operator_ S _on_ ℓ_p^n, _there exists a subset_ τ _of_ $\{1,2,\ldots,n\}$ _such that_ $|\tau| \geq n/2$ _and_

$$\|R_\tau SR_\tau\|_2 \leq 4K_G\|S\|_p,$$

where K_G _denotes Grothendieck's constant._

In the next section, we shall present a direct proof of a generalization of Proposition 3.4 for spaces with a symmetric or even unconditional basis.

In order to deduce Theorem 3.1 from Theorem 3.3, we first notice that the assertion of Theorem 3.1 is self-dual and, therefore, it suffices to consider the case $1 < p < 2$. If S is a linear operator on ℓ_p^n; $1 < p < 2$, and its matrix relative to the unit vector basis has 0's on the diagonal then, by Proposition 3.4 and Theorem 2.2, one can find a subset τ_1 of $\{1,2,\ldots,n\}$ so that

$$|\tau_1| \geq \delta_2\left(\rho(\varepsilon,p)/4K_G\|S\|_p\right)n/2$$

and

$$\|R_{\tau_1} SR_{\tau_1}\|_2 \leq \rho(\varepsilon,p),$$

where $\rho(\varepsilon,p)$ is the constant appearing in the statement of Theorem 3.3. Then, by applying this theorem, we conclude the existence of a subset η of τ_1 such that $|\eta| = [\varepsilon|\tau_1|]$ and

$$\|R_\eta S R_\eta\|_p \leq C(p)\varepsilon^{1/p}(1 + \|S\|_p).$$

This completes the argument provided ε is chosen small enough. As one can easily see from this proof, the difficulty in finding sharp estimates for the function $\delta_p(\varepsilon)$, appearing in the statement of Theorem 3.1, lies in the fact that we do not have such good estimates for $\delta_2(\varepsilon)$. For an operator on ℓ_p^n; $1 < p < 2$, which has small norm on ℓ_2^n (i.e. $\|S\|_2 \leq \rho(\varepsilon,p)$), one clearly obtains $\delta_p(\varepsilon) \sim \varepsilon^p$.

Contrary to the case of operators on euclidean spaces, one cannot expect to prove a general restricted invertibility theorem for matrices with "large" rows which act on ℓ_p^n-spaces, $1 \leq p \neq 2$. While such a result is still true for $p > 2$, in the case $1 < p < 2$ even the algebraic rank of a matrix with "large" rows need not be proportional to n. For instance, if H_p is a well-complemented subspace of ℓ_p^n; $1 < p < 2$, of maximal dimension which is, say, 2-isomorphic to a Hilbert space (i.e. $\dim H_p = [n^{2/p'}]$) then the orthogonal projection P on H_p has "large" rows. This fact is discussed in detail in [7] Section 3.

In the case $1 < p < 2$ it is still possible, however, to prove a restricted invertibility result if we require that the rows of the matrix have a rather "large" Rademacher average. We summarize these results, as follows.

THEOREM 3.5. <u>For every</u> $p \geq 1$ <u>and</u> $M < \infty$, <u>there exists a constant</u> $c = c(p,M) > 0$ <u>such that</u>, <u>whenever</u> T <u>is a linear operator on</u> ℓ_p^n <u>of norm</u> $\|T\|_p \leq M$ <u>for which</u>

$$\|Te_i\|_p = 1;\ 1 \leq i \leq n, \quad \text{if} \quad p \geq 2 \quad \text{or}$$

$$\int \|\sum_{i=1}^n \varepsilon_i Te_i\|_p d\varepsilon \geq n^{1/p}, \quad \text{if} \quad 1 \leq p < 2,$$

<u>then one can find a subset</u> σ of $\{1,2,\ldots,n\}$ <u>so that</u>

$$|\sigma| \geq cn$$

<u>and</u>

$$\|\sum_{i\in\sigma} a_i Te_i\|_p \geq c(\sum_{i\in\sigma} |Te_i|^p)^{1/p},$$

<u>for all</u> $\{a_i\}_{i\in\sigma}$.

Theorem 3.5 for $p \neq 2$ is actually a direct consequence of Corollary 3.2. Indeed, the conditions imposed on T in the statement of Theorem 3.5 yield that the expressions $(\sum_{i=1}^n |Te_i|^2)^{1/2}$ and $(\sum_{i=1}^n |Te_i|^p)^{1/p}$ have both

norms of order of magnitude $n^{1/p}$. Thus, by a simple use of Holder's inequality, one deduces that also the expression $\max_{1\le i\le n} |Te_i|$ has norm of the same order of magnitude. This easily implies the existence of a permutation π of the integers $\{1,2,\ldots,n\}$ such that the matrix $\pi^{-1}T$ has a "large" diagonal.

We return now to the study of matrices which have 1's on the diagonal. As we have already pointed out, the norm of such a matrix on ℓ_p^n should be ≥ 1. In the remarkable case when the norm does not exceed the number 1 by a large amount, one can prove a new invertibility result which requires no restrictions on the domain or range of the matrix.

THEOREM 3.6. *For every* $1 \le p < \infty$, $p \ne 2$, *there exists a constant* $C_p < \infty$ *so that, whenever* S *is a linear operator on* ℓ_p^n *whose matrix relative to the unit vector basis has* 0's *on the diagonal and the operator* $T = I + S$ *satisfies the condition* $\|T\|_p < 1 + \varepsilon$, *for some* $0 < \varepsilon < 1$, *then* $\|S\|_p \le C_p \varepsilon^{2-p^*}$, *where* $p^* = \min\{p,p'\}$. *Consequently, if* $0 < \varepsilon < \varepsilon_p = (1/C_p)^{1/(2-p^*)}$ *then* T *is invertible and*

$$\|T^{-1}\|_p \le \left(1 - C_p\varepsilon^{2-p^*}\right)^{-1}.$$

Proof. Since the assertion is self-dual it suffices to study only the case $1 \le p < 2$. The case $p = 1$ has already been considered in Proposition 1.5 and, therefore, we may assume that $1 < p < 2$.

The argument that we intend to present here consists of a averaging procedure which allows to exploit the essential fact that S has 0's on the diagonal. To this end, fix a vector $x = \sum_{i=1}^n a_i e_i \in \ell_p^n$ of norm one, let $\{\xi_i\}_{i=1}^n$ be a sequence of $\{0,1\}$-valued independent random variables of mean $\delta = \varepsilon^{p-1}$ over some probability space (Ω,Σ,μ) and, for $\omega \in \Omega$, put

$$\sigma(\omega) = \{1 \le i \le n;\ \xi_i(\omega) = 1\}.$$

In order to average over the random sets $\sigma(\omega)$; $\omega \in \Omega$, we need first some preliminary estimates for such a subset $\sigma = \sigma(\omega)$. A simple calculation shows that

$$(1 + \varepsilon)^p\|R_\sigma x\|_p^p > \|T\|_p^p\|R_\sigma x\|_p^p \ge \|TR_\sigma x\|_p^p = \|R_\sigma x + R_\sigma SR_\sigma x\|_p^p + \|R_{\sigma^c} S R_\sigma x\|_p^p \ge$$

$$\ge \left|\|R_\sigma x\|_p - \|R_\sigma S R_\sigma x\|_p\right|^p + \left|\|S R_\sigma x\|_p - \|R_\sigma S R_\sigma x\|_p\right|^p.$$

Hence, by using the inequality

$$|\alpha - \beta|^p \ge \alpha^p - p\alpha^{p-1}\beta,$$

which is valid for any pair $\alpha,\beta > 0$, we conclude that

$$4\varepsilon\|R_\sigma x\|_p^p \geq \|SR_\sigma x\|_p^p - 8\|R_\sigma x\|_p^{p-1}\|R_\sigma SR_\sigma x\|_p.$$

Now, by replacing σ with $\sigma(\omega)$, integrating and then using Holder's inequality, it follows that

$$4\varepsilon\int_\Omega \|R_{\sigma(\omega)}x\|_p^p d\mu(\omega) \geq \int_\Omega\|SR_{\sigma(\omega)}x\|_p^p d\mu(\omega) -$$

$$- 8\Big(\int_\Omega\|R_{\sigma(\omega)}SR_{\sigma(\omega)}x\|_p^p d\mu(\omega)\Big)^{1/p}\Big(\int_\Omega\|R_{\sigma(\omega)}x\|_p^p d\mu(\omega)\Big)^{1/p'}.$$

We shall calculate next the expressions appearing in the above inequality. First, observe that

$$\text{(i)} \qquad \int_\Omega\|R_{\sigma(\omega)}x\|_p^p d\mu(\omega) = \int_\Omega \sum_{i=1}^n |a_i|^p\xi_i(\omega)d\mu(\omega) = \delta.$$

Then if the matrix associated to S is defined by $Se_i = \sum_{j=1}^n c_{i,j}e_j$; $1 \leq i \leq n$, and $c_{i,i} = 0$; $1 \leq i \leq n$, by our assumption, it follows that

$$\text{(ii)} \quad \int_\Omega\|R_{\sigma(\omega)}SR_{\sigma(\omega)}x\|_p^p\, d\mu(\omega) = \int_\Omega \sum_{j=1}^n \xi_j(\omega)\Big|\sum_{\substack{i=1\\ i\neq j}}^n a_i c_{i,j}\xi_i(\omega)\Big|^p d\mu(\omega) =$$

$$= \delta\int_\Omega\|SR_{\sigma(\omega)}x\|_p^p d\mu(\omega).$$

Hence, the expression

$$\Lambda = \Big(\int_\Omega\|SR_{\sigma(\omega)}x\|_p^p d\mu(\omega)\Big)^{1/p},$$

satisfies the inequality

$$4\varepsilon\delta \geq \Lambda^p - 8\delta\Lambda$$

from which it easily follows that

$$\Lambda \leq \max\{(8\varepsilon\delta)^{1/p}, (16\delta)^{1/(p-1)}\}.$$

However, since

$$\Lambda \geq \int_\Omega\|SR_{\sigma(\omega)}x\|_p d\mu(\omega) \geq \|S(\int_\Omega R_{\sigma(\omega)}x)\|_p = \delta\|Sx\|_p$$

we conclude, in view of the value assigned to δ, that

$$\|S\|_p \leq C_p\varepsilon^{2-p},$$

for a suitable constant $C_p < \infty$ and all $0 < \varepsilon < 1$. □

REMARK. Theorem 3.6 fails for $p = 2$. Indeed, for each n, consider the orthogonal projection P_n in ℓ_2^n whose range is the one-dimensional subspace of ℓ_2^n generated by the vector $x_0 = \sum_{i=1}^n e_i$. The operator $S_n = I/n - P_n$ has clearly 0's on the diagonal and $\|I + S_n\|_2 \leq 1 + 1/n$. On the other hand, $\|S_n\|_2 \geq 1 - 1/n$, for all n.

We pass now to some applications to the geometry of Banach spaces. As a rule, the results presented above are useful to study finite dimensional subspaces of an L_p-space which have extremal euclidean distance. The case $1 < p \neq 2$ is considerably more difficult than $p = 1$, which was briefly discussed in Section 1, and the matrix method seems to be well suited for this kind of problems.

The starting point of this discussion is the well known and relatively trivial fact that the euclidean distance of an ℓ_p^n-space is equal to $n^{|1/p-1/2|}$, for any $1 \leq p \leq \infty$, while, by a result of D.R. Lewis [21], the euclidean distance of an arbitrary subspace of L_p of dimension n is $\leq n^{|1/p-1/2|}$, for $1 \leq p \leq \infty$. This maximality property leads naturally to several questions regarding those subspaces of L_p having extremal euclidean distance. These problems can be considered from an isometric, isomorphic or almost isometric point of view. Naturally, the point of view affects the nature of the questions. Results of an isometric and isomorphic type have already been discussed in [7] Section 4 and they will be only reviewed here. The main tool in studying isomorphic questions is Corollary 3.2. For problems of an almost isometric nature, Theorem 3.6 seems to be quite useful.

The main isometric question is whether the ℓ_p^n-spaces are the only subspaces of L_p; $1 < p < \infty$, whose euclidean distance is equal to $n^{|1/p-1/2|}$. This problem has a positive solution for $1 < p < 2$ but it is still unsolved in the case $p > 2$. The case $p = 1$ was discussed in Section 1.

THEOREM 3.7. For every $1 < p < 2$ and every integer n, any n-dimensional subspace of an L_p-space, whose euclidean distance is equal to $n^{1/p-1/2}$, is isometric to ℓ_p^n.

This result is proved in detail in [7] Section 4.

More interesting is, of course, the isomorphic case in which one considers n-dimensional subspaces X of an L_p-space that satisfy the condition

$$d_X \geq cn^{|1/p-1/2|},$$

for some constant $c > 0$, independent of n. In this case, it is quite clear that X need not be well isomorphic to ℓ_p^n. However, one might hope to prove that such an X contains, in turn, a subspace Y of dimension k proportional to n which is well isomorphic to ℓ_p^k. For $p = 1$, this is true and was proved in [18] and [5]. On the other hand, examples of random subspaces on which the L_1 and L_2-norms are equivalent (cf. [11]) show that, for $p > 2$, L_p has an n-dimensional subspace X_n of maximal euclidean distance in the above sense which contains subspaces well-isomorphic to ℓ_p^m only for $m \leq Cn^{2/p'}$, with C being a fixed constant. The case $1 < p < 2$ is still open.

The situation changes radically when we consider subspaces of L_p which, in addition to being of extremal euclidean distance in the isomorphic sense, are also well-complemented. For instance, W.B. Johnson and G. Schechtman [18] found a quite complicated proof to the fact that such a space X of dimension equal to n contains, for every $\varepsilon > 0$, a well complemented subspace Y of dimension $k \geq n^{1-\varepsilon}$ which is also well isomorphic to ℓ_p^k; $1 < p < \infty$. They raised the question if this result holds also with Y of dimension proportional to n. It turns out that, with the aid of Corollary 3.2, one can provide a positive solution to their problem.

THEOREM 3.8. <u>For every</u> $1 < p < \infty$ <u>and</u> $c > 0$, <u>there exists a constant</u> $d = d(p,c) > 0$ <u>so that, whenever</u> X <u>is a</u> c^{-1}<u>-complemented subspace of an</u> L_p<u>-space which satisfies the condition</u>

$$d_X \geq cn^{|1/p-1/2|},$$

<u>then one can find a</u> d^{-1}<u>-complemented subspace</u> Y <u>of</u> X <u>of dimension</u> $k \geq dn$ <u>such that</u>

$$d(Y,\ell_p^k) \leq d^{-1}.$$

The proof of Theorem 3.8 is based on an adaptation of Corollary 3.2 to the case of functions in L_p.

PROPOSITION 3.9. <u>For every</u> $1 < p < \infty$ <u>and</u> $M < \infty$, <u>there exists a constant</u> $c = c(p,M) > 0$ <u>so that, whenever</u> $\{g_i\}_{i=1}^m$ <u>and</u> $\{h_i\}_{i=1}^m$ <u>are functions in</u> L_p, <u>respectively in</u> $L_{p'}$, <u>for which</u>

(i) $\|\sum_{i=1}^m a_i g_i\|_p \leq M(\sum_{i=1}^m |a_i|^p)^{1/p}$, for all $\{a_i\}_{i=1}^m$,

(ii) $\|\sum_{i=1}^m b_i h_i\|_{p'} \leq M(\sum_{i=1}^m |b_i|^{p'})^{1/p'}$, for all $\{b_i\}_{i=1}^m$, and

(iii) $\langle g_i, h_i \rangle = 1$, for any $1 \leq i \leq m$,

<u>then one can find a subset</u> σ <u>of</u> $\{1,2,\dots,m\}$ <u>with</u> $|\sigma| \geq cm$ <u>and a projection</u> R <u>from</u> L_p <u>onto its subspace</u> $[g_i]_{i\in\sigma}$ <u>such that</u> $\|R\|_p \leq c^{-1}$ <u>and</u>

$$\|\sum_{i\in\sigma} a_i g_i\|_p \geq c(\sum_{i\in\sigma} |a_i|^p)^{1/p},$$

<u>for any choice of</u> $\{a_i\}_{i\in\sigma}$.

Proposition 3.9 is easily proved by applying Corollary 3.2 to the matrix $(\langle g_i,h_j\rangle)_{i,j=1}^m$ or, more precisely, to the operator T on ℓ_p^m, defined by

$$Te_i = \sum_{j=1}^m \langle g_i,h_j\rangle e_j \ ; \ 1 \le i \le m.$$

In order to deduce Theorem 3.8 from Proposition 3.9, we may assume that $1 < p < 2$ since the assertion of Theorem 3.8 is self-dual. Then, by a standard argument already presented in [18], one constructs a set $\{y_i\}_{i=1}^m$ of norm one functions in X which have "large" mutually disjoint supports $\{A_i\}_{i=1}^m$ with m proportional to $n = \dim X$. This means that there is a constant $a > 0$, independent of X, so that $\|y_i\chi_{A_i}\|_p \ge a$, for all $1 \le i \le m$. Next, if P denotes a projection from L_p onto X of norm $\le c^{-1}$ and we set $h_i = P^*(|y_i|^{p-1}\operatorname{sgn}\chi_{A_i})$; $1 \le i \le m$, then the functions $\{y_i\}_{i=1}^m$ and $\{h_i\}_{i=1}^m$ will satisfy the conditions of Proposition 3.9 in an essential manner with the exception of the fact that $\{y_i\}_{i=1}^m$ does not have a good upper p-estimate i.e. (i) fails to hold for this system. However, with some additional work one can replace this system by another one which does satisfy all the conditions of Proposition 3.9. Additional details are given in [7] Section 4.

Proposition 3.9 can be interpreted also as a factorization theorem in the following way. Given an operator $T : \ell_p^m \to L_p$ of norm $\le M$, for some $M < \infty$, such that

$$\|(\sum_{i=1}^m |Te_i|^2)^{1/2}\|_p \ge m^{1/p}$$

one can find an integer k proportional to m and a linear operator $R : L_p \to \ell_p^k$ for which the identity I on ℓ_p^k factors through T as $I = RTJ$, where J is the formal identity map from ℓ_p^k onto a subspace of ℓ_p^m generated by a certain subset of k unit vectors. In the statement of Proposition 3.9, the operator T is defined by $g_i = Te_i$; $1 \le i \le m$, and the fact that the above condition on the square function holds follows from (ii) and (iii) there. In this formulation, Proposition 3.9 is an improvement of a previous result from [12] which asserts the factorization of the identity on ℓ_p^k through T in the weaker form $I = \int R_\varepsilon T J_\varepsilon d\varepsilon$, with $\{R_\varepsilon\}$ and $\{J_\varepsilon\}$ being families of operators having the same properties as R, respectively J, above.

Proposition 3.9 serves also to give a positive solution to another problem of W.B. Johnson and G. Schechtman from [18].

THEOREM 3.10. _For every_ $1 < p < \infty$ _and_ $M < \infty$, _there exists a constant_ $d = d(p,M) > 0$ _such that, whenever_ $\{f_i\}_{i=1}^n$ _is a sequence of functions in_ L_p _satisfying_

$$M^{-1}\Big(\sum_{i=1}^{n}|a_i|^p\Big)^{1/p} \le \Big\|\sum_{i=1}^{n}a_if_i\Big\|_p \le M\Big(\sum_{i=1}^{n}|a_i|^p\Big)^{1/p},$$

for any choice of $\{a_i\}_{i=1}^n$, *then one can find a subset* σ *of* $\{1,2,\ldots,n\}$ *of cardinality* $|\sigma| \ge dn$ *so that* $[f_i]_{i\in\sigma}$ *is* d^{-1}*-complemented in* L_p.

It is well known that, for every $1 \le p \ne 2$, there exist isomorphic copies of ℓ_p^n in L_p which are not well complemented i.e. Theorem 3.10 cannot be improved as to yield that $[f_i]_{i=1}^n$ is already d^{-1}-complemented. However, in the case when the constant M, appearing in the statement, is close to 1 the situation is completely different. The extreme case $M = 1$, i.e. when the functions $\{f_i\}_{i=1}^n$ span an isometric copy of ℓ_p^n in L_p is trivial. Then $\{f_i\}_{i=1}^n$ are disjointly supported if $1 \le p \ne 2$ and, thus, their closed linear span is 1-complemented in L_p. It turns out that this property is stable under a small perturbation of M. In the case $p = 1$, which is much easier, the complementation of almost isometric copies of ℓ_1^n in L_1 was proved by L.E. Dor [10] while the general case $p \ne 2$ was solved by G. Schechtman [33]. Their results can be stated together in the following way.

THEOREM 3.11. *For every* $1 \le p < \infty$, *there exist a number* $\varepsilon_p > 0$ *and a function* $M_p(\varepsilon)$, *defined for* $0 < \varepsilon < \varepsilon_p$, *such that* $\lim_{\varepsilon\to 0} M_p(\varepsilon) = 1$ *and, whenever* $\{f_i\}_{i=1}^n$ *is a system of functions in an* L_p*-space satisfying*

$$(1-\varepsilon)\Big(\sum_{i=1}^{n}|a_i|^p\Big)^{1/p} \le \Big\|\sum_{i=1}^{n}a_if_i\Big\|_p \le (1+\varepsilon)\Big(\sum_{i=1}^{n}|a_i|^p\Big)^{1/p},$$

for all $\{a_i\}_{i=1}^n$ *and some* $0 < \varepsilon < \varepsilon_p$, *then* $[f_i]_{i=1}^n$ *is* $M_p(\varepsilon)$*-complemented in* L_p.

Both Dor's result for $p = 1$ and Schechtman's theorem for $1 < p \ne 2$ are proved by using the following disjointness result of L.E. Dor [10].

THEOREM 3.12. *For any* $1 \le p \ne 2$, *there exists a function* $\tau_p(\varepsilon)$ *such that* $\lim_{\varepsilon\to 0}\tau_p(\varepsilon) = 0$ *and, whenever* $\{f_i\}_{i=1}^n$ *is a normalized sequence in an* L_p*-space satisfying*

$$(1-\varepsilon)\Big(\sum_{i=1}^{n}|a_i|^p\Big)^{1/p} \le \Big\|\sum_{i=1}^{n}a_if_i\Big\|_p \le (1+\varepsilon)\Big(\sum_{i=1}^{n}|a_i|^p\Big)^{1/p},$$

for all $\{a_i\}_{i=1}^n$ *and some* $\varepsilon > 0$, *then there are mutually disjoint subsets* $\{A_i\}_{i=1}^n$ *of the underlying measure space for which*

$$\|\chi_{A_i^c} f_i\|_p \le \tau_p(\varepsilon); \quad 1 \le i \le n.$$

Besides the original proof of Theorem 3.12 given by L.E. Dor in the Appendix of his paper [10], a simple separation argument was discovered independently by R. Kaufman (see [10]) and B. Maurey [27].

The deduction of Theorem 3.11 from Theorem 3.12 in the case $p = 1$ is immediate. On the other hand, Schechtman's proof that Theorem 3.12 implies Theorem 3.11 in the general case $1 < p \neq 2$ is quite complicated. We propose here instead an almost obvious argument which uses the invertibility result for matrices proved in Theorem 3.6.

Proof of Theorem 3.11. Fix $1 < p \neq 2$ and let $\{f_i\}_{i=1}^n$ be a sequence of elements in an L_p-space which satisfies the condition

$$(1 - \varepsilon)\left(\sum_{i=1}^n |a_i|^p\right)^{1/p} \le \left\|\sum_{i=1}^n a_i f_i\right\|_p \le (1 + \varepsilon)\left(\sum_{i=1}^n |a_i|^p\right)^{1/p},$$

for all $\{a_i\}_{i=1}^n$ and some sufficiently small $\varepsilon > 0$. We can assume without loss of generality that $\|f_i\|_p = 1$; $1 \le i \le n$.

Let now $\tau_p(\varepsilon)$ and $\{A_i\}_{i=1}^n$ be given by Theorem 3.12 so that

$$\|\chi_{A_i^c} f_i\|_p \le \tau_p(\varepsilon),$$

for all $1 \le i \le n$. We also assume that ε was chosen small enough as to ensure that $\tau_p(\varepsilon) < 1$. This condition implies that the functions

$$g_i = \chi_{A_i} |f_i|^{p-1} \operatorname{sgn} f_i / \textstyle\int_{A_i} |f_i|^p d\mu; \quad 1 \le i \le n,$$

belong to $L_{p'}$ and $\|g_i\|_{p'} = \|\chi_{A_i} f_i\|_p^{-1} \le (1 - \tau_p(\varepsilon))^{-1}$, for all $1 \le i \le n$.

Next, consider the operator $T : \ell_p^n \to \ell_p^n$, defined by

$$Te_i = \sum_{j=1}^n \langle f_i, g_j \rangle e_j; \quad 1 \le i \le n,$$

and notice that the matrix associated to T has 1's on the diagonal (since $\langle f_i, g_i \rangle = 1$, for all $1 \le i \le n$). Furthermore,

$$\left\|T\left(\sum_{i=1}^n a_i e_i\right)\right\|_p = \left(\sum_{j=1}^n \left|\left\langle \sum_{i=1}^n a_i f_i, g_j \right\rangle\right|^p\right)^{1/p} \le \left(\sum_{j=1}^n \left\|\chi_{A_j} \sum_{i=1}^n a_i f_i\right\|_p^p \|g_j\|_{p'}^p\right)^{1/p}$$

$$\le (1 - \tau_p(\varepsilon))^{-1} \left\|\sum_{i=1}^n a_i f_i\right\|_p \le (1 + \varepsilon)(1 - \tau_p(\varepsilon))^{-1}\left(\sum_{i=1}^n |a_i|^p\right)^{1/p},$$

for any choice of $\{a_i\}_{i=1}^n$, i.e.

$$\|T\|_p \le (1 + \varepsilon)(1 - \tau_p(\varepsilon))^{-1}.$$

Since $(1 + \varepsilon)(1 - \tau_p(\varepsilon))^{-1} \to 1$, as $\varepsilon \to 0$, it follows from Theorem 3.6 that

$$\|I - T\|_p \le \rho_p(\varepsilon),$$

for some function $\rho_p(\varepsilon)$ which satisfies the condition

$$\lim_{\varepsilon\to 0} \rho_p(\varepsilon) = 0.$$

In order to prove that $[f_i]_{i=1}^n$ is well-complemented in L_p, we first define the operator $Q : L_p \to L_p$, by setting,

$$Qf = \sum_{j=1}^n \langle f, g_j\rangle f_j,$$

for $f \in L_p$. Since

$$\|Qf\|_p \le (1 + \varepsilon)\Big(\sum_{j=1}^n |\langle f, g_j\rangle|^p\Big)^{1/p} \le (1 + \varepsilon)(1 - \tau_p(\varepsilon))^{-1}\|f\|_p;\ f \in L_p,$$

it follows that

$$\|Q\|_p \le (1 + \varepsilon)(1 - \tau_p(\varepsilon))^{-1}.$$

Furthermore, if $f = \sum_{i=1}^n a_i f_i$, for some choice of $\{a_i\}_{i=1}^n$, then

$$\begin{aligned}\|(I - Q)f\|_p &= \Big\|\sum_{i=1}^n a_i f_i - \sum_{i=1}^n a_i \sum_{j=1}^n \langle f_i, g_j\rangle f_j\Big\|_p \\ &\le (1 + \varepsilon)\Big(\sum_{j=1}^n \Big|a_j - \sum_{i=1}^n a_i\langle f_i, g_j\rangle\Big|^p\Big)^{1/p} = (1 + \varepsilon)\Big\|(I - T)\Big(\sum_{i=1}^n a_i e_i\Big)\Big\|_p \\ &\le (1 + \varepsilon)\rho_p(\varepsilon)\Big(\sum_{i=1}^n |a_i|^p\Big)^{1/p},\end{aligned}$$

which clearly yields that

$$\|(I - Q)_{|[f_i]_{i=1}^n}\|_p \le \tilde{\tau}_p(\varepsilon),$$

for some function $\tilde{\tau}_p(\varepsilon)$ satisfying the condition

$$\lim_{\varepsilon\to 0} \tilde{\tau}_p(\varepsilon) = 0.$$

Then the desired projection R from L_p onto $[f_i]_{i=1}^n$ is constructed by putting

$$R = \big(Q_{|[f_i]_{i=1}^n}\big)^{-1} Q$$

provided that ε is sufficiently small as to ensure that $\tilde{\tau}_p(\xi) < 1$. Notice that in this case

$$\|R\|_p \leq (1+\varepsilon)(1+\tau_p(\varepsilon))^{-1}(1-\tilde{\tau}_p(\varepsilon))^{-1} = M_p(\varepsilon),$$

where

$$\lim_{\varepsilon\to 0} M_p(\varepsilon) = 1. \qquad []$$

4. GENERALIZATIONS TO SPACES WITH AN UNCONDITIONAL OR SYMMETRIC BASIS.

The invertibility results for matrices with 1's on the diagonal acting on ℓ_p^n-spaces; $1 \leq p \leq \infty$, which were presented in the previous sections, can be extended to certain classes of spaces with an unconditional or symmetric basis. However, some restrictions are needed since, contrary to the cases described in Theorems 1.2, 2.2 and 3.1, the invertibility of a submatrix of a matrix with 1's on the diagonal cannot be always achieved by proving a supression result for the off-diagonal part. This fact is well illustrated by the following example.

Example 4.1. For each integer n, there exist a Banach space X_n with a 1-symmetric normalized basis $\{e_i\}_{i=1}^n$ and a linear operator $S_n : X_n \to X_n$ of norm $\|S_n\|_{X_n} \leq 3$, whose matrix relative to $\{e_i\}_{i=1}^n$ has 0's on the diagonal, such that, whenever σ is a subset of the integer $\{1,2,\ldots,n\}$ of cardinality $|\sigma| \geq n^{2/3}$, then

$$\|R_\sigma S_n R_\sigma\|_{X_n} \geq 1 - n^{-2/3}.$$

In other words, the operator S_n has the feature that no supression theorem for a submatrix of rank k can be proved unless $k < n^{2/3}$.

For sake of simplicity, we shall construct the spaces X_n and the operators S_n only for values of n of the form 2^h. Fix such an integer n, let $m = [n^{2/3}]$ and define the norm in X_n of a vector $x = \sum_{i=1}^n a_i e_i$, by setting

$$\|x\| = \max\{(\sum_{i=1}^n |a_i|^2)^{1/2}, \sum_{i\in\tau} |a_i|; \tau \subset \{1,2,\ldots,n\}, |\tau| = m\}.$$

This definition clearly yields that the unit vectors $\{e_i\}_{i=1}^n$ form a 1-symmetric normalized basis in the space X_n.

Let now $W = W_n$ be the operator on X_n defined by the usual Walsh matrix, normalized in ℓ_∞^n (i.e. such that all its entries are of the form ± 1), and put

$$T = W/m.$$

In order to estimate the norm of T in X_n, notice first that, for any $x = \sum_{i=1}^{n} a_i e_i$ in X_n of norm one, we have that,

$$\|Tx\|_2 = m^{-1}\|Wx\|_2 \leq m^{-1}n^{1/2}\|x\|_2 \leq m^{-1}n^{1/2},$$

since $W/n^{1/2}$ is a unitary matrix. Furthermore, for any subset τ of $\{1,2,\ldots,n\}$ of cardinality $|\tau| = m$, we get that

$$\|R_\tau Tx\|_1 \leq \|R_\tau TR_{\tau_x} x\|_1 + \|R_\tau T(x - R_{\tau_x} x)\|_1,$$

where

$$\tau_x = \{1 \leq i \leq n;\ |a_i| \geq m^{-1}\}.$$

The fact that $\|x\| = 1$ implies that $|\tau_x| < m$ and, thus, the operator $R_\tau TR_{\tau_x}$ is doubly-stochastic. Hence,

$$\|R_\tau TR_{\tau_x} x\|_1 \leq \|R_{\tau_x} x\|_1 \leq \|x\| = 1.$$

On the other hand, we also have that

$$\|R_\tau T(x - R_{\tau_x} x)\|_1 \leq m^{1/2}\|R_\tau T(x - R_{\tau_x} x)\|_2 \leq m^{-1/2}\|W(x - R_{\tau_x} x)\|_2$$

$$\leq m^{-1/2}n^{1/2}\|x - R_{\tau_x} x\|_2 \leq m^{-3/2}n.$$

By combining these estimates, it follows that

$$\|Tx\| = \max\{\|Tx\|_2,\ \|R_\tau Tx\|_1;\ \tau \subset \{1,2,\ldots,n\},\ |\tau| = m\}$$

$$\leq \max\{m^{-1}n^{1/2},\ 1 + m^{-3/2}n\}$$

i.e.

$$\|T\| \leq 2.$$

Consider now a subset σ of $\{1,2,\ldots,n\}$ which has cardinality $|\sigma| \geq n^{2/3}$. Then, for any subset τ of σ of cardinality exactly equal to m and $i \in \tau$, we have

$$\|R_\sigma TR_\sigma e_i\| \geq \|R_\tau Te_i\|_1 = 1$$

since all the entries of T are of the form $\pm m^{-1}$. This implies that

$$\|R_\sigma TR_\sigma\| \geq 1$$

and the construction would be completed provided T had only 0's on the diagonal. Since this is not the case we consider instead the operator

$$S = T - E(T)$$

i.e. T from which we remove the diagonal. However,

$$\|E(T)\| < m^{-1}$$

and, therefore, $\|S\| < 3$ and $\|R_\sigma S R_\sigma\| > 1 - m^{-1}$, as desired. []

In spite of Example 4.1, a supression theorem for matrices acting on a space with a symmetric basis can be proved provided some restrictions on the space are imposed. In order to be able to discuss the nature of the restrictions and state the result, we shall recall the notion of Boyd indices which is commonly used in interpolation theory for rearrangement invariant (r.i.) spaces. Their definition is related to the so-called dilation operator D_s which is defined, as follows.

For $0 < s < \infty$ and f being a measurable function on $[0,\infty)$, the dilation $D_s f$ of f is defined by the formula,

$$(D_s f(t) = f(t/s);\ 0 < t < \infty.$$

The operator D_s dilates the graph of f by the ratio $s:1$ and its norm on an arbitrary r.i. function space X on $[0,\infty)$ can be computed by using only non-increasing functions.

We shall adopt the notation from [23] (rather the original one from [9]) and define the Boyd indices p_X and q_X of a r.i. function space X, by setting,

$$p_X = \lim_{s\to\infty} \frac{\log s}{\log\|D_s\|_X} = \sup_{s>1} \frac{\log s}{\log\|D_s\|_X} \quad \text{and} \quad q_X = \lim_{s\to 0^+} \frac{\log s}{\log\|D_s\|_X} = \inf_{0<s<1} \frac{\log s}{\log\|D_s\|_X}$$

if $\|D_s\|_X \neq 1$ for $s \neq 1$. If $\|D_s\|_X = 1$, for some and hence all $s > 1$, we put $p_X = \infty$. Similarly, if $\|D_s\|_X = 1$, for some $0 < s < 1$, we put $q_X = \infty$. These indices can be defined as well in the case of a space X_n with a normalized 1-symmetric basis $\{e_i\}_{i=1}^n$. One way is to extend by linear interpolation any vector $x = \sum_{i=0}^n a_i e_i \in X_n$ with $a_1 \geq a_2 \geq \ldots \geq a_n \geq 0$ to a non-increasing function $x(t)$ such that $x(t) = 0$ for $t \geq n + 1$. The norm of D_s on X_n can then be calculated as the supremum of $\|\sum_{i=1}^n (D_s x)(i) e_i\|_{X_n}$ over all vectors x, as above, which have norm in X_n equal to one. Another possibility is to define the dilation operator D_s only for s being an integer or the reciprocal of an integer. Then the Boyd indices can be defined by the limits considered above except that $s \to \infty$ or $s \to 0^+$ via the integers, respectively, reciprocals of the integers. For the proofs given in the sequel, it is easier to adopt the former approach. Additional details about Boyd indices and their properties can be found in [23] 2.b.

The Boyd indices of a space X can be obviously interpreted in the following manner: p_X is the supremum of all the numbers $p > 1$ for which there exists a constant $K > \infty$ so that

$$\|D_s\|_X \leq Ks^{1/p},$$

for all $s > 1$, and q_X is the infimum of all $q > \infty$ (if any) for which there exists a constant $K < \infty$ so that

$$\|D_s\|_X \leq Ks^{1/q}$$

for all $0 < s < 1$. We shall use this quantitative formulation of Boyd's indices in the statement of our main result.

THEOREM 4.2. _For every_ $1 < p \leq q < 2$, $K < \infty$ _and_ $\varepsilon > 0$, _there exists a constant_ $c = c(p,q,K,\varepsilon) > 0$ _so that, whenever_ $n \geq c^{-1}$, X _is a space with a normalized basis_ $\{e_i\}_{i=1}^n$ _for which_

$$\|D_s\|_X \leq Ks^{1/p}, \text{ if } s \geq 1, \text{ and } \|D_s\|_X \leq Ks^{1/q}, \text{ if } 0 < s \leq 1.$$

and S _is a linear operator on_ X _whose matrix relative to_ $\{e_i\}_{i=1}^n$ _has_ 0's _on the diagonal, then one can find a subset_ σ _of the integers_ $\{1,2,\dots,n\}$ _so that_ $|\sigma| \geq cn$ _and_

$$\|R_\sigma S R_\sigma\|_X \leq \varepsilon \|S\|_X.$$

It is not clear at all if, in the statement of Theorem 4.2, one has to impose the condition $q < 2$. The proof requires $q < 2$ because of some use of the square function but it is likely that the result remains true whenever the afore condition holds for $1 < p < q < \infty$. This is exactly the condition which is not satisfied by the spaces $\{X_n\}_{n=1}^\infty$, constructed in Example 4.1.

In order to interpret Theorem 4.2 as an interpolation result, we recall first Boyd's theorem [8], [9] (see e.g. also [23] 2.b.3) which aserts that, for any $1 \leq p_0 < p \leq q < q_0 < \infty$ and any $K < \infty$, there exists a constant $C = C(p,q,p_0,q_0,K) < \infty$ so that if X is a space with a normalized 1-symmetric basis satisfying the conditions

$$\|D_s\|_X \leq Ks^{1/p}, \text{ for all } s \geq 1, \text{ and } \|D_s\|_X \leq Ks^{1/q}, \text{ for all } 0 < s \leq 1,$$

and T is a linear operator on X, then

$$\|T\|_X \leq C \max(\|T\|_{p_0}, \|T\|_{q_0}),$$

i.e. T interpolates between $\ell_{p_0}^n$ and $\ell_{q_0}^n$, where $n = \dim X$.

Let us now focus our attention on Theorem 3.3 but first, for an operator S on $\mathbb{R}^n$, denote its graph by $\Gamma(S)$ i.e. put

$$\Gamma(S) = \{(x,Sx);\ x \in \mathbb{R}^n\}.$$

The graph $\Gamma(S)$ will be denoted by $\Gamma_p(S)$ when it is considered as a subspace of $\ell_p^n \oplus \ell_p^n$. Also, let π_2 stand for the projection from $\mathbb{R}^n \oplus \mathbb{R}^n$ onto the second coordinate i.e. $\pi_2(x,y) = y$. With this notation, Theorem 3.3 can be reinterpreted as asserting that, under the conditions appearing in its statement, $R_\eta\pi_2$, considered as an operator from $\Gamma_p(S)$ into ℓ_p^n, is bounded by $C(p)\varepsilon^{1/p}$, for all $1 < p < 2$. In this formulation, Theorem 4.2 could be deduced from Theorem 3.3 by Boyd's interpolation theorem mentioned above except that one encounters two difficulties in the attempt to interpolate the operator $R_\eta\pi_2$. The first concerns the condition required in Theorem 3.3 that S have a "small" norm as an operator from ℓ_2^n into itself. As we shall see below, we can overcome this problem without additional restrictions on the underlying space with a symmetric basis. The other problem one encounters is the fact that $\Gamma_p(S)$ is rather a subspace of $\ell_p^n \oplus \ell_p^n$ and not an ℓ_p^n-space in itself. In general, interpolation theorems do not apply for subspaces of L_p-spaces. In order to overcome this difficulty, too, we shall present in the next section an argument that combines ideas from Boyd's proof together with an exhaustion argument.

As we have already mentioned, the first fact needed in order to proof Theorem 4.2 is the restricted boundedness on ℓ_2^n of an operator acting on a space with a symmetric basis. Since this result, which extends Proposition 3.4, is of interest in itself we shall state it in a more general form.

PROPOSITION 4.3. _Let_ X _be a Banach space with a normalized_ 1-_unconditioned basis_ $\{e_i\}_{i=1}^n$. _Then there exists a sequence_ $\{\lambda_i\}_{i=1}^n$ _of positive reals such that, whenever_ S _is a linear operator on_ X, _one can find a subset_ σ _of_ $\{1,2,\dots,n\}$ _for which_

$$|\sigma| > n/2$$

and

$$\|R_\sigma M_\lambda S M_{\lambda^{-1}} R_\sigma\|_2 < 16\ K_G \|S\|_X,$$

where M_λ _and_ $M_{\lambda^{-1}}$ _denote the operators "multiplication" by the sequence_ $\lambda = \{\lambda_i\}_{i=1}^n$, _respectively_ $\lambda^{-1} = \{\lambda_i^{-1}\}_{i=1}^n$. _Moreover, in the case when_ $\{e_i\}_{i=1}^n$ _is actually a_ 1-_symmetric basis then we have_

$$\|R_\sigma S R_\sigma\|_2 < 16\ K_G \|S\|_X.$$

Proof. Let $\{e_i^*\}_{i=1}^n$ denote the sequence of the functional biorthogonal to $\{e_i\}_{i=1}^n$. By [24] (see also [14], [15] and [30]), one can find positive reals $\{\lambda_i\}_{i=1}^n$ such that

$$\|\sum_{i=1}^n \lambda_i e_i^*\|_{X^*} \cdot \|\sum_{i=1}^n \lambda_i^{-1} e_i\|_X = n.$$

In the case when $\{e_i\}_{i=1}^n$ is actually 1-symmetric, the step is redundant and one can just take $\lambda_i = 1$, for all $1 \le i \le n$.

Suppose now that the assertion of Proposition 4.3 is false for any subset σ of $\{1,2,\ldots,n\}$ of cardinality $|\sigma| \ge n/2$ and some linear operator S on X. Then one can find a vector $y_1 = \sum_{j=1}^n b_{1,j}e_j \in \ell_2^n$ with $\|y_1\|_2 = 1$ and

$$\|M_\lambda S M_{\lambda^{-1}} y_1\|_2 > 16\, K_G \|S\|_X.$$

Hence, there exists an element $y_1^* = \sum_{j=1}^n c_{1,j}e_j \in \ell_2^n$ such that $\|y_1^*\|_2 = 1$ but

$$y_1^* M_\lambda S M_{\lambda^{-1}} y_1 > 16\, K_G \|S\|_X.$$

We continue now by induction. If the vectors $y_i = \sum_{j=1}^n b_{i,j}e_j$ and $y_i^* = \sum_{j=1}^n c_{i,j}e_j$ have already been chosen for $1 \le i \le \ell$ so that $\|y_i\|_2 = \|y_i^*\|_2 = 1$ and

$$y_i^* M_\lambda S M_{\lambda^{-1}} y_i > 16\, K_G \|S\|_S.$$

for all $1 \le i \le \ell$, then we put

$$\tau_{\ell+1} = \{1 \le j \le n;\ \sum_{i=1}^{\ell} |b_{i,j}|^2 < 1,\ \sum_{i=1}^{\ell} |c_{i,j}|^2 < 1\}$$

and if $|\tau_{\ell+1}| < n/2$ we stop the procedure. On the other hand, if $|\tau_{\ell+1}| \ge n/2$ then, by our assumption, there exists a vector $y_{\ell+1} = \sum_{j\in\tau_{\ell+1}} b_{\ell+1,j}e_j$ with $\|y_{\ell+1}\|_2 = 1$ so that

$$\|R_{\tau_{\ell+1}} M_\lambda S M_{\lambda^{-1}} y_{\ell+1}\|_2 > 16\, K_G \|S\|_X.$$

Therefore, one can choose a vector $y_{\ell+1}^* = \sum_{j\in\tau_{\ell+1}} c_{\ell+1,j}e_j$ with $\|y_{\ell+1}^*\|_2 = 1$, for which

$$y_{\ell+1}^* M_\lambda S M_{\lambda^{-1}} y_{\ell+1} > 16\, K_G \|S\|_X.$$

Suppose that this construction stops at step m. Then, with the convention that $b_{i,j} = c_{i,j} = 0$, whenever these indices are not defined, we obtain

$$2m = \sum_{i=1}^{m}\sum_{j=1}^{n}(|b_{i,j}|^2 + |c_{i,j}|^2) \geq \sum_{j\in\tau_{m+1}^c}\sum_{i=1}^{m}(|b_{i,j}|^2 + |c_{i,j}|^2)$$

$$\geq |\tau_{m+1}^c| \geq n/2$$

i.e.

$$m \geq n/4.$$

We also have, by using [22] 1.d.2 (ii) and 1.f.14, that

$$16\, K_G \|S\|_{X^m} \leq \sum_{i=1}^{m} y_i^* M_\lambda S M_{\lambda^{-1}} y_i \leq \left\langle \left(\sum_{i=1}^{m}|M_\lambda^* y_i^*|^2\right)^{1/2}, \left(\sum_{i=1}^{m}|SM_{\lambda^{-1}}y_i|^2\right)^{1/2}\right\rangle$$

$$\leq K_G \|S\|_X \left\|\left(\sum_{i=1}^{m}|M_\lambda^* y_i^*|^2\right)^{1/2}\right\|_{X^*} \left\|\left(\sum_{i=1}^{m}|M_{\lambda^{-1}}y_i|^2\right)^{1/2}\right\|_X$$

$$= K_G \|S\|_X \left\|\sum_{j=1}^{n}\lambda_j\left(\sum_{i=1}^{m}|c_{i,j}|^2\right)^{1/2} e_j^*\right\|_{X^*} \left\|\left(\sum_{j=1}^{m}|b_{i,j}|^2\right)^{1/2} e_j\right\|_X .$$

However, our construction ensures that both expressions $\left(\sum_{i=1}^{m}|b_{i,j}|^2\right)^{1/2}$ and $\left(\sum_{i=1}^{m}|c_{i,j}|^2\right)^{1/2}$ are $\leq \sqrt{2}$, for all $1 \leq j \leq n$. Hence,

$$4n \leq (\sqrt{2})^2 \left\|\sum_{j=1}^{n}\lambda_j e_j^*\right\|_{X^*}\left\|\sum_{j=1}^{n}\lambda_j^{-1}e_j\right\|_X = 2n$$

which is contradictory. □

The interpolation argument needed to complete the proof of Theorem 4.2 will be given only in the next section after the proof of Theorem 3.3 on which is based. Meanwhile, we would like to state the following restricted invertibility result which is an immediate consequence of Theorem 4.2.

COROLLARY 4.4. <u>For every</u> $1 < p \leq q < 2$, $K < \infty$ <u>and</u> $\varepsilon > 0$, <u>there exists a constant</u> $c = c(p,q,K,\varepsilon) > 0$ <u>such that, whenever</u> $n \geq c^{-1}$, X <u>is an</u> n-<u>dimensional Banach space with a</u> 1-<u>symmetric basis for which</u>

$$\|D_s\|_X \leq Ks^{1/p}, \ \underline{\text{if}}\ s \geq 1, \ \underline{\text{and}}\ \|D_s\|_X \leq Ks^{1/2}, \ \underline{\text{if}}\ 0 < s < 1,$$

<u>and</u> T <u>is a linear operator on</u> X <u>of norm</u> $\leq K$ <u>with a matrix having</u> 1's <u>on the diagonal, then one can find a subset</u> σ <u>of the intgers</u> $\{1,2,\ldots,n\}$ <u>of cardinality</u> $|\sigma| \geq cn$ <u>so that</u> $R_\sigma T R_\sigma$ <u>is invertible and</u>

$$\|(R_\sigma T R_\sigma)^{-1}\|_X \leq (1-\varepsilon)^{-1}.$$

As we have already pointed out, there is a good chance that condition $q < 2$ could be relaxed to $q < \infty$. However, in the method of proof of Theorem 4.2 it is essential that X have a symmetric basis rather than an unconditional basis.

For spaces with an unconditional basis, there is a completely different approach which can be used to prove the invertibility of a "large" submatrix of a matrix with 1's on the diagonal. The theorem can be stated, as follows:

THEOREM 4.5. _For every_ $1 < r < \infty$, $c_r > 0$, $\varepsilon > 0$ _and_ $M < \infty$, _there exists a constant_ $C = C(r,c_r,\varepsilon,M) < \infty$ _such that, whenever_ $n \geq C$, X _is a Banach space with a normalized_ 1-_unconditional basis_ $\{e_i\}_{i=1}^n$ _which satisfies the condition_

$$\|\sum_{i=1}^n a_i e_i\| \geq c_r(\sum_{i=1}^n |a_i|^r)^{1/r},$$

for all $\{a_i\}_{i=1}^n$, _and_ T _is a linear operator on_ X _of norm_ $\leq M$ _whose matrix relative to_ $\{e_i\}_{i=1}^n$ _has_ 1's _on the diagonal, then one can find a subset_ σ _of the integers_ $\{1,2,\ldots,n\}$ _of cardinality_ $|\sigma| \geq n^{1-\varepsilon}$ _for which the operator_ $T_\sigma TR_\sigma$ _is invertible and_

$$\|(R_\sigma TR_\sigma)^{-1}\|_X \leq C.$$

The requirement that $\{e_i\}_{i=1}^n$ satisfy a lower r-estimate is not very restrictive and, for instance, it holds whenever the underlying space X has some non-trivial cotype. There are, however, two essential differences between Theorem 4.5 and all the other restricted invertibility results for matrices with 1's on the diagonal, discussed in the paper. The first difference is that "large" in the present context means $n^{1-\varepsilon}$ rather than cn and we do not know if this is best possible. The second difference is that Theorem 4.5, contrary to all the other invertibility results of this type, is not the consequence of a supression theorem for the off-diagonal part. Actually, it is very interesting to compare Theorem 4.5 with Example 4.1. First, notice that the basis of the space X_n, defined in the Example 4.1, satisfies a lower 2-estimate with constant $c_2 = 1$; this explains the role of the expression $(\sum_{i=1}^n |a_i|^2)^{1/2}$ in the definition of the norm in X_n. Therefore, Theorem 4.5 applies to the operators $I + S_n$ and, for every $\varepsilon > 0$, yields a submatrix of $I + S_n$ of rank $n^{1-\varepsilon}$ which is well invertible. On the other hand, the off diagonal part of $I + S_n$, namely S_n, can be supressed only by passing to submatrices of rank $\leq n^{2/3}$.

The proof of Theorem 4.5 is given in [7] Section 6, in detail, and, therefore, we do not reproduce it here.

5. PROOFS.

The object of this section is to supply the proofs of Theorems 3.3 and 4.2 with complete details.

We begin with Theorem 3.3 whose proof consists of a more efficient presentation of the arguments used to prove Theorem 3.1 from [7]. We need first some preliminary results.

PROPOSITION 5.1. There exists a constnat $C < \infty$ with the property that, for every $0 < \varepsilon < 1/4e^2$ and $1 < r < 2$, one can find an integer $n(\varepsilon,r)$ such that, whenever $n > n(\varepsilon,r)$, S is a linear operator on ℓ_r^n and $\{\xi_i\}_{i=1}^n$ is a sequence of $\{0,1\}$-valued independent random variables of mean 4ε over some probability space (Ω,ε,μ), then, with the notation $\tau(\log 1/\tau) = e^2(4\varepsilon)^{r'}$ and

$$\sigma(\omega) = \{1 \leq i \leq n;\ \xi_i(\omega) = 1\},$$

we have

$$J = \int_\Omega \max\{\|R_{\sigma(\omega)}Sx\|_1/\|Sx\|_r;\ x \in \ell_r^n;\ |\text{supp } x| \leq \tau n\}d\mu(\omega) \leq C\varepsilon n^{1/r'}.$$

Proof. Put $h = [\tau n]$ and, for each subset σ of $\{1,2,\dots,n\}$ of cardinality h, select a 1/2-net $G(\sigma)$ in the unit sphere of $[Se_i]_{i\in\sigma}$, considered as a subspace of ℓ_r^n, such that

$$|G(\sigma)| \leq 4^h.$$

Then, by Stirling's formula, the set

$$G = \cup\{G(\sigma);\ \sigma \subset \{1,2,\dots,n\},\ |\sigma| = h\}$$

has cardinality

$$|G| \leq 4^h\binom{n}{h} \leq (4en/h)^h$$

from which it easily follows that

$$|G|^{1/m} \leq B,$$

for $m = [\tau(\log 1/\tau)n]$ and some constant $B < \infty$, independent of n, provided n is sufficiently large. On the other hand,

$$\begin{aligned}
J &\leq \int_\Omega \max\{\|R_{\sigma(\omega)}y\|_1;\ y \in G\}d\mu(\omega) \\
&= \int_\Omega \max\{\sum_{i=1}^n |c_i|\xi_i(\omega);\ y = \sum_{i=1}^n c_ie_i \in G\}d\mu(\omega) \\
&\leq \Big(\Sigma\{\|\sum_{i=1}^n |c_i|\xi_i\|_{L_m}^m;\ y = \sum_{i=1}^n c_ie_i \in G\}\Big)^{1/m} \\
&\leq |G|^{1/m} \max\{\|\sum_{i=1}^n |c_i|\xi_i\|_{L_m};\ y = \sum_{i=1}^n c_ie_i \in G\}.
\end{aligned}$$

Thus, by the estimate above for the cardinality of G and [7] Proposition 1.8, we obtain that

$$J \le BA \left(\frac{m}{\log \frac{\tau(\log 1/\tau)}{(4\varepsilon)^{r'}}} \right)^{1/r'} \le 4e\, BA\varepsilon\, n^{1/r'}.$$

provided n is large enough. This concludes the proof. []

The main argument needed to prove Theorem 3.3, which is given below, consists of an estimate from ℓ_r^n into ℓ_1^n.

PROPOSITION 5.2. <u>There exists a constant</u> $D < \infty$ <u>with the property that, for every</u> $0 < \varepsilon < 1/4e^2$ <u>and</u> $1 < r < 2$, <u>there exist an integer</u> $n(\varepsilon,r)$ <u>and a constant</u> $\rho(\varepsilon,r) > 0$ <u>such that, whenever</u> $n \ge n(\varepsilon,r)$ <u>and</u> S <u>is a linear operator on</u> ℓ_2^n <u>for which</u> $\|S\|_2 \le \rho(\varepsilon,r)$, <u>then one can find a subset</u> σ <u>of</u> $\{1,2,\ldots,n\}$ <u>of cardinality</u> $|\sigma| = [2\varepsilon n]$ <u>so that</u>

$$\|R_\sigma Sx\|_1 \le D\varepsilon n^{1/r'}(\|x\|_r + \|Sx\|_r),$$

<u>for all</u> $x \in \ell_r^n$.

Proof. By Proposition 5.1, if n is sufficiently large one can find a subset σ of $\{1,2,\ldots,n\}$ of cardinality $|\sigma| = [2\varepsilon n]$ so that

$$\|R_\sigma Sx\|_1 \le C\varepsilon n^{1/r'}\|Sx\|_r,$$

for any $x \in \ell_r^n$ whose support is $\le \tau n$ with τ satisfying the condition $\tau(\log 1/\tau) = e^2(4\varepsilon)^{r'}$. The problem is to pass from vectors with "small" support, i.e. support of cardinality $\le \tau n$, to general vectors in ℓ_r^n. To this end, fix a vector $x = \sum_{i=1}^n a_i e_i \ne 0$ and put

$$\tau(x) = \{1 \le i \le n,\ |a_i| < \|x\|_r/(\tau n)^{1/r}\}.$$

$$y = R_{\tau(x)}x \quad \text{and} \quad z = x - y.$$

Since

$$\|x\|_r^r \ge \|z\|_r^r \ge \|x\|_r^r |\tau(x)^c|/\tau n$$

we can conclude that the vector z, which contains the peak part of x, has support satisfying the condition

$$|\operatorname{supp} z| = |\tau(x)^c| \le \tau n.$$

Hence,

$$\|R_\sigma Sz\|_1 \le C\varepsilon n^{1/r'}\|Sz\|_r$$

which yields that

$$\|R_\sigma Sx\|_1 \leq \|R_\sigma Sy\|_1 + \|R_\sigma Sz\|_1 \leq |\sigma|^{1/2}\|R_\sigma Sy\|_2 + C\varepsilon n^{1/r'}\|Sz\|_r$$
$$\leq (2\varepsilon n)^{1/2}\|S\|_2\|y\|_2 + C\varepsilon n^{1/r'}(\|Sx\|_r + \|Sy\|_r)$$
$$\leq (2\varepsilon n)^{1/2}\|S\|_2\|y\|_2 + C\varepsilon n^{1/r'}(\|Sx\|_r + n^{1/r-1/2}\|S\|_2\|y\|_2)$$
$$\leq C\varepsilon n^{1/r'}\|Sx\|_r + (C\varepsilon^{1/2} + 2)(\varepsilon n)^{1/2}\|S\|_2\|y\|_2.$$

However,

$$\|y\|_2^2 = \sum_{i\in\tau(x)} |a_i|^2 < \|x\|_r^{2-r} \sum_{i\in\tau(x)} |a_i|^r/(\tau n)^{2/r-1} \leq \|x\|_r^2/(\tau n)^{2/r-1}$$

i.e.

$$\|y\|_2 \leq \|x\|_r/(\tau n)^{1/r-1/2}.$$

It follows that

$$\|R_\sigma Sx\|_1 \leq C\varepsilon n^{1/r'}\|Sx\|_r + (C + 2)\varepsilon^{1/2} n^{1/r'} \tau^{1/2-1/r}\|S\|_2\|x\|_r$$

and the proof is completed provided we ensure that

$$\|S\|_2 \leq \rho(\varepsilon,r) = \varepsilon^{1/2}\tau^{1/r-1/2}. \qquad []$$

Proof of Theorem 3.3. Fix $\varepsilon > 0$ and $1 < p < 2$, and choose an $r = r(p)$ such that $1 < r < p$. For instance, take $r = (1 + p)/2$.

Then let D, $n(\varepsilon,r)$ and $\rho(\varepsilon,r)$ be given by Proposition 5.2. It follows that, whenever $n \geq n(\varepsilon,r)$ and S is a linear operator on ℓ_2^n which satisfies the condition $\|S\|_2 \leq \rho(\varepsilon,r)$, then one can find a subset σ of $\{1,2,\ldots,n\}$ for which $|\sigma| = [2\varepsilon n]$ and

$$\|R_\sigma Sx\|_1 \leq D\varepsilon n^{1/r'}(\|x\|_r + \|Sx\|_r).$$

for $x \in \ell_r^n$.

It turns out that it is easier to handle the proof in the terminology of function spaces, i.e. we prefer to work with the spaces L_p^n rather than with ℓ_p^n. For a vector $x = \sum_{i=1}^n a_i e_i$, the norms of x in these two spaces, are related by the obvious identity

$$\|x\|_{L_p^n} = \|x\|_p/n^{1/p}.$$

We shall also recall the notation discussed in the previous section: π_2 stands for the projection from the direct sum $\mathbb{R}^n \oplus \mathbb{R}^n$ onto the second $\mathbb{R}^n$ and $\Gamma(S)$ for the graph of S. When $\Gamma(S)$ is considered as a subspace of $L_p^n \oplus L_p^n$ we use the notation $\Gamma_p(S)$ to stress this point.

With these conventions, the above inequality can be reinterpreted as asserting that the operator $W = R_\sigma\pi_2$, considered as acting from $\Gamma_r(S)$

into $L_1^{|\sigma|}$, is bounded by D (one should be careful with the difference in normalization between the norms in $L_1^{|\sigma|}$ and L_1^n).

The main part of the proof of Theorem 3.3 consists of showing that W, considered as an operator acting from $\Gamma_p(S)$ into $L_1^{|\sigma|}$, factorizes through $L_p^{|\sigma|}$ in a suitable manner. This factorization can be proved either directly in a way similar to the proof of Proposition 3.7 from [7] or e.g., by a method suggested by G. Pisier based on the fact that any linear operator from an L_r-space into an L_1-space is p-stable if $1 < r < p \le 2$. We present here a slightly different approach. First, we notice that W, considered as an operator from $\Gamma_p(S)$ into $L_1^{|\sigma|}$, is p-convex in the sense that

$$\|(\sum_{i=1}^m |Wu_i|^p)^{1/p}\|_{L_1^{|\sigma|}} \le D_1(\sum_{i=1}^m \|u_i\|^p_{L_p^{2n}})^{1/p},$$

for any choice of $\{u_i\}_{i=1}^m$ in $\Gamma_p(S)$ and some constant D_1, depending only on D and p. In fact, this property holds for any operator from a subspace of an L_r-space into an L_1-space. In order to verify this assertion in our case, let $\{u_i\}_{i=1}^m$ be a sequence of elements in $\Gamma_p(S)$ and $\{\phi_i\}_{i=1}^m$ a sequence of p-stable independent random variable over some probability space (Ω,Σ,μ), which are normalized in $L_1(\Omega,\Sigma,\mu)$. Then

$$\|(\sum_{i=1}^m |Wu_i|^p)^{1/p}\|_{L_1^{|\sigma|}} = \|\int_\Omega|\sum_{i=1}^m \phi_i(\omega)Wu_i|d\mu(\omega)\|_{L_1^{|\sigma|}}$$

$$\le D\int_\Omega\|\sum_{i=1}^m \phi_i(\omega)u_i\|_{L_r^{2n}}d\mu(\omega) \le D\|(\int_\Omega|\sum_{i=1}^m \phi_i(\omega)u_i|^r)^{1/r}d\mu(\omega)\|_{L_r^{2n}}$$

$$\le D\|\phi_1\|_{L_r}\|(\sum_{i=1}^m |u_i|^p)^{1/p}\|_{L_r^{2n}} \le D\|\phi_1\|_{L_r}(\sum_{i=1}^m \|u_i\|^p_{L_p^{2n}})^{1/p},$$

i.e. $D_1 = D\|\phi_1\|_{L_r}$. The fact that W is p-convex implies that its adjoint W^* is p'-concave with the same constant D_1 when is considered as an operator from $L_\infty^{|\sigma|}$ into $\Gamma_p(S)^*$. Thus, by a result of B. Maurey [26] (see also [23] 1.d.10),

$$\pi_{p'}(W^*: L_\infty^{|\sigma|} \to \Gamma_p(S)^*) \le D_1$$

which further implies, by Pietsch's factorization theorem [29], that there exist an operator U from $L_{p'}^{|\sigma|}$ into $\Gamma_p(S)^*$ of norm equal to one and a diagonal operator V from $L_\infty^{|\sigma|}$ into $L_{p'}^{|\sigma|}$ of norm $\le D$, such that $W^* = UV$.

Once this factorization diagram is proved, the rest of the argument is standard. By dualizing this diagram, we conclude that $W = VU^*$ since V

is clearly self-adjoint. Suppose now that V, which acts from $L_p^{|\sigma|}$ into $L_1^{|\sigma|}$, is defined by $Ve_i = \lambda_i e_i$; $i \in \sigma$. Then, in view of the function space normalization, we get that

$$\|V\|_{L_p^{|\sigma|} \to L_1^{|\sigma|}} = \Big(\sum_{i\in\sigma} |\lambda_i|^{p'}/|\sigma|\Big)^{1/p'} \leq D_1$$

i.e.

$$\sum_{i\in\sigma} |\lambda_i|^{p'} \leq D_1^{p'}|\sigma|.$$

Hence, if we put

$$\eta = \{i \in \sigma;\ |\lambda_i| < 2D_1\}$$

then $|\eta| \geq |\sigma|/2 = [\varepsilon n]$. Furthermore, since

$$R_\eta W = R_\eta \pi_2 = R_\eta VU^*$$

it easily follows that

$$\|R_\eta \pi_2\|_{\Gamma_p(S)\to L_p^{|\sigma|}} \leq (\max_{i\in\eta}|\lambda_i|)\|U^*\|_{\Gamma_p(S)\to L_p^{|\sigma|}} \leq 2D_1$$

i.e.

$$\|R_\eta Sx\|_{L_p^{|\sigma|}} \leq 2D_1\big(\|x\|_{L_p^n} + \|Sx\|_{L_p^n}\big),$$

for all $x \in L_p^n$. The desired inequality is then obtained by switching back to the sequence type notation. □

We pass now to the proof of Theorem 4.2. The main part of the proof consists of a Calderon type distributional inequality which is stated below separately. The statment of this result uses the notion of decreasing rearrangement x^* of an element $x \in \mathbb{R}^n$. A priori, x^* is defined only on the integers $\{1,2,\ldots,n\}$. However, for our purpose, it is more convenient to extend the definition of the decreasing rearrangement x^* to the whole interval $[0,\infty)$, e.g. by linear interpolation, in such a way that $x^*(t) = x^*(1)$, for $0 \leq t \leq 1$, and $x^*(t) = 0$, for $t \geq n+1$.

We caution the reader that in the rest of this section the functions x^*, $(Sx)^*$ and so on, denote always decreasing rearrangements and not linear functionals, as in the previous sections.

PROPOSITION 5.3. For every $\varepsilon > 0$ and $1 < p < 2$, there exist an integer $n(\varepsilon,p)$ and reals $\rho(\varepsilon,p) > 0$ and $D(p) < \infty$ so that, whenever $n \geq n(\varepsilon,p)$ and S is a linear operator on ℓ_2^n of norm $\|S\|_2 \leq \rho(\varepsilon,p)$, then one can find a subset σ of $\{1,2,\ldots,n\}$ of cardinality $|\sigma| = [\varepsilon n]$ such that

$$(R_\sigma Sx)^*(t) \leq D(p)\varepsilon^{1/2}\Phi(x,t),$$

<u>for all</u> $t > 0$ <u>and</u> $x \in \mathbb{R}^n$, <u>where</u>

$$\Phi(x,t) = \int_0^1 (x^* + (Sx)^*)(tu)u^{-1/p'}du + \int_1^\infty (x^* + (Sx)^*)(tu)u^{-1/2}du.$$

Proof. The starting point is Theorem 3.3 which, for $\varepsilon > 0$ and $1 < p < 2$, yields constants $n(\varepsilon,p)$, $\rho(\varepsilon,p)$ and $C(p)$, and, for S satisfying its hypotheses, yields a subset η of $\{1,2,\dots,n\}$ such that $|\eta| = [\varepsilon n]$ and

$$\|R_\eta Sx\|_p \leq C(p)\varepsilon^{1/p}(\|x\|_p + \|Sx\|_p);\ x \in \ell_p^n.$$

We shall prove by exhaustion that there exists a subset σ of η of cardinality $|\sigma| \geq |\eta|/2$ so that

$$(R_\sigma Sx)^*(t) \leq 16C(p)\varepsilon^{1/2}\Phi(x,t),$$

for all $x \in \mathbb{R}^n$ and $t > 0$. Indeed, suppose that this assertion is false, for some $\varepsilon > 0$, $1 < p < 2$ and some operator S. Then there exist an element $x_0 \in \mathbb{R}^n$ and an integer $t_0 \geq 1$ so that, with the notation $\eta_0 = \eta$, we have that

$$(R_{\eta_0} Sx)^*(t_0) > 16C(p)\varepsilon^{1/2}\Phi(x_0,t_0).$$

Since this inequality is homogeneous we can assume without loss of generality that

$$\int_1^\infty (x_0^* + (Sx)^*)(tu)u^{-1/2}du = 1.$$

Moreover, the inequality satisfied by $(R_{\eta_0} Sx_0)^*(t_0)$ yields a subset τ_0 of η_0 of cardinality $|\tau_0| = t_0$ so that

$$|Sx_0|(i) > 16C(p)\varepsilon^{1/2}\Phi(x_0,t_0),$$

for all $i \in \tau_0$. Here, the notation $|Sx|(i)$ refers to the absolute value of the ith coordinate of the vector Sx_0.

Next, we consider the set $\eta_1 = \eta_0 \sim \tau_0$ and if $|\eta_1| < |\eta|/2$ then we stop this construction. On the other hand, if $|\eta_1| \geq |\eta|/2$ then we repeat the procedure with η_1 instead of η_0. In this way, we construct vectors $\{x_j\}_{j=0}^m$, integers $\{t_j\}_{j=0}^m$ and mutually disjoint subsets $\{\tau_j\}_{j=0}^m$ of η such that, for all $1 \leq j \leq m$, we have

(i) $|\tau_j| = t_j$

(ii) $|Sx_j|(i) > 16C(p)\varepsilon^{1/2}\Phi(x_j,t_j);\ i \in \tau_j$,

(iii) $\int_1^\infty (x_j^* + (Sx_j)^*)(tu)u^{-1/2}du = 1$

(iv) $|\eta|/2 \leq \sum_{j=1}^m t_j \leq |\eta|$.

We shall average now the expressions $\|\sum_{j=1}^{m} \varepsilon_j R_\eta S x_j\|_p$, over all choices of signs $\varepsilon_j = \pm 1$, $j = 1,2,\ldots,m$, and obtain in this way a contradiction. Since $\{\tau_j\}_{j=1}^m$ are mutually disjoint subsets we get, by (ii), (iii), and (iv), that

$$\begin{aligned}\int \|\sum_{j=1}^{m} \varepsilon_j R_\eta S x_j\| d\varepsilon &\geq 2^{-1/2} \|(\sum_{j=1}^{m} |R_\eta S x_j|^2)^{1/2}\|_p \\ &= 2^{-1/2} \|\sum_{j=1}^{m} R_{\tau_j} |Sx_j|\|_p > 2^{-1/2}\ 16C(p)\varepsilon^{1/2}(\sum_{j=1}^{m} t_j \Phi(x_j,t_j)^p)^p \\ &\geq 4C(p)\varepsilon^{1/2}[(\sum_{j=1}^{m} t_j \Phi(x_j,t_j)^p)^{1/p} + (2^{-1}\varepsilon n)^{1/p}] \\ &\geq 4C(p)\varepsilon^{1/2}(\sum_{j=1}^{m} t_j \Phi(x_j,t_j)^p)^{1/p} + 2C(p)\varepsilon^{1/2+1/p}\, n^{1/p}.\end{aligned}$$

On the other hand, it follows from Theorem 3.3 that

$$\int \|\sum_{j=1}^{m} \varepsilon_j R_\eta S x_j\|_p d\varepsilon \leq C(p)\varepsilon^{1/p} \int (\|\sum_{j=1}^{m} \varepsilon_j x_j\|_p + \|\sum_{j=1}^{m} \varepsilon_j S x_j\|_p) d\varepsilon.$$

Moreover, if we denote

$$\sigma_j = \{1 \leq i \leq n;\ x_j(i) > x_j^*(t_j)\};\ 1 \leq j \leq m,$$

then we have that

$$\begin{aligned}\int \|\sum_{j=1}^{m} \varepsilon_j x_j\|_p d\varepsilon &\leq \int (\|\sum_{j=1}^{m} \varepsilon_j x_j \chi_{\sigma_j}\|_p + \|\sum_{j=1}^{m} \varepsilon_j x_j \chi_{\sigma_j^c}\|_p) d\varepsilon \\ &\leq (\sum_{j=1}^{m} \|x_j \chi_{\sigma_j}\|_p^p)^{1/p} + n^{1/p-1/2}(\sum_{j=1}^{m} \|x_j \chi_{\sigma_j^c}\|_2^2)^{1/2}.\end{aligned}$$

Each of the expressions appearing in the right hand side of the above inequality can be further estimated by using the inequalities

$$\|x\|_p \leq \|x\|_{p,1} \quad \text{and} \quad \|x\|_2 \leq \|x\|_{2,1}$$

concerning vectors x in the Lorentz spaces $L_{p,1}$ and $L_{2,1}$. It follows that, for $1 \leq j \leq m$, we have

$$\|x_j \chi_{\sigma_j}\|_p \leq \int_0^{t_j} x_j^*(t) t^{-1/p'} dt = t_j^{1/p} \int_0^1 x_j^*(tu) u^{1/p'} du$$

and also

$$\|x_j \chi_{\sigma_j^c}\|_2 \int_{t_j}^\infty x_j^*(t) t^{-1/2} dt = t_j^{1/2} \int_1^\infty x_j^*(tu) u^{-1/2} du \leq t_j^{1/2},$$

in view ot (iii). Hence,

$$\int \|\sum_{j=1}^{m} \varepsilon_j x_j\|_p d\varepsilon \le \Big(\sum_{j=1}^{m} t_j \Phi(x_j,t_j)^p\Big)^{1/p} + n^{1/p-1/2}\Big(\sum_{j=1}^{m} t_j\Big)^{1/2}$$

$$\le \Big(\sum_{j=1}^{m} t_j \Phi(x_j,t_j)^p\Big)^{1/p} + \varepsilon^{1/2} n^{1/p}.$$

Since a similar estimate holds also for the average $\int \|\sum_{j=1}^{m} \varepsilon_j Sx_j\|_p d\varepsilon$ we get that

$$\int \|\sum_{j=1}^{m} \varepsilon_j R_\eta Sx_j\|_p \le 2C(p)\varepsilon^{1/p}\Big(\sum_{j=1}^{m} t_j \Phi(x_j,t_j)^p\Big)^{1/p} + 2C(p)\varepsilon^{1/p+1/2} n^{1/p}.$$

A comparison between the estimate from the above and that from below yields clearly a contradiction. □

Proof of Theorem 4.2. Fix $1 < p \le q < 2$ and $K < \infty$, and choose p_0 so that $1 < p_0 < p$. For $\varepsilon < 0$ and this real number p_0, let $n(\varepsilon,p_0)$, $\rho(\varepsilon,p_0) > 0$ and $D(p_0) < \infty$ be the constants given by Proposition 5.3. Furthermore, let X be a space with a normalized 1-symmetric basis $\{e_i\}_{i=1}^n$ such that

$$\|D_s\|_X \le Ks^{1/p}, \quad \text{for } s \ge 1, \quad \text{and} \quad \|D_s\|_X \le Ks^{1/q}, \quad \text{for } 0 < s < 1,$$

and fix a linear operator S on X with $\|S\|_X \le 1$, whose matrix relative to $\{e_i\}_{i=1}^n$ has 0's on the diagonal. The dimension of X i.e. n will be subject to a bound from below which is described in the sequel.

By Proposition 4.3, there exists a subset σ_0 of $\{1,2,\dots,n\}$ of cardinality $|\sigma_0| \ge 2$ so that

$$\|R_{\sigma_0} SR_{\sigma_0}\|_2 \le 16K_G\|S\|_X \le 16K_G.$$

Thus, by Theorem 2.2, one can find a subset σ_1 of σ_0 such that

$$|\sigma_1| \ge \delta_2(\rho(\varepsilon,p_0)/16K_G)|\sigma_0|$$

and

$$\|R_{\sigma_1} SR_{\sigma_1}\|_2 \le \rho(\varepsilon,p_0).$$

Suppose now that we have taken n sufficiently large as to have

$$\delta_2(\rho(\varepsilon,p_0)/16K_G)n \ge 2n(\varepsilon,p_0).$$

Then $|\sigma_1| \ge n(\varepsilon,p_0)$ and, thus, by Proposition 5.3 applied to the operator $R_{\sigma_1} SR_{\sigma_1}$, acting on $\ell_2^{|\sigma|}$, we conclude the existence of a subset σ of σ_1 such that $|\sigma| = [\varepsilon|\sigma_1|]$ and

$$(R_\sigma SX)^*(t) \leq D(p_0)\varepsilon^{1/2}\Phi_1(x,t),$$

for all $t > 0$ and $x \in [e_i]_{i\in\sigma_1}$, where in the present case

$$\Phi_1(x,t) = \int_0^1 \big(x^* + (R_{\sigma_1} SR_{\sigma_1} x)^*\big)(tu)u^{-1/p_0'}du + \int_1^\infty \big(x^* + (R_{\sigma_1} SR_{\sigma_1} x)^*\big)(tu)u^{-1/2}du.$$

Fix $x \in [e_i]_{i\in\sigma_1}$ and choose a non-increasing vector g in the dual X^* of X such that $\|g\|_{X^*} = s$. Let $g(t)$ denote the extension, by linear interpolation, of g to a function on $[0,\infty)$, as we did in the proof of Proposition 5.3. Then

$$g(R_\sigma Sx) = \int_0^\infty (R_\sigma Sx)(t)g(t)dt \leq \int_0^\infty (R_\sigma Sx)^*(t)g(t)dt$$

$$\leq D(p_0)\varepsilon^{1/2} \int_0^\infty \Phi_1(x,t)g(t)dt.$$

However,

$$\int_0^\infty \Phi_1(x,t)g(t)dt = \int_0^1 \Big\{\int_0^\infty \big(D_{u^{-1}}x^* + D_{u^{-1}}(R_{\sigma_1} SR_{\sigma_1} x)^*\big)(t)g(t)dt\Big\}u^{1/p_0'}du$$

$$+ \int_0^\infty \Big\{\int_0^\infty \big(D_{u^{-1}}x^* + D_{u^{-1}}(R_{\sigma_1} SR_{\sigma_1} x)^*\big)(t)g(t)dt\Big\}u^{-1/2}du$$

$$\leq \Big(\int_0^1 \|D_{u^{-1}}\|_X\, u^{-1/p_0'}du + \int_1^\infty \|D_{u^{-1}}\|_X\, u^{-1/2}du\Big)(\|x\|_X + \|Sx\|_X)$$

$$\leq K\Big(\int_0^1 u^{-1/p-1/p_0'}du + \int_1^\infty u^{-1/q-1/2}du\Big)(\|x\|_X + \|Sx\|_X)$$

$$\leq K\big(1/p_0 - 1/p + (1/q - 1/2)^{-1}\big)(\|x\|_X + \|Sx\|_X).$$

Hence, there is a constant $M = M(p,\varepsilon,K) < \infty$ with the property that

$$\|R_\sigma Sx\|_X \leq M\varepsilon^{1/2}(\|x\|_X + \|Sx\|_X),$$

for all $x \in [e_i]_{i\in\sigma}$. This completes the proof since ε can be taken as small as we desire. □

References

1. J. Anderson, Extensions, restrictions and representations of states on C^*-algebras, Trans. Amer. Math. Soc., 249 (1979), 303-329.

2. K. Ball, Private communication.

3. K. Berman, H. Halpern, V. Kaftal and G. Weiss, Matrix norm inequalities and the relative Diximer property, Integral Eq. and Oper. Theory J., to appear.

4. J. Bourgain, New classes of L_p-spaces, Lecture Notes in Math., 889, Springer-Verlag, Berlin 1981.

5. J. Bourgain, A remark on finite dimensional P_λ-spaces, Studia Math., 72 (1981), 87-91.

6. J. Bourgain, A counterexample to a complementation problem, Compositio Math., 43 (1981) 133-144.

7. J. Bourgain and L. Tzafriri, Invertibility of "large" submatrices with applications to the geometry of Banach spaces and harmonic analysis, Israel J. Math., 57 (1987), 137-224.

8. D. W. Boyd, The spectral radius of averaging operators, Pacific J. Math., 24 (1968), 19-28.

9. D. W. Boyd, Indices of function spaces and their relationship to interpolation, Canadian J. Math., 21 (1969), 1245-1254.

10. L. E. Dor, On projections in L_1, Annals of Math., 102 (1975), 463-474.

11. T. Figiel and W. B. Johnson, large subspaces of ℓ_n^∞ and estimates of the Gordon-Lewis constant, Israel J. Math., 37 (1980), 92-112.

12. T. Figiel, W. B. Johnson and G. Schechtman, Random sign - embeddings from ℓ_r^n; $2 < r < \infty$, Proc. Amer. Math. Soc., to appear.

13. T. Figiel, J. Lindenstrauss and V. Milman, The dimension of almost spherical sections of convex bodies, Acta Math., 139 (1977), 53-94.

14. T. A. Gillespie, Factorization in Banach function spaces, Indag. Math., 43 (1981), 287-300.

15. K. Gregson, Thesis, University of Aberdeen, 1986.

16. R. E. Jamison and W. H. Ruckle, Factoring absolutely convergent series, Math. Ann., 224 (1976), 143-148.

17. W. B. Johnson and L. Jones, Every L_p-operator is an L_2-operator, Proc. Amer. Soc., 72 (1978), 309-312.

18. W. B. Johnson and G. Schechtman, On subspaces of L_1 with maximal distances to Euclidean space, Proc. Research Workshop on Banach Space Theory, Univ. of Iowa (Bor - Luh - Lin, ed.), 1981, 83-96.

19. R. Kadison and I. Singer, Extensions of pure states, Amer. J. Math., 81 (1959), 547-564.

20. B. S. Kashin, Some properties of matrices bounded operators from space ℓ_2^n to ℓ_2^m, Izvestiya Akademii Nauk Armyanoskoi SSR, Mathematika, 15 (1980), 379-394.

21. D. R. Lewis, Finite dimensional subspaces of L_p, Studia Math., 63 (1978), 207-212.

22. J. Lindenstrauss and L. Tzafriri, Classical Banach Spaces I, Sequence Spaces, Springer-Verlag, Berlin 1977.

23. J. Lindenstrauss and L. Tzafriri, Classical Banach Spaces II, Function Spaces, Springer-Verlag, Berlin 1979.

24. G. Ya. Lozanovskii, On some Banach lattices, Siberian Math. J., 10 (1969), 419-431 (English translation).

25. B. Maurey, Thoremes de factorisations pour les operateures a valeurs dans un espace L_p, Asterisque 11, Soc. Math., France 1974.

26. B. Maurey, Type et cotype dans les espaces munis de structure locales inconditionnelles, Seminaire Maurey-Schwartz 1973-74, Exposes 24-25, Ecole Polyt., Paris.

27. B. Maurey, Projections dans L^1 d'apres L. Dor, Seminaire Maurey-Schwartz 1974-75, Expose 21, Ecole Polyt., Paris.

28. Y. Peres, A combinatorial application of the maximal ergodic theorem.

29. A. Pietsch, Absolute p-summierende Abbildugen in normierten Raumen, Studia Math., 28 (1967), 333-353.

30. S. Reisner, On two theorems of Lozanovskii concerning intermediate Banach lattices.

31. I. Z. Ruzsa, On difference sets, Studia Sci. Math. Hungar., 13 (1978), 319-326.

32. N. Sauer, On the density of families of sets, J. Combinatorial Theory Ser. A 13 (1972), 145-147.

33. G. Schechtman, Almost isometric L_p subspaces of $L_p(0,1)$, J. London Math. Soc., 20 (1979), 516-528.

34. S. Shelah, A combinatorial problem: stability and order for models and theories in infinitary languages, Pacific J. Math., 41 (1972), 247-261.

35. E. Szemeredi, On sets of integers containing no k elements in arithmetic progression, Acta Arith., 27 (1975), 199-245.

36. V. N. Vapnik and A. Ya. Cervonenkis, On uniform convergence of the frequencies of events to their probabilities, SIAM Theory of Prob. and Its Appl. 16 (1971), 264-280.

I.H.E.S. and The University of Illinois

The Hebrew University of Jerusalem

THE COMMUTING B.A.P. FOR BANACH SPACES

by

Peter G. Casazza

1. Introduction. In light of Enflo's famous counterexample to the approximation problem [3], the study of "weaker structures" has gained added importance. The most fruitful of these has been the bounded approximation property (B.A.P.) (see section 2 for the definitions), the π_λ-property, and the finite dimensional decomposition property (F.D.D.P.). Johnson, Rosenthal and Zippin [11] examined certain relationships between these weaker structures. Since 1970, this paper has been the standard reference for people working in the area. Essentially no further positive progress has been made on the important problem of finding general conditions which imply the F.D.D.P. for a Banach space X.

Enflo's example [3] was the first in a long series of important counterexamples in the area. Figiel and Johnson [4] then showed the existence of a Banach space which has the approximation property (A.P.) but fails the B.A.P.. Lindenstrauss (see [14]) found a Banach space X with a basis so that X* is separable and fails the A.P. Recently, S. J. Szarek [18] constructed a Banach space with the F.D.D.P. which fails to have a basis.

In the sequel, we will see that the much ignored concept of commuting B.A.P. (C.B.A.P.) plays a central role in passing from "weaker structures" to a F.D.D. for a separable Banach space. In section 2 we give the definitions and review the work to date on these problems. Section 3 is a study of C.B.A.P. and what conditions imply its existence. In section 4 we prove the main results and examine their consequences.

2. Definitions and known results.

We will work with the standard five types of structures for a Banach space determined by the following properties:

1) The approximation property (A.P. for short). A Banach space X is said to have the **approximation property** (A.P. for short) if for every compact subset K of X and for every $\varepsilon > 0$, there is a finite rank operator $T : X \to X$ so that $\| Tx - x \| \leq \varepsilon$, for every $x \varepsilon K$. That is, the identity operator on compact subsets of X can be approximated by finite rank operators.

2) The bounded approximation property (B.A.P., in short). Let $\lambda \geq 1$. A Banach space X is said to have the **λ-metric approximation property** (λ-M.A.P. in short) if for every finite dimensional subspace E of X and every $\varepsilon > 0$, there is a finite rank operator $T : X \to X$ such that $\| T \| \leq \lambda$ and $\| Tx - x \| \leq \varepsilon \| x \|$ for all $x \varepsilon E$. The space X is said to have the **bounded approximation property** if it has the λ-M.A.P. for some $\lambda \geq 1$. If X has λ-M.A.P. for $\lambda = 1$, we say X has the **metric approximation property** (M.A.P. for short).

3) The π property. Let $\lambda \geq 1$. A Banach space X is called a **π_λ-space** if there is a net (directed by inclusion) of finite dimensional subspaces $\{E_\alpha\}_{\alpha \varepsilon A}$ of X, whose union is dense in X, and there is a projection P_α from X onto E_α, with $\| P_\alpha \| \leq \lambda$, for all $\alpha \varepsilon A$. The space X is called a **π-space** if it is a π_λ-space for some $\lambda \geq 1$.

4) The finite dimensional decomposition property (F.D.D.P. for short). A Banach space X is said to have the F.D.D.P. if there is a sequence $\{F_n\}$ of finite dimensional subspaces of X such that each $x \varepsilon X$ has a unique representation

$$x = \sum_{n=1}^{\infty} P_n x$$

with $P_n x \varepsilon F_n$ for all n. The sequence $\{F_n\}$ is called the finite dimensional decomposition for X and we write $X = \sum \oplus F_n$. The functions P_n are bounded linear projections on X and $P_n P_m = \delta_{n,m} P_m$, for all n, m. Moreoever, for each n, the operators

$$Q_n = \sum_{k=1}^{n} P_k$$

is a projection from X onto

$$E_n = [\bigcup_{k=1}^{n} P_k X],$$

$Q_n \to I$ strongly on X, and $Q_n Q_m = Q_{min(n,m)}$ for all n,m. These Q_n's will be called the natural projections of the decomposition.

(5) The basis property (B.P. for short). A sequence $\{x_n\}$ of elements of X is called a basis if each $x \varepsilon X$ has a unique representation

$$x = \sum_{n=1}^{\infty} a_n x_n$$

where $\{a_n\}$ are scalars. A Banach space is said to have the basis property if it has a basis. If $\{x_n\}$ is a basis of X, then the 1-dimensional subspaces $F_n = [x_n]$ form a F.D.D. for X and so we use the notation and terminology from (4) here except that the Q_n's are now called the natural projections of the basis.

It is immediate that: *B.P.* $\Rightarrow$ *A.P.*. Figiel and Johnson [4] have shown that *A.P.* $\not\Rightarrow$ *B.A.P.*. However, an old result of Grothendiek [5] asserts that for a separable dual space, A.P. implies M.A.P. (and hence B.A.P.) It is an important open problem whether a space with the B.A.P. can be renormed to have the M.A.P. It is not known if the B.A.P. implies the π_λ-property. The problem here seems to be that there is no general method for constructing projections from bounded operators at this time. It is known that the "metric" π_λ-property implies the F.D.D.P. for separable Banach spaces. This is a result of Johnson [6]. By the "metric" π_λ-property we mean that X has π_λ-property for every $\lambda > 1$. It is unknown if the π_λ-property implies the F.D.D.P. The strongest result in this direction is due to Johnson, Rosenthal and Zippin [11]:

Theorem A: Let X be a separable Banach space. Then X has the F.D.D.P. if any of the following hold:

(a) X is a π space and X^* has the B.A.P.

(b) X^* is a π space

(c) X is a π space and X is isomorphic to a conjugate Banach space.

A much stronger result is contained in section 4.

An important result classifying separable spaces with the B.A.P. was proved independently by Johnson, Rosenthal and Zippin [11] and Pelczynski [16]:

Theorem B: Let X be a separable Banach space. Then X has the B.A.P. if and only if X embeds complementably into a space with a basis.

S. J. Szarek [18] has given an example of a Banach space with the F.D.D.P. which fails to have a basis. Combined with Theorem B, this says there is a Banach space with a basis which has a complemented subspace which fails to have a basis.

There is an important technique of W. B. Johnson [9] which is the only method we have at this time for producing projections from bounded operators. This technique has the drawback that it requires attaching another space to the original one. We will use this technique in the proof of Theorem 4 below.

If X is a Banach space, let $\{E_n\}_{n=1}^{\infty}$ be a sequence of finite dimensional subspaces of X which is dense in the Banach-Mazur distance in all finite dimensional subspaces of X.

$$\text{For } 1 \le p < \infty,\ C_p(X) = \Big(\sum_{n=1}^{\infty} \oplus E_n\Big)_{\ell_p}, \text{ and } C_0(X) = \Big(\sum_{n=1}^{\infty} \oplus E_n\Big)_{C_0}.$$

The Theorem of Johnson [9] now states:

Theorem C: If X is separable and has the B.A.P., then $X \oplus C_p(X)$ is a π_λ-space.

Even in this setting, we do not know if $X \oplus C_p(X)$ has the F.D.D.P. If $\{E_n\}_{n=1}^{\infty}$ is dense in the class of all finite dimensional Banach spaces, we write just C_p.

(3) Commuting B.A.P.

It follows easily from the definition that a separable Banach space X has the B.A.P. if and only if there is a sequence $\{T_n\}$ of finite rank operators on X so that $T_n \to I$ strongly. (i.e. $T_n(x) \to x$, for every $x \varepsilon X$.) It is also known (see [11]) we may assume that $T_m T_n = T_n$ for all $m > n$. We say that a separable Banach space X has the **commuting B.A.P.** if there is a sequence $\{T_n\}$ of finite rank operators on X so that $T_n \to I$ strongly and $T_n T_m = T_m T_n$, for all m,n.

Clearly commuting B.A.P. is implied by the F.D.D.P. and implies the B.A.P. It is not known if a space with the B.A.P. has commuting B.A.P. It is also unknown if the M.A.P. implies commuting B.A.P. It is not known if the commuting B.A.P. implies X is a π space or vice-versa. However, Johnson [7] has shown that a separable space with the commuting B.A.P. can be renormed to have the metric commuting B.A.P. But it is not known if the metric commuting B.A.P. implies X is a π space or X has the F.D.D.P.

To work with the commuting B.A.P., we need first some weak conditions which imply it. This is the thrust of our next Theorem.

Theorem 1: For a separable Banach space X, the following are equivalent:

(1) There are finite rank operators $T_n : X \to X$ so that $T_n \to I$ strongly and $T_n T_m = T_{min\ (n,m)}$ for all $n \neq m$.

(2) X has the commuting B.A.P.

(3) There are finite rank operators $T_n : X \to X$ so that $T_n \to I$ strongly and $T_n T_m = T_n$ for all $m > n$.

(4) There are finite rank operators

$$T_n : X \to X \text{ so that } T_n \to I \text{ strongly and } \lim_{m \to \infty} \| T_n|_{(I-T_m)X} \| = 0$$

for all n.

(5) There are finite rank operators

$$T_n : X \to X \text{ so that } T_n \to I \text{ strongly and } \lim_{m \to \infty} \| T_n(I - T_m) \| = 0$$

for all n.

(6) There are finite rank operators $T_n : X \to X$ with $T_n \to I$ strongly and finite rank operators $S_n : X^* \to Y = [T_n^* X^*]$ so that $S_n|_Y \to I|_Y$ strongly and $\sup_{n\geq 1} \| S_n \| < \infty$.

Proof: It is known that (1) $\Leftrightarrow$ (2)(see [7]). It is immediate that (1) $\Rightarrow$ (3) $\Rightarrow$ (4) $\Rightarrow$ (5). Also, if we have (5), then

$$\lim_{m\to\infty} \| (I - T_m^*)T_n^* \| = 0$$

implies (since T_n^* is finite rank) $T_m^* \to I$ strongly on $T_n^* X^*$, for all n. Hence, $T_m^*|_Y \to I|_Y$ strongly and we have (6).

To prove that (6) $\Rightarrow$ (1) we will need several results from [11]. For the first result, it is really the "proof" we need so we will state the lemma and sketch the proof from [11]:

Lemma D: Let T be an operator from a Banach space X onto an n-dimensional subspace E of X. Let $k \leq n$ and F be a k-dimensional subspace of X such that $\| T|_F - I|_F \| < \varepsilon < 1$, where $(1-\varepsilon)^{-1}\varepsilon k < 1$. Then

(1) There is an operator S from X onto an n-dimensional subspace of X such that $S|_F = I|_F, \| S - T \| < (1-\varepsilon)^{-1}\varepsilon k \| T \|$, $S^* X^* = T^*(X^*)$, and range $S \subset [E, F]$,

(2) If, in addition, T is a projection then S can be chosen to be a projection and $\| S|_{T(X)} - I|_{T(X)} \| < (1-\varepsilon)^{-1}\varepsilon k$.

Proof: By assumption, if $U = T|_F : F \to T(F)$ then $\| U \| < 1+\varepsilon$ and $\| U^{-1} \| < (1-\varepsilon)^{-1}$. Let $P : E \to T(F)$ be a projection onto of norm $\leq k$. Let $V = U^{-1}P + I_E - P$ and let $S = VT$. Then $\| V - I_E \| = \| U^{-1}P - P \| = \| (U^{-1} - I|_{T(F)})P \| < k\varepsilon(1-\varepsilon)^{-1}$. So V is one to one and T and S have the same null space. Since the null space of T has finite codimension in X, it follows that S^* and T^* have the same range. The rest follows easily from the definition of S. ∎

We also need the following lemma from [11]:

Lemma E: Let X and Y be Banach spaces with $\dim Y < \infty$. Let F be a finite dimensional subspace of X^*, let R be an operator from X^* into Y and let $\varepsilon > 0$. Then there is a w^* continuous operator S from X^* to Y such that

(1) $S|_F = R|_F$

(2) $\| S \| \leq \| R \| (1+\varepsilon)$

(3) $S^*y^* = R^*y^*$ whenever R^*y^* belongs to X (regarding $X \subset X^{**}$)

(4) If $Y \subset X^*$ then there is an operator T on X with T^*=S.

Finally, we need a lemma which is essentially copied from [11] except we use bounded operators in the place of projections and make some minor alterations in the proof. Therefore, we will sketch the proof only.

Lemma F: Suppose there are finite rank operators $T_n : X \to X$ with $T_n \to I$ strongly and finite rank operators $S_n : X^* \to Y = [T_n^* X^*]$ so that $S_n|_Y \to I|_Y$ strongly and

$$\sup_{n\geq 1} \| S_n \| < \infty.$$

Then for all finite dimensional subspaces $E \subset X$ and $F \subset Y$ There is a finite rank operator $L : X \to X$ satisfying:

(a) $Lx = x$ for all $x \varepsilon E$,

(b) $L^*f = f$ for all $f \varepsilon F$,

(c) $L^*X^* \subset Y$,

(d) $\| L \| \leq 2\lambda + 2K + 4\lambda K$ where

$$\lambda = \sup_{n\geq 1} \| T_n \|, \text{ and } K = \sup_{n\geq 1} \| S_n \| .$$

Proof: Choose δ so that $\delta \dim F < 1/4$ and m so large that $\| S_m f - f \| < \delta \| f \|$ for every $f \varepsilon F$. By lemma D, there is a finite rank operator $T : X^* \to X^*$ so that $Tf = f$ for all $f \varepsilon F$, $\| T \| \leq \frac{3}{2}K$, and $TX^* \subset [F, S_m X^*] \subset Y$. By lemma E there is a w^* continuous

$R: X^* \to Y$ so that $rng\, R \subset rng\, T$ and $Rf = f$ for all $f \varepsilon F$. Also by lemma E, there is a operator $S: X \to X$ so that $S^* = R$(so, $rng\, S^* \subset Y$), $S^* f = Rf = f$ for all $f \varepsilon F$, and $\| S \| \leq 2K$. Let $G = [E, rng\, S]$ and choose $\gamma dim\, G < \varepsilon/4$ and n so large that $\| T_n g - g \| < \gamma \| g \|$ for every $g \varepsilon G$. Again by lemma D, there is a finite rank operator Q on X so that $Qg = g$ for all $g \varepsilon G$, and $\| Q \| \leq \frac{3}{2}\lambda$, and $Q^* X^* = T_n^* X^* \subset Y$. Now let $L = S + Q - SQ$. It is easily seen that L is the required operator. ∎

Proof of (6) ⇒ (1) of Theorem 1: This also follows the outline of the proof of Theorem 4.1 from [11]. Let $\{x_n\}$ and $\{y_n\}$ be dense in X, Y respectively, $x_1 = y_1, = 0$. We proceed by induction. Let $Q_1 = T_1$, let $k \geq 1$ and suppose $Q_1, \ldots, Q_k$ have been constructed to satisfy:

(a) $Q_i Q_j = Q_j Q_i = Q_{\min(i,j)}$, for all $1 \leq i, j \leq k, i \neq j$,

(b) $Q_i(X) \supset [x_1, \ldots, x_i]$ for all $1 \leq i \leq k$,

(c) $Y \supset Q_i^* X^* \supset [y_1, \ldots, y_i]$ for all $1 \leq i \leq k$,

(d) $\| Q_i \| \leq 2\lambda + 2K + 2\lambda K$ for all $1 \leq i \leq k$,

Now let $E = [x_{k+1}, Q_k(X)]$ and $F = [y_{k+1}, Q_k^* X^*]$. Let Q_{k+1} be the operator given by lemma F. Then Q_{k+1} has property (d) above by lemma F. Properties (b) and (c) above follow from properties (a) and (b) of lemma F. To see (a), let $i < k$ and note that $Q_{k+1}|_E = I$ implies $Q_{k+1} Q_k = Q_k$. Also, $Q_{k+1} Q_i = Q_{k+1} Q_k Q_i = Q_k Q_i = Q_i$. Similarly, $Q_{k+1}^* Q_i^* = Q_i^*$ for all $1 \leq i \leq k$. Hence, $Q_i Q_{k+1} = Q_i$. ∎

Now we can see where the properties which allow us to get a F.D.D. in Theorem A actually come from. We will group these into a single theorem but note that (1) is due to Johnson [8] and (2) follows from [11]:

Theorem G: Let X be a separable Banach space. Then each of the following implies that X has the commuting B.A.P.:

(1) X^* has the B.A.P.

(2) X has the B.A.P. and X is isomorphic to a dual space.

This theorem will be important in the next section. For now, it gives us conditions which guarantee that a space has the commuting B.A.P.

A recent result of Zippin [19] asserts that a Banach space X has a separable dual X^* if and only if X embeds into a Banach space with a shrinking basis. Recall that a basis $\{x_n\}$ for a Banach space X is said to be shrinking if its dual functionals (given by $x_n^*(x_m) = \delta_{nm})\{x_n^*\}$ form a basis for X^*. Comparing this result to Theorem B, we should ask when X embeds complementably into a space with a shrinking basis. The space X having the B.A.P. is not enough, since X^* may fail A.P. [14] and so X cannot embed complementably into a space with a shrinking basis. Our next Theorem will give the complemented version of Zippin's result.

Theorem 2: For a Banach space X, the following are equivalent:

(1) X^* is separable and has the B.A.P.,

(2) X embeds complementably into a space with a shrinking basis,

(3) X has the shrinking commuting B.A.P. (That is, there is a sequence $\{T_n\}$ of finite rank operators on X so that $T_n \to I$ strongly, $T_nT_m = T_mT_n$ for all m,n and $T_n^* \to I$ strongly on X^*).

Proof: (1) $\Rightarrow$ (3) This follows easily from Theorem 1 and lemmas E,F, applied to X^*. (3) $\Rightarrow$ (1) Since T_n is finite rank so is T_n^*. So $T_n^* \to I$ strongly on X^* implies X^* is separable and has the B.A.P. (2) $\Rightarrow$ (1) This is also quick. Since X embeds complementably in Y and Y has a shrinking basis, X^* embeds complementably in Y^* and Y^* has a basis. Hence, X^* is separable and has the B.A.P. (3) $\Rightarrow$ (2) This is a result of Johnson [6]. ∎

4. Commuting B.A.P. and the F.D.D.P. In this section we will show how the commuting B.A.P. comes into play when constructing F.D.D.'s for a space X. Our first Theorem explains the conditions which appeared in Theorem A of Johnson, Rosenthal and Zippin

[11]. Combined with Theorem G, it shows that in passing from the π_λ-property to the F.D.D.P., they were really checking for the presence of the commuting B.A.P.

Theorem 3: Let X be a separable Banach space. Then X has the F.D.D.P. if and only if X is a π space and X has the commuting B.A.P.

Proof: The "only if" part is immediate since the natural projections of the F.D.D. form a commuting family of projections on X which converge strongly to the identity.

Now assume there is a sequence of finite rank projections $\{\pi_n\}$ on X so that $\pi_n \to I$ strongly and $\pi_m \pi_n = \pi_n$ for all $m \geq n$, and that there is a sequence of finite rank operators $\{T_n\}$ on X so that $T_n \to I$ strongly and $T_n T_m = T_{\min(n,m)}$ for all $n \neq m$. Let $Y = [T_n^* X^*]$. Choose T_{n_1} so that

$$\| T_{n_1}|_{rng\pi_1} - I|_{rng\pi_1} \| < \varepsilon < 1,$$

where

$$(1-\varepsilon)^{-1}\varepsilon(rnk\pi_1)\lambda < \frac{1}{2},$$

and

$$\lambda = (\sup_{n\geq 1} \| T_n \|)(\sup_{n \geq 1} \| \pi_n \|).$$

By Theorem D, there is an operator S_1 on X so that $S_1|_{rng\pi_1} = I|_{rng\pi_1}, \| S_1 - T_{n_1} \| \leq \frac{1}{2}$ and $S_1^* X^* = T_{n_1}^* X^* \subset Y$. We can continue to pick $n_1 < n_2 < \ldots$ and finding $S_1, S_2, \ldots$ so that

(a) $S_i \pi_i = \pi_i$ for all $i = 1, 2, \ldots$

(b) $\| S_i - T_{n_i} \| < \frac{1}{i}$, for all $i = 1, 2, \ldots,$

(c) $S_i^* X^* = T_{n_i}^* X^* \subset Y$ for all $i = 1, 2, \ldots.$

Let $Q_i = \pi_i S_i$ for all i. Then $\pi_j \pi_i = \pi_i$ for all $i \leq j$ implies $S_j \pi_i = \pi_i$ for all $i \leq j$. Hence, $Q_i^2 = (\pi_i S_i)(\pi_i S_i) = \pi_i \pi_i S_i = \pi_i S_i = Q_i$ for all i. That is, the Q_i are finite rank projections on X. Also, $(\pi_i S_i)^* X^* = S_i^* \pi_i^* X^* \subset T_{n_i}^* X^* \subset Y$. Next, fix $x \varepsilon X$ and m. For all $i > m$, $Q_i \pi_m x = \pi_i S_i \pi_m x = \pi_m x$. Thus, $Q_i \to I$ strongly on $E = \bigcup_{m=1}^{\infty} \pi_m X$. Since

E is dense in X, we have $Q_i \to I$ strongly on X. Finally $T_nT_m = T_{\min(n,m)}$ for all $n \neq m$ implies for any $x^* \in X^*$,

$$\lim_{m\to\infty} T_n^*T_m^*(x^*) = T_m^*(x^*).$$

Since T_m^* is finite rank, this implies $T_n^* \to I$ strongly on a dense subset of Y and hence on Y. Now, By Theorem 4.1 of [11], X has a F.D.D. ∎

Theorem B asserts that a space with the B.A.P. embeds complementably into a space with a basis. But both proofs of this result $\Big($[11] and [16]$\Big)$ involve attaching an artificial space to X as a direct summand. This has the inharent problem that properties of the space X may not be carried over to the new space which contains it and has a basis. In individual cases, it was shown that the new space could be chosen to have the properties of X: i.e. reflexivity, type, cotype and so on. But, these did not handle, for example, the case where X fails to contain an embedded copy of C_0 or ℓ_p, for $1 \leq p < \infty$ as well as many others. Our next result shows that in the presence of commuting B.A.P., we can attach a subspace of X to X to get a F.D.D. Therefore, X will be complemented in a space Y with a F.D.D. and Y will have all properties of X which pass to finite direct sums and are subspace properties.

Theorem 4: If X is separable and has the commuting B.A.P., then there is a subspace Y of X so that both Y and $X \oplus Y$ have finite dimensional decompositions.

Proof: Choose finite rank operators T_n on X so that $T_n \to I$ strongly and $T_nT_m = T_{\min(n,m)}$ for all $n \neq m$. For each n let $E_n = T_{2n}(I - T_{2n})X$. Let $Y = [\{E_n\}_{n=1}^{\infty}]$. We will show that $\{E_n\}$ is a F.D.D. for Y and that $X \oplus Y$ has a F.D.D. For any $n \geq i, T_{2n+1}T_{2i} = T_{2i}$, so $T_{2n+1}|_{E_i} = I|_{E_i}$. Also, for any $n \leq i-1, T_{2n+1}(I - T_{2i}) = T_{2n+1} - T_{2n+1}T_{2i} =$

$T_{2n+1} - T_{2n+1} = 0$. Hence, for any $x_i \in E_i$,

$$\begin{aligned} \| \sum_{i=1}^{n} x_i \| &= \| \sum_{i=1}^{n} T_{2n+1} x_i \| \\ &= \| \sum_{i=1}^{n+m} T_{2n+1} x_i \| \\ &\leq \| T_{2n+1} \| \| \sum_{i=1}^{n+m} x_i \| \\ &\leq K \| \sum_{i=1}^{n+m} x_i \| \end{aligned}$$

for all $m = 1, 2, \ldots$, and $K = \sup_{n \geq 1} \| T_n \|$. It is well known (see [14]) that this implies $\{E_n\}$ is a F.D.D. for Y. Now we use a technique of Johnson [9] to find a commuting sequence of finite rank projections π_n on $X \oplus Y$ so that $\pi_n \to I$ strongly. This will give that $X \oplus Y$ has a F.D.D. Let P_n be the natural projection of Y onto E_n and

$$Q_n = \sum_{i=1}^{n} P_i.$$

To simplify the notation, we will replace T_{2n} with T_n, so now $E_n =$ Range $(T_n(I - T_n))$. Let $L_n :$ Range $(T_n(I - T_n)) \to E_n$ be the injection from a subspace of X to a subspace E_n of Y. Now define $\pi_n : X \oplus Y \to X \oplus Y$ by :

$$\pi_n(x, y) = \Big(T_n(x) + L_n^{-1} P_n y, L_n(I - T_n) T_n(x) + L_n(I - T_n) L_n^{-1} P_n(y) + Q_n(y) \Big).$$

It is messy to check that these π_n's have the required properties, so we will break it into steps.

Step I: $\pi_n \to I$ strongly on $X \oplus Y$. Let $m < n$. If $x \in$ RangeT_m, then $T_n(x) = x$ and $(I - T_n)T_n(X) = 0$, so $\pi_n(x, 0) = (x, 0)$. Also, if $y \in Q_m(Y)$, then $Q_m(y) = y$ and $P_n(y) = 0$, so $\pi_n(0, y) = (0, y)$. That is, π_n is the identity on $T_m X$ and on $Q_m(Y)$. So $\pi_n \to I$ strongly on $X \oplus Y$.

For the next three steps, we need to consider $\pi_n \pi_m(x,y)$. We will first write out this expression and then examine the individual parts separately. For any $(x,y) \in X \oplus Y$, let

(1) $x_o = T_m(x) + L_m^{-1}\ P_m\ (y)$ and

(2) $y_o = L_m(I - T_m)T_m(x) + L_m(I - T_m)L_m^{-1}P_m(y) + Q_m(y)$.

Then,

(3) $\pi_n \pi_m(x,y) = \pi_n(x_o, y_o) =$
$$\Big(T_n(x_o) + L_n^{-1}P_n(y_o), L_n(I - T_n)T_n(x_o) + L_n(I - T_n)L_n^{-1}P_n(y_o) + Q_n(y_o)\Big).$$

Step 2: For each n, π_n is a projection. For n=m, and $(x, 0) \in X \oplus Y$, (1) & (2) becomes $x_o = T_n(x)$ and $y_o = L_n(I - T_n)T_n(x)$. So,

$$\begin{aligned}\pi_n(x_o, y_o) &= \Big(T_nT_n(x) + (I - T_n)T_n(x), L_n(I - T_n)T_n(T(x)) + \\ &\quad L_n(I - T_n)(I - T_n)T_n(x)\Big) \\ &= \Big(T_n(x), L_n(I - T_n)T_n(x)\Big) \\ &= \Big(x_o, y_o\Big).\end{aligned}$$

For n=m, and $(o, y) \in X \oplus Y$, (1) and (2) becomes $x_o = L_n^{-1}P_n(y)$ and $y_o = L_n(I - T_n)L_n^{-1}P_n(y) + Q_n(y)$. So,

$$\begin{aligned}\pi_n(x_o, y_o) &= \Big(T_nL_n^{-1}P_n(y) + (I - T_n)L_n^{-1}P_n(y), \\ &\quad L_n(I - T_n)T_nL_n^{-1}P_n(y) + L_n(I - T_n)(I - T_n)L_n^{-1}P_n(y) + Q_nQ_n(y)\Big) \\ &= \Big(L_n^{-1}T_n(y), L_n(I - T_n)L_n^{-1}P_n(y) + Q_n(y)\Big) \\ &= (x_o, y_o).\end{aligned}$$

Hence, $\pi_n = \pi_n^2$ is a projection.

Step III: For $m > n, \pi_n \pi_m = \pi_n$. By assumption, $T_mT_n = T_n = T_nT_m$. So, for $m > n$,

(4) $T_nL_m^{-1} = 0$,

(5) $P_nL_m = 0$,

(6) $Q_nL_m = 0$,

(7) $P_nQ_m = P_n$,

(8) $Q_nQ_m = Q_n$.

Therefore, from (1)–(8),

(9) $T_n(x_o) = T_nT_m(x) + T_nL_m^{-1}P_m(y) = T_n(x)$

(10) $L_n^{-1}P_n(y_o) = L_n^{-1}P_n\Big(L_m(I - T_m)T_m(x)\Big)+$
$L_n^{-1}P_n\Big(L_m(I - T_m)L_m^{-1}P_m(y)\Big) + L_n^{-1}P_nQ_m(y)$
$= 0 + 0 + L_n^{-1}P_n(y) = L_n^{-1}P_n(y)$.

(11) $L_n(I - T_n)T_n(x_o) = L_n(I - T_n)T_nT_m(x) + L_n(I - T_n)T_n\Big(L_m^{-1}P_m(y)\Big) = L_n(I - T_n)T_n(x) + 0 = L_n(I - T_n)T_n(x)$.

(12) $L_n(I - T_n)L_n^{-1}P_n(y_o) = L_n(I - T_n)L_n^{-1}P_n\Big(L_m(I - T_m)T_m(x)\Big)+$
$L_n(I - T_n)L_n^{-1}P_n\Big(L_m(I - T_m)L_m^{-1}P_m(y)\Big) + L_n(I - T_n)L_n^{-1}P_n\Big(Q_m(y)\Big)$
$= 0 + 0 + L_n(I - T_n)L_n^{-1}P_n(y) = L_n(I - T_n)L_n^{-1}P_n(y)$.

(13) $Q_n(y_o) = Q_n\Big(L_m(I - T_m)T_m(x)\Big) + Q_n\Big(L_m(I - T_m)L_m^{-1}P_m(y)\Big) + Q_nQ_m(y)$
$= 0 + 0 + Q_n(y)$
$= Q_n(y)$.

Combining (9)-(13) gives, for $m > n$,

$\pi_n\pi_m(x, y) = \pi_n(x_o, y_o) = \pi_n(x, y)$.

Step IV: For $m < n$, $\pi_n\pi_m = \pi_m$. Since $m < n$, we have:

(14) $P_nL_m = 0$,

(15) $T_nL_m^{-1} = L_m^{-1}$,

(16) $P_nQ_m = 0$,

(17) $T_nT_m = T_m$,

(18) $Q_nL_m = L_m$

(19) $Q_nQ_m = Q_m$.

Therefore, by (1), (2) and (14)–(19) we have,

(20) $T_n(x_o) = T_nT_m(x) + T_nL_m^{-1}P_m(y) = T_m(x) + L_m^{-1}P_m(y)$.

(21) $L_n^{-1}P_n(y_o) = L_n^{-1}P_n\Big(L_m(I - T_m)T_m(x)\Big)+$
$L_n^{-1}P_n\Big(L_m(I - T_m)L_m^{-1}P_m(y)\Big) + L_n^{-1}P_nQ_m(y)$
$= 0 + 0 + 0 = 0.$

(22) $L_n(I - T_n)T_n(x_o) = L_n(I - T_n)T_nT_m(x) + L_n(I - T_n)T_nL_m^{-1}P_m(y) =$

$L_n(I-T_n)T_m(x)+L_n(I-T_n)L_m^{-1}P_m(y) = L_n(T_m-T_m)(x)+L_n(L_m^{-1}-L_m^{-1})P_m(y) = 0.$

(23) $L_n(I - T_n)L_n^{-1}P_n(y_o) = L_n(I - T_n)L_n^{-1}P_n\Big(L_m(I - T_m)T_m(x)\Big)+$
$L_n(I - T_n)L_n^{-1}P_n\Big(L_m(I - T_m)L_m^{-1}P_m(y)\Big) + L_n(I - T_n)L_n^{-1}P_n\Big(Q_m(y)\Big)$
$= 0 + 0 + 0 = 0.$

(24) $Q_n(y_o) = Q_n\Big(L_m(I - T_m)T_m(x)\Big) + Q_n\Big(L_m(I - T_m)L_m^{-1}P_m(y)\Big) + Q_nQ_m(y) =$
$L_m(I - T_m)T_m(x) + L_m(I - T_m)L_m^{-1}P_m(y) + Q_m(y).$

Numbers (20)–(24) yield for $m < n$, $\pi_n\pi_m = \pi_m$.

Steps 1–4 complete the proof of Theorem 4. ∎

Note that the proof of Theorem 4 shows that $X \oplus C_p(X)$ and $X \oplus C_p$ have F.D.D.'s as long as X has the commuting B.A.P. Actually, $X \oplus Cp$ has a basis. By [11], if X^* is separable, we may choose Y in Theorem 4 to have a shrinking F.D.D. If X^* is separable and has the B.A.P., then we may choose Y so that both Y and $X \oplus Y$ have shrinking F.D.D.'s. Finally, note that we have actually shown that for

$$Y = \sum \oplus E_n$$

in Theorem 4,

$$X \oplus \sum_{i=1}^{\infty} \oplus E_{n_i}$$

has the F.D.D.P. uniformly for all $n_1 < n_2 < \dots$. Our next corollary now follows from Theorem 4 and standard F.D.D. techniques (see [1]).

Corollary 5: If X is reflexive and embeds complementably into a space with a unconditional F.D.D., then there is a subspace Y of X so that Y has a unconditional F.D.D. and $X \oplus Y$ has a F.D.D.

It would be important to discover if we could get $X \oplus Y$ to also have a unconditional F.D.D. but our techniques do not show this.

We conclude by considering some of the consequences of Theorem 4. Pisier [17] and Milman and Pisier [15] introduced the notion of a weak Hilbert space.

Definition: (1) A Banach space X is said to be a **weak cotype 2 space** if there is a $0 < \delta < 1$ and a $C_\delta > 1$ so that for every finite dimensional subspace E of X there is a subspace $F \subset X$ with $\dim F \geq \delta \dim E$ and $d(F, \ell_2^{\dim F}) \leq C_\delta$.

(2) A Banach space X is said to be a **weak type 2 space** if there is a $0 < \delta < 1$ and a $C_\delta > 1$ so that for every finite dimensional subspace $E \subset X$ and for every operator $T : E \to \ell_2^n$ (for any n) there is an orthogonal projection $P : \ell_2^n \to \ell_2^n$ with rank $P \geq \delta n$ and an extension $\widetilde{PT} : X \to \ell_2^n$ so that $\| \widetilde{PT} \| \leq C_\delta \| T \|$ and $\widetilde{PT}|_E = PT$.

(3) A Banach space X is a **weak Hilbert space** if X is both a weak type 2 space and a weak cotype 2 space.

Pisier [17] showed that all weak Hilbert spaces have the B.A.P. Since the weak type 2 and weak cotype 2 properties are stable under taking subspaces and finite direct sums, Theorem 4 yields:

Corollary 5: Every separable weak Hilbert space embeds complementably into a weak Hilbert space with a F.D.D.

A major unsolved problem in weak Hilbert space Theory is whether every separable weak Hilbert space has a basis. Corollary 5 is a step in the direction of proving this.

W. B. Johnson [10] proved that convexified Tsirelson's space $T^{(2)}$ (see [2]) has, uniformly, the finite basis property. That is, there is a constant $K > 1$ so that for any finite dimensional subspace $E \subset T^{(2)}$, there is a basis $\{x_i\}_{i=1}^n$ for E whose basis constant is $\leq K$. Recall that the basis constant of $\{x_i\}_{i=1}^n$ is the sup of the norms of the natural projections of the basis (see definition (5) of section 2). We say a Banach space X is crudely finitely representable in a Banach space Y if there is a $K > 1$ so that for every finite dimensional $E \subset X$there is an operator $T: E \to Y$ so that $\| T \| \| \left(T|_{TE}\right)^{-1} \| \leq K$. Since any space finitely representable in $T^{(2)}$ is a weak Hilbert space and hence has the B.A.P., Theorem 4 now yields:

Corollary 7: If a separable Banach space X is crudely finitely representable in $T^{(2)}$ then there is a subspace Y of X so that both Y and $X \oplus Y$ have bases.

Finally, we can give a partial answer to an old question of Johnson and Zippin [12], [13]. Recall that for Banach spaces X_n,

$$\left(\sum_{n=1}^{\infty} \oplus X_n\right)_{\ell p} = \{x = \{x_n\} | x_n \in X_n \text{ and } \| x \| = \left(\sum_{n=1}^{\infty} \| x_n \|_{X_n}^p\right)^{1/p} < \infty\}.$$

By applying the theorem of Johnson and Zippin [12] and Theorem 4 we have:

Corollary 8: If X is a subspace of a quotient space of

$$\left(\sum_{n=1}^{\infty} \oplus E_n\right)_{\ell p} \left(or \left(\sum_{n=1}^{\infty} \oplus E_n\right)_{C_o}\right)$$

where dim $E_n < \infty$, and if X has the B.A.P., then there is a subspace $Y \subset X$ so that

$$Y \approx \left(\sum_{n=1}^{\infty} \oplus F_n\right)_{\ell p}$$

for some dim $F_n < \infty$, and

$$X \oplus Y \approx \left(\sum_{n=1}^{\infty} \oplus G_n\right)_{\ell p}$$

for some dim $G_n < \infty$ (resp. For C_o-sums).

Theorem 4 gives us a technique for constructing F.D.D.'s from "inside" the space. We conjecture that this is "best possible" in the following sense:

Conjecture: There is a Banach space X with the F.D.D.P. so that for any subspace $Y \subset X$, the space $X \oplus Y$ fails to have a basis.

That is, although we can construct F.D.D.'s for X from "inside" of X, we conjecture that we cannot construct bases from inside of X. There is some strong evidence in support of this conjecture. Namely, P. Mankiewicz pointed out to us that a variation of the construction of Szarek [18] yields a Banach space X with the F.D.D.P. so that for every **complemented** subspace Y of X, $X \oplus Y$ fails to have a basis. In particular, $X \oplus X$ fails to have a basis.

We conclude by mentioning that in light of these results, the major question now is whether metric commuting B.A.P. implies the π_λ property. For if it does, than we would have for a separable space X that the commuting B.A.P. implies the F.D.D.P.. Such a result would have important applications throughout Banach space Theory.

BIBLIOGRAPHY

1. P. G. Casazza, *Finite Dimensional Decompositions in Banach spaces*, Contemp. Math. (**52**) 1986, pp. 1–32.
2. P. G. Casazza and T. Shura, *Tsirelson's Space*, To appear in Springer lecture notes.
3. P. Enflo, *A counterexample to the approximation property in Banach Spaces*, Acta. Math. (**130**) 1973, pp. 309–317.
4. T. Figiel and W. B. Johnson, *The approximation property does not imply the bounded approximation property*, Proc. Amer. Math. Soc. (**41**) 1973, pp. 197–200.
5. A. Grothendieck, *Produits Tensoriels Topologigues et espaces nucléaires*, Mem. Amer. Math. So. No.**16** (1955).
6. W. B. Johnson, *Finite-dimensional Schauder decompositions in* π_λ *and dual* π_λ *spaces*, Illinois J. Math. (**14**) 1970, pp. 642–647.
7. W. B. Johnson, *A complementably universal conjugate Banach space and its relation to the approximation property*, Israel J. Math (**13**) 1972, pp. 301–310.
8. W. B. Johnson, *On the existence of strongly series summable Markuschevich bases in Banach spaces*, Trans. Am. Math. Soc. (**157**) 1971, pp. 481–486.
9. W. B. Johnson, *Factoring Compact operators*, Israel J. Math (**9**) 1971, pp. 337–345.
10. W. B. Johnson, *Banach spaces all of whose subspaces have the approximation property*, special topics of Appl. Math. (1980) North Holland, PP. 15–26.
11. W. B. Johnson, H. P. Rosenthal, and M. Zippin, *On Bases, finite dimensional decompositions and weaker structures in Banach spaces*, Israel J. Math. (**9**) 1971, pp. 488–506.
12. W. B. Johnson and M. Zippin, *On subspaces of quotients of* $(\sum G_n)_{\ell p}$ *and* $(\sum G_n)_{Co}$, Israel J. Math (**13**) 1972, pp. 311–316.
13. W. B. Johnson and M. Zippin, *On subspaces and quotient spaces of* $(\sum G_n)_{\ell p}$ *and* $(\sum G_n)_{Co}$, Israel J. Math. (**17**) 1974, pp. 50–55.

14. J. Lindenstrauss and L. Tzafriri, *Classical Banach spaces I*, Ergebnisse de Mathematik, No. **92**, 1977, Springer Verlag.

15. V. Milman and G. Pisier, *Banach spaces with a weak cotype 2 property*, Israel J. Math. (**54**) 1986, pp. 139–158.

16. A. Pelczynski, *Any separable Banach space with the bounded approximation property is a complemented subspace of a Banach space with a basis*, Studia Math. (**40**) 1971, pp. 239–242.

17. G. Pisier, *Weak Hilbert Spaces*, (preprint).

18. S. J. Szarek, *A Banach space without a basis which has the bounded approximation property*, Acta. Math. (**159**) 1987, pp. 81–98.

19. M. Zippin, *Banach spaces with separable duals*, (Preprint).

The research in this paper was supported by NSF DMS 8500938.

The Minimal Normal Extension of a Function of a Subnormal Operator

by

John B Conway
Indiana University
Bloomington, IN 47405

This research was partially supported by National Science Foundation Grant MCS 83-204-26.

Let S be a subnormal operator on a separable Hilbert space $\mathcal{H}$ and let N be its minimal normal extension acting on $\mathcal{K}$. If μ is a scalar valued spectral measure for N and $P^\infty(\mu)$ is the weak* closure of the polynomials in $L^\infty(\mu)$, then for every φ in $P^\infty(\mu)$ the normal operator $\varphi(N)$ leaves $\mathcal{H}$ invariant. This permits the definition of a subnormal operator $\varphi(S)$ as $\varphi(N)|\mathcal{H}$. This functional calculus was studied by the author and R F Olin in [3] (also see [4]). Various other authors have extended some of the results of [3] to the functional calculus resulting from the weak* closure in $L^\infty(\mu)$ of the rational functions with poles off the spectrum of S (for example, see [5] and [6]).

It is easy to see that $\varphi(S)$ is a subnormal operator whenever φ is in $P^\infty(\mu)$ and one of the basic questions answered in [3] is "When is $\varphi(N)$ the minimal normal extension of $\varphi(S)$?" The purpose of this note is to give a new proof of this result. This new proof uses a technique of the author from a forthcoming paper in which a functional calculus for an n- tuple of subnormal operators is studied. The basic ingredient in this proof is the use of the disintegration of measures.

In order to phrase the statement of the theorem and give its proof, it is necessary to understand something about Sarason's Theorem [7] characterizing $P^\infty(\mu)$. For any compactly supported measure μ on $\mathbb{C}$, let

$$G(\mu) = \{ \hat{z}(\rho) : \rho \text{ is a weak* continuous multiplicative linear functional on } P^\infty(\mu) \}.$$

Here, $\hat{z}$ is the Gelfand transform of the polynomial z. A statement of Sarason's Theorem is as follows.

1 THEOREM For any compactly supported measure μ on $\mathbb{C}$ there are mutually singular measures μ_1 and μ_0 such that $\mu = \mu_1 + \mu_0$, the set $G(\mu_1)$ is an open subset of $\mathbb{C}$, and and the following properties hold.

(a) μ_0 is supported on $\mathbb{C} \backslash G(\mu_1)$, $\mu_1 | \partial G(\mu_1)$ is absolutely continuous with respect to harmonic measure for $G(\mu_1)$, and μ_1 is supported on the closure of $G(\mu_1)$.

(b) Each bounded analytic function φ on $G(\mu_1)$ can be assigned boundary values on $\partial G(\mu_1)$ almost everywhere with respect to harmonic measure for $G(\mu_1)$ in such a way that the identity map on polynomials extends to an isometric isomorphism of $H^\infty(G(\mu_1))$ onto $P^\infty(\mu_1)$. Under this identification, a bounded sequence in $P^\infty(\mu_1)$ converges weak* to 0 if and only if the corresponding sequence in $H^\infty(G(\mu_1))$ converges uniformly on compact subsets of $G(\mu_1)$.

(c) The identity map on polynomials extends to an isometric isomorphism of $P^\infty(\mu)$ onto $P^\infty(\mu_1) \oplus L^\infty(\mu_0)$.

(d) The spectrum of z in $P^\infty(\mu_1)$ is $\text{cl}G(\mu_1)$ and $R(\text{cl}G(\mu_1))$ is a Dirichlet algebra.

The proof of this theorem as well as all the necessary background information can also be found in [4].

The statement in part (b) of the preceding theorem that each φ in $H^\infty(G(\mu))$ can be assigned boundary values on $\partial G(\mu)$ a.e. $[\mu_1]$ is one that should command the reader's attention for an extra moment. Since $\mu_1|\partial G(\mu_1)$ is absolutely continuous with respect to harmonic measure for $G(\mu_1)$, this is analogous to the fact that a bounded analytic function on $\mathbb{D}$ has radial limits on $\partial\mathbb{D}$ a.e. with respect to Lebesgue measure. Indeed, if $G(\mu_1) = \mathbb{D}$, the two statements are identical.

Another consequence of part (b) is that $P^\infty(\mu_1)$ and $H^\infty(G(\mu_1))$ can be identified. In fact, Theorem 1 can and will be abbreviated by saying that $P^\infty(\mu) = H^\infty(G(\mu_1)) \oplus L^\infty(\mu_0)$. In particular (this is important in reading the results and proofs below), functions in $P^\infty(\mu_1)$ will be considered as defined and analytic on $G(\mu_1)$ and, conversely, any bounded analytic function on $G(\mu_1)$ will be considered as an element of $P^\infty(\mu_1)$.

The main contribution of this note is a new proof of Theorem 6.1 of [3] whose statement is as follows.

2 THEOREM Suppose that S is a subnormal operator on $\mathcal{H}$ with minimal normal extension N acting on $\mathcal{K}$ and let μ be a scalar valued spectral measure for N. If $\varphi \in P^\infty(\mu)$ and φ is not constant on any component of $G(\mu_1)$, then $\varphi(N)$ is the minimal normal extension of $\varphi(S)$.

The proof of Theorem 2 is postponed until some necessary background is filled in. The first stroke in the painting of this background is the next lemma which is a consequence of Theorem 1. If $\varphi \in H^\infty(\mathbb{D})$ and $\zeta \in \mathbb{C}$, then extend φ to $\partial\mathbb{D}$ by letting $\varphi(e^{i\theta})$ be the value of the radial limit whenever it exists and letting $\varphi(e^{i\theta}) = 0$ if the radial limit does not exist. Thus $\varphi^{-1}(\zeta)$ consists of the points $\{a_n\}$ in $\mathbb{D}$ where $\varphi(z) = \zeta$ (and so $\{a_n\}$ is a Blaschke sequence) together with the points on $\mathbb{D}$ where $\varphi(e^{i\theta}) = \zeta$.

3 LEMMA If $\varphi \in H^\infty(\mathbb{D})$ and φ is not constant, then for any ζ in $\mathbb{C}$ and any measure μ that is carried by $\varphi^{-1}(\zeta)$, $P^\infty(\mu) = L^\infty(\mu)$.

Proof. Let K be the (compact) support of μ and form the polynomially convex hull of K, $\hat{K}$. It is not difficult to see that the hypothesis implies that K consists of the closure of the set $\{ e^{i\theta} : \varphi(e^{i\theta}) = \zeta \}$ together with $\{ z \in \mathbb{D} : \varphi(z) = \zeta \}$ and its limit points on $\partial\mathbb{D}$. Hence K either contains $\partial\mathbb{D}$ or K fails to separate the plane. In the first case $\hat{K} = \text{cl}\mathbb{D}$ and in the second case $\hat{K} = K$. If $\hat{K} = K$, K is a polynomially convex set that does not separate the plane. Thus $P(K) = C(K)$ by Lavrentiev's Theorem so $P^\infty(\mu) = L^\infty(\mu)$.

If $\hat{K} = \text{cl}\mathbb{D}$, the Sarason process will be used to determine $P^\infty(\mu)$ (see [7] or [4], page 405). This process tells us to consider $\mu|\partial\mathbb{D}$ and take the absolutely continuous part of μ with respect to Lebesgue measure on $\partial\mathbb{D}$. But $\mu|\partial\mathbb{D}$ is carried by $\varphi^{-1}(\zeta) \cap \partial\mathbb{D}$, a set of Lebesgue measure 0. So $\mu|\partial\mathbb{D}$ is singular. Put $\mu_0 = \mu|\partial\mathbb{D}$ and $\mu_1 = \mu - \mu_0$. The next step of the process is to examine the compact set $K_1 = \text{cl}\{ z \in \mathbb{D} : |f(z)| \le \|f\|_{L^\infty(\mu_1)}$

for every f in $H^\infty(\mathbb{D})$ }. But μ_1 is carried by the Blaschke sequence $\varphi^{-1}(\zeta) \cap \mathbb{D}$. So taking f in the definition of K_1 to be the corresponding Blaschke product shows that $K_1 = \mathrm{cl}[\, \varphi^{-1}(\zeta) \cap \mathbb{D}]$, a set without interior. Thus the Sarason process stops and $P^\infty(\mu) = L^\infty(\mu)$. ∎

The next result is a special case of the Theorem 2, and will be used to prove it. Indeed, this is the heart of the new proof. For a compactly supported measure on the plane, $P^2(\mu)$ denotes the closure of the polynomials in $L^2(\mu)$.

4 PROPOSITION Let μ be a measure supported on $\mathrm{cl}\mathbb{D}$ such that $G(\mu) = \mathbb{D}$ and $P^\infty(\mu) = H^\infty(\mathbb{D})$. If φ is a non- constant function in $P^\infty(\mu)$, then the minimal normal extension of multiplication by φ on $P^2(\mu)$ is multiplication by φ on $L^2(\mu)$.

Proof. Let $S: P^2(\mu) \to P^2(\mu)$ and $N: L^2(\mu) \to L^2(\mu)$ be defined by $Sh = \varphi h$ and $Nf = \varphi f$ for h in $P^2(\mu)$ and f in $L^2(\mu)$, respectively. Define

$$\text{(5)} \qquad \mathcal{Q} = \text{linear span}\{\, \bar{\varphi}^n p : n \geq 0 \text{ and } p \text{ is a polynomial} \,\}$$

and let $\mathcal{Q}^2(\mu)$ denote the closure of $\mathcal{Q}$ in $L^2(\mu)$. N is the minimal normal extension of S if and only if $\mathcal{Q}^2(\mu) = L^2(\mu)$ ([4], page 128).

Since $\varphi \in P^\infty(\mu) \subset L^\infty(\mu)$ and is therefore defined a.e. $[\mu]$, $\nu \equiv \mu \circ \varphi^{-1}$ is a well - defined regular Borel measure whose support is the μ-

essential range of φ. Let $\mu = \int \lambda_\zeta \, d\nu(\zeta)$ be the disintegration of μ determined by the function φ ([2], page 58). Thus each λ_ζ is a probability measure carried by $\varphi^{-1}(\zeta)$ and if $g \in L^1(\mu)$, then the restriction of g to $\varphi^{-1}(\zeta)$ belongs to $L^1(\lambda_\zeta)$ a.e. $[\nu]$, the function $\zeta \mapsto \int g(z) \, d\lambda_\zeta(z)$] is in $L^1(\nu)$, and

$$\int g \, d\mu = \int \left[\int g(z) \, d\lambda_\zeta(z) \right] d\nu(\zeta). \tag{6}$$

In particular,

$$\int |f(z)|^2 \, d\mu = \int \left[\int |f(z)|^2 \, d\lambda_\zeta(z) \right] d\nu(\zeta) \tag{7}$$

for every f in $L^2(\mu)$.

Let $\mathcal{R} = \{ f \in L^2(\mu) : f|\varphi^{-1}(\zeta) \in P^2(\lambda_\zeta) \text{ a.e. } [\nu] \}$. It follows from (7) that $\mathcal{R}$ is closed in $L^2(\mu)$. In fact, if $\{f_n\} \subset \mathcal{R}$ and $f_n \to f$, then (7) implies there is a subsequence $\{f_{n_k}\}$ and a set Δ with $\nu(\Delta) = 0$ such that $\int |f(z) - f_{n_k}(z)|^2 \, d\lambda_\zeta(z) \to 0$ for $\zeta \notin \Delta$. Moreover the definition of $\mathcal{R}$ is such that there is a Borel set Δ_1 containing Δ with $\nu(\Delta_1) = 0$ and $f_{n_k}|\varphi^{-1}(\zeta)$ in $P^2(\lambda_\zeta)$ for $\zeta \notin \Delta_1$ and for all n_k. Thus $f|\varphi^{-1}(\zeta) \in P^2(\lambda_\zeta)$ for ζ

$\notin \Delta_1$ and so $f \in \mathfrak{R}$.

If p is a polynomial and $n \geq 0$, then $\overline{\varphi}^n p | \varphi^{-1}(\zeta) = \overline{\zeta}^n p \in P^2(\lambda_\zeta)$ for every ζ ; hence $\mathfrak{Q} \subset \mathfrak{R}$. Since $\mathfrak{R}$ is closed, $\mathfrak{Q}^2(\mu) \subset \mathfrak{R}$.

Claim: $\mathfrak{Q}^2(\mu) = \mathfrak{R}$.

Indeed, suppose $f \in \mathfrak{R} \ominus \mathfrak{Q}^2(\mu)$. So for every polynomial p and for $n \geq 0$,

$$
\begin{aligned}
0 &= \langle \overline{\varphi}^n p \, , f \rangle \\
&= \int \left[\int_{\varphi^{-1}(\zeta)} \overline{\varphi}^n p \overline{f} \; d\lambda_\zeta \right] d\nu(\zeta) \\
&= \int \left[\overline{\zeta}^n \int_{\varphi^{-1}(\zeta)} p \overline{f} \; d\lambda_\zeta \right] d\nu(\zeta) .
\end{aligned}
$$

Since $\varphi^m p \in P^2(\mu) \subset \mathfrak{Q}^2(\mu)$ for every $m \geq 0$, this equality remains valid if $\varphi^m p$ is substituted for p above. Thus

$$
\begin{aligned}
0 &= \int \left[\overline{\zeta}^n \int_{\varphi^{-1}(\zeta)} \varphi^m p \overline{f} \; d\lambda_\zeta \right] d\nu(\zeta) \\
&= \int \left[\overline{\zeta}^n \zeta^m \int_{\varphi^{-1}(\zeta)} p \overline{f} \; d\lambda_\zeta \right] d\nu(\zeta) .
\end{aligned}
$$

For a polynomial p let $F_p(\zeta) \equiv \int p\overline{f} \; d\lambda_\zeta$. The preceding equation thus becomes $\int \overline{\zeta}^n \zeta^m F_p(\zeta) \, d\nu(\zeta) = 0$ for all $n, m \geq 0$. By the Stone-

Weierstrass Theorem, $F_p(\zeta) = 0$ a.e. $[\nu]$. Let $\mathbf{P}$ be the set of all polynomials with complex - rational coefficients. Since $\mathbf{P}$ is countable, there is a Borel set Δ with $\nu(\Delta) = 0$ such that $F_p(\zeta) = 0$ whenever $\zeta \notin \Delta$ and $p \in \mathbf{P}$. By taking uniform limits, this implies that $\int p\bar{f}\, d\lambda_\zeta = 0$ whenever $\zeta \notin \Delta$ and p is any polynomial. Thus $f \perp P^2(\lambda_\zeta)$ for $\zeta \notin \Delta$. But $f|\varphi^{-1}(\zeta) \in P^2(\lambda_\zeta)$ a.e. $[\nu]$ since $f \in \mathbf{R}$. Hence $f|\varphi^{-1}(\zeta) = 0$ a.e. $[\nu]$; from (7) it follows that $f = 0$ in $L^2(\mu)$. This establishes the claim.

But Lemma 3 implies that $P^2(\lambda_\zeta) = L^2(\lambda_\zeta)$ for every ζ. Hence $\mathbf{Q}^2(\mu) = L^2(\mu)$ and the proof is complete. ∎

The remainder of this note is devoted to showing how the proof of Theorem 2 can be reduced to Proposition 4. There is no originality in this and it is included for the sake of completeness.

A standard tool in studying minimal normal extension problems is the following result of Ball, Olin, and Thomson [1] which permits a reduction of the problem to the case of a cyclic subnormal operator. Recall that if (X, Ω, μ) is any measure space, then $L^\infty(\mu) = L^\infty(\nu)$ for any measure ν that is equivalent to μ (that is, ν and μ are mutually absolutely continuous). So if $\mathbf{W} \subset L^\infty(\mu)$, $\mathbf{W}$ can also be considered as a subset of $L^\infty(\nu)$. In fact, $\mathbf{W}^\infty(\nu)$, the weak* closure of $\mathbf{W}$ in $L^\infty(\nu)$, is equal to $\mathbf{W}^\infty(\mu)$ whenever ν and μ are mutually absolutely continuous. Let $\mathbf{W}^2(\nu)$ denote the closure of $\mathbf{W}$ in $L^2(\nu)$.

8 THEOREM If (X, Ω, μ) is a finite measure space and $\mathcal{W}$ is a linear subspace of $L^\infty(\mu)$, then

$$\mathcal{W}^\infty(\mu) = \bigcap \{ \mathcal{W}^2(\nu) \cap L^\infty(\nu) : \nu \equiv \mu \} .$$

Proof of Theorem 2. The theorem will be proved by showing through a series of reductions that the general problem can be answered by the special case shown in Proposition 4.

To show that $\varphi(N)$ acting on $\mathcal{K}$ is the minimal normal extension of $\varphi(S)$, it must be shown that

$$\mathcal{K} = \bigvee \{ \varphi(N)^{*n} h : n \geq 0 \text{ and } h \in \mathcal{H} \}.$$

If $\mathcal{Q}$ is defined as in (5) and if $\mathcal{Q}$ is weak* dense in $L^\infty(\mu)$, then $N^* \in$ the weak operator topology closure of $\{ g(N) : g \in \mathcal{Q} \}$. Thus $\bigvee \{ g(N)h : g \in \mathcal{Q} \}$ is a reducing subspace for N that contains $\mathcal{H}$ and hence equals $\mathcal{K}$ by the minimality of N. From here it follows that $\varphi(N)$ is the minimal normal extension of $\varphi(S)$. Thus it suffices to show that

$$\mathcal{Q}^\infty(\mu) = L^\infty(\mu) .$$

To do this, Theorem 8 implies that it suffices to prove that if μ is a compactly supported measure, $\varphi \in P^\infty(\mu)$ and φ is not constant on any component of $G(\mu)$, and if $\mathfrak{A}$ is defined as in (5), then $\mathfrak{A}^2(\mu) = L^2(\mu)$. This is what will be done.

If μ_0 and μ_1 are as in Theorem 1, then $P^\infty(\mu_1) \oplus L^\infty(\mu_0) \subset \mathfrak{A}^2(\mu)$. Since $\mathfrak{A}$ contains the polynomials, it follows that $\mathfrak{A}^2(\mu) = L^2(\mu)$ if and only if $\mathfrak{A}^2(\mu_1) = L^2(\mu_1)$. So it may be assumed that the measure μ_0 in Theorem 1 is 0. That is, it may be assumed that $\mu = \mu_1$ and so $P^\infty(\mu) = H^\infty(G)$, where $G = G(\mu)$.

Let $G_1, G_2, \ldots$ be the components of G and let $\chi_1, \chi_2, \ldots$ be their characteristic functions. Because $\chi_j \in H^\infty(G)$, $H^\infty(G) = H^\infty(G_1) \oplus H^\infty(G_2) \oplus \ldots$. Since $P^\infty(\mu) = H^\infty(G)$, there is a corresponding decomposition of μ as the sum of measures $\mu_1 + \mu_2 + \ldots$ such that $P^\infty(\mu) = P^\infty(\mu_1) \oplus P^\infty(\mu_2) \oplus \ldots$ and $P^\infty(\mu_j) = H^\infty(G_j)$. Since $P^\infty(\mu) \subset \mathfrak{A}^2(\mu)$, $\mathfrak{A}^2(\mu) = \mathfrak{A}^2(\mu_1) \oplus \mathfrak{A}^2(\mu_2) \oplus \ldots$ and $\mathfrak{A}^2(\mu_j)$ contains $P^\infty(\mu_j) = H^\infty(G_j)$. So to show that $\mathfrak{A}^2(\mu) = L^2(\mu)$ is equivalent to showing that $\mathfrak{A}^2(\mu_j) = P^2(\mu_j)$ for all $j \geq 1$. That is, it can be assumed that G is connected.

Since $R(\mathrm{cl}G)$ is a Dirichlet algebra, G is simply connected and if τ is the Riemann map of G onto $\mathbb{D}$, then τ is a weak* generator of $H^\infty(G)$ and τ^{-1} is a weak* generator of $H^\infty(\mathbb{D})$ (see [4], page 409). Thus $\tau \in P^\infty(\mu)$ and is hence defined a.e. $[\mu]$. If $\eta = \mu \circ \tau^{-1}$, a measure on $\mathrm{cl}\mathbb{D}$, then $f \mapsto f \circ \tau$ defines a unitary from $L^2(\eta)$ onto $L^2(\mu)$. This same formula defines an isometric isomorphism of $L^\infty(\eta)$ onto $L^\infty(\mu)$ that is a weak* homeomorphism. Since τ is a weak* generator of $H^\infty(G)$, $f \mapsto f\circ\tau$ maps $P^\infty(\eta) = H^\infty(\mathbb{D})$ onto $P^\infty(\mu) = H^\infty(G)$. So if $n \geq 0$, $(\overline{\varphi^n})\circ\tau^{-1} = \overline{\psi^n}$, where $\psi = \varphi\circ\tau^{-1} \in H^\infty(\mathbb{D})$. These statements imply that the map $f \mapsto f\circ\tau$ defines a unitary from $\mathfrak{A}_1^2(\eta)$ onto $\mathfrak{A}^2(\mu)$, where $\mathfrak{A}_1^2(\eta)$ is the closed linear span in $L^2(\eta)$ of $\{ \overline{\psi^n}p : n \geq 0$ and p is a polynomial $\}$. So to show that $\mathfrak{A}^2(\mu) = L^2(\mu)$ is equivalent to showing that $\mathfrak{A}_1^2(\eta) = L^2(\eta)$. But ψ is a non-constant bounded analytic function on $\mathbb{D}$ and so Proposition 4 implies $\mathfrak{A}_1^2(\eta) = L^2(\eta)$. ∎

Bibliography

[1] J A Ball, R F Olin, and J E Thomson, "Weakly closed algebras of subnormal operators," Ill J Math **22** (1978) 315 - 326.

[2] N Bourbaki, Éléments de Mathématique, Livre VI, Intégration, Chap. 6, Intégration Vectorielle, Hermann et Cie, Paris (1959).

[3] J B Conway and R F Olin, "A functional calculus for subnormal operators, II," Memoirs Amer. Math. Soc. **184** (1977).

[4] J B Conway, Subnormal Operators, Pitman Publ. Co. London (1981).

[5] J Dudziak, "Spectral mapping theorems for subnormal operators," J. Funct. Anal. **56** (1984) 360 - 387.

[6] J Dudziak, "The minimal normal extension problem for subnormal operators," J. Funct. Anal. **65** (1986) 314 - 338.

[7] D Sarason, "Weak - star density of polynomials," J Reine Angew Math **252** (1972) 1 - 15.

Two $C^\star$–Algebra Inequalities

GUSTAVO CORACH, HORACIO PORTA[1] AND LÁZARO RECHT[1]

Dedicated to Mischa Cotlar on his 75th birthday.

The purpose of this note is to prove the following inequalities

$$\|\eta\| \leq \|c\eta a \pm b\eta^* b\| \tag{1}$$

$$\|c\eta b^* \pm b\eta^* c\| \leq K \, \|c\eta a \pm b\eta^* b\| \tag{2}$$

where η and b are elements of a $C^\star$–algebra $\mathcal{A}$, a and c are the positive square roots of $1 + b^*b$ and $1 + bb^*$ respectively, and $K = 2\|b\|\sqrt{1+\|b\|^2}/(1+2\|b\|^2)$.

Inequality (1) has a geometric interpretation that can be briefly described as follows. Denote by Q and P the subsets of $\mathcal{A}$ formed by all projections (=idempotent elements) and all self-adjoint projections. The polar decomposition allows us to define a retraction $\pi : Q \to P$ from Q to P and inequality (1) applies to conclude that the tangent map $d\pi : TQ \to TP$ is a contraction. Inequality (2) has similar application to the study of hyperbolic unitary groups. Suppose $p \in P$ and $B(x, y) = \langle (2p - I)x, y \rangle$ (here we assume that $\mathcal{A}$ is represented in a Hilbert space). Denote by U the group of B–unitary elements of $\mathcal{A}$ and by V the subset of positive elements of U. The "bending" of V in U can be partially described by the following fact: for $Z \in TV_v$ decompose $v^{-1}Z = Z_0 + Z_1$ where Z_0 commutes with p and Z_1 anticommutes with p; then $\|Z_0\| < \|Z_1\|$. This follows from inequality (2). For details and complete proofs we refer to our forthcoming paper *The structure of projections in a $C^\star$–algebra.*

Some abbreviations will be helpful: $r = \|b\|$, $T = ba^{-1} = c^{-1}b$, $D = b^*b(1+b^*b)^{-1} = T^*T$, $F = bb^*(1+bb^*)^{-1} = TT^*$. Define also real-linear maps from $\mathcal{A}$ into itself by $\Delta(\eta) = T\eta^*T$, $\Psi(\eta) = \eta - \Delta(\eta)$, $\Phi(\eta) = c\eta a - b\eta^* b$. Calculations give $\Delta^{2n}(\eta) = F^n \eta D^n$, $\Delta^{2n+1}(\eta) = F^n T\eta^* T D^n$ for $n = 0, 1, 2, \ldots$ Hence using $\|D\| = \|F\| = r^2/(1+r^2)$, $\|T\| = r/\sqrt{1+r^2}$ we get

$$\|\Delta^{2n}(\eta)\| \leq (r^2/(1+r^2))^{2n}\|\eta\|$$
$$\|\Delta^{2n+1}(\eta)\| \leq (r^2/(1+r^2))^{2n+1}\|\eta\|$$

so that $\sum_{k=0}^{\infty} \Delta^k$ converges to the inverse of $\Psi = I - \Delta$. This means that Φ is also invertible, and

$$\Phi^{-1}(\xi) = \Psi^{-1}(c^{-1}\xi a^{-1}) = \sum_{k=0}^{\infty} \Delta^k(c^{-1}\xi a^{-1})$$

[1]Partially supported by CONICET (Argentina) and Universidad de Buenos Aires.

whence

$$\|\Phi^{-1}(\xi)\| \leq \Big(\sum_{k=0}^{\infty}(\frac{r^2}{1+r^2})^k\Big)\, \|c^{-1}\|\, \|a^{-1}\|\, \|\xi\|.$$

But $\|a^{-1}\| = \|c^{-1}\| = (1+r^2)^{-1/2}$ implies

$$\Big(\sum_{k=0}^{\infty}(\frac{r^2}{1+r^2})^k\Big)\|c^{-1}\|\, \|a^{-1}\| = \frac{1}{1-(r^2/(1+r^2))} \cdot \frac{1}{1+r^2} = 1$$

and so $\|\Phi^{-1}(\xi)\| \leq \|\xi\|$. Setting $\xi = c\eta a - b\eta^* b$ gives (1) with $-$ sign. Changing η into $i\eta$ gives the $+$ sign.

To prove (2) we will calculate:

$$\begin{aligned} &c\Phi^{-1}(\xi)b^* - b\big(\Phi^{-1}(\xi)\big)^* c \\ &= \sum_{k=0}^{\infty}\big(c\Delta^k(c^{-1}\xi a^{-1})b^* - b\big(\Delta^k(c^{-1}\xi a^{-1})\big)^* c\big). \end{aligned}$$

The term of order $2n$ reads

$$\begin{aligned} &c\Delta^{2n}(c^{-1}\xi a^{-1})b^* - b\big(\Delta^{2n}(c^{-1}\xi a^{-1})\big)^* c \\ &= cF^n c^{-1}\xi a^{-1} D^n b^* - bD^n a^{-1}\xi^* c^{-1} F^n c. \end{aligned}$$

The identities $cF = Fc$, $Db^* = b^*F$, $T = ba^{-1}$, and $T^* = a^{-1}b^*$ reduce this expression to $F^n(\xi T^* - T\xi^*)F^n$. Similarly for terms of odd order $2n+1$:

$$\begin{aligned} &c\Delta^{2n+1}(c^{-1}\xi a^{-1})b^* - b\big(\Delta^{2n+1}(c^{-1}\xi a^{-1})\big)^* c \\ &= cF^n T a^{-1}\xi^* c^{-1} T D^n b^* - bD^n T^* c^{-1}\xi a^{-1} T^* F^n c, \end{aligned}$$

and using now also the identities $cTa^{-1} = T$, $c^{-1}Tb^* = bT^*c^{-1} = TT^* = F$, and $bD = Fb$ we obtain the simpler expression $F^n(T\xi^*F - F\xi T^*)F^n$. Thus combining the terms of order $2n$ and $2n+1$ the series has the form

$$\begin{aligned} &c\Phi^{-1}(\xi)b^* - b\big(\Phi^{-1}(\xi)\big)^* c \\ &= \sum_{n=0}^{\infty} F^n\big((1-F)\xi T^* - T\xi^*(1-F)\big)F^n. \end{aligned}$$

From $\|1-F\| = \|(1+bb^*)^{-1}\| = (1+r^2)^{-1}$ and $\|T\| = r/\sqrt{1+r^2}$ we get

$$\|(1-F)\xi T^* - T\xi(1-F)\| \leq 2r(1+r^2)^{-3/2}\|\xi\|$$

and therefore

$$\begin{aligned} &\|c\Phi^{-1}(\xi)b^* - b\big(\Phi^{-1}(\xi)\big)^* c\| \\ &\leq \Big(\sum_{n=0}^{\infty}\|F\|^{2n} 2r(1+r^2)^{-3/2}\Big)\|\xi\| \\ &= \left(\sum_{n=0}^{\infty}\Big(\frac{r^2}{1+r^2}\Big)^{2n} \cdot 2r(1+r^2)^{-3/2}\right)\|\xi\|. \end{aligned}$$

This proves (2) with − signs; changing η into $i\eta$ produces the + signs, so the proof is complete.

GUSTAVO CORACH, Instituto Argentino de Matemática, Buenos Aires.
HORACIO PORTA, University of Illinois, Urbana.
LÁZARO RECHT, Universidad Simón Bolívar, Caracas.

GENERALIZED BOCHNER THEOREM IN ALGEBRAIC SCATTERING SYSTEMS

Mischa Cotlar
and
Cora Sadosky[1]

0. INTRODUCTION. The Bochner integral representation of positive definite functions is a central result in harmonic analysis. Its extension to a *generalized Bochner theorem*, GBT, first proved in [19], [20], has applications to another central theory, that of the Hilbert transform; moreover, it was shown in [12], [23], [4], [27] and subsequent papers how it provides new extensions and refinements of classical results in one and several dimensions. The purpose of this survey article is to summarize those papers and outline new results.

The classical Bochner theorem in $\mathbb{R}$ says that if $K(x,y)$ is a positive definite continuous Toeplitz kernel, giving rise to a positive sesquilinear form $\langle f,g\rangle = \int K(x,y)f(x)\overline{g(y)}dxdy$, there exists a positive measure μ in $\mathbb{R}$, such that

$$(0.1)\qquad K(x,y) = \int e_t(x)\overline{e_t(y)}d\mu(t) \quad \text{and} \quad \langle f,g\rangle = \int e_t(f)\overline{e_t(g)}d\mu(t)$$

where $e_t(x) = \exp(itx)$ and $e_t(f)$, are, for $t \in \mathbb{R}$, eigenvalues and eigenfunctionals of the operator $D = i\, d/dx$. Moreover,

1980 *Mathematics Subject Classification* (1985 *Revision*). 47A20, 47B35, 42A50, 42A32.

[1] Research supported in part by the National Science Foundation, Grants DMS-8613265 and DMS-8709934.

(0.2) $\langle Df,g\rangle = \langle f,Dg\rangle$

and (0.1) can be rewritten as

(0.3) $K(x,y) = \langle U_{x-y}1,1\rangle$

where (U_t), for $(U_tf)(x) = e_t(x)f(x)$, is a unitary group in $L^2(\mu)$, showing that U_{x-y} is a unitary dilation of $K(x,y)$.

A general method of M. G. Krein [40], developed by Iu. M. Berezanski in [13], extends the representation (0.1) to other positive sesquilinear forms and operators, satisfying symmetry conditions such as (0.2), and having a family of eigenfunctionals "large enough" to fulfill certain separation conditions (for related ideas see [58], [59] and [45]).

The representation (0.3) led to a general unitary dilation theorem of M. A. Naimark and to the Nagy unitary dilation of contraction semigroups, followed by the Nagy-Foias, N-F, lifting theorem as well as the general dilation theory in [47].

The N-F theory provides a powerful approach to the spectral theory of dissipative operators, clarifying the notion of scattering and characteristic functions, while the N-F lifting theorem also includes the Nehari interpolation theorem, closely related to the BMO spaces in $\mathbf{T}$.

On the other hand, the N-F theory appears to be only remotely related to the theorems of Bochner and Krein. It is natural to ask whether there exists a deeper connection between them, and whether the N-F theorem is also related to other BMO and prediction theory results.

Here we consider another development of Bochner's theory, the GBT, which arose outside the N-F theory, from a suggestion in [16] concerning the possible use of Krein's method in Hilbert transform theory, but which provides a positive answer to the above questions, thus widening the range of the N-F theory and leading towards new directions.

In [30] the GBT was presented as a special line of development of the N-F theory. Here we discuss the relation of the GBT with the method of Krein, and emphasize the features that go beyond the framework of the Nagy-Foias and the Krein theories.

In Section 1 the GBT is given in the simplest context of the one-dimensional trigonometric case. Some of its applications to harmonic analysis, including the theorems of Nehari and Helson-Szegö for one and two weights, are summarized in Section 2. In Section 3 the GBT is extended to general (algebraic) scattering systems. The lifting and dilation properties, as well as their extensions to multiparametric evolution groups, are given in Section 4. That section also includes the statements of the bidimensional generalizations of the Nehari and Helson-Szegö theorem to Helson-Lowdenslager halfplanes.

Section 5 deals with the GBT in function algebras, where integral representation and elementary expansion are no longer equivalent. The Grothendieck Inequality for GTK and the Helson-Szegö theorem for $p \neq 2$ follow. Sections 6 and 7 summarize the connections of the GBT with the Krein method and the theories of local semigroups and stochastic processes. Section 8 is a short account of work in progress.

1. TOEPLITZ-HANKEL TRIPLETS IN $\mathbb{Z}$. In this section V will stand for the vector space of all trigonometric polynomials, $f(t) = \sum \hat{f}(n)e_n(t)$, $e_n(t) = \exp(int)$, defined in the unit circle $\mathbb{T} \sim [0,2\pi)$. We set $\mathbb{Z}_1 = \{n \in \mathbb{Z}: n \geq 0\}$, $\mathbb{Z}_2 = \{n \in \mathbb{Z}: n < 0\}$ and $W_1 = \{f \in V: \hat{f}(n) = 0 \text{ for } n \in \mathbb{Z}_2\}$, $W_2 = \{f \in V: \hat{f}(n) = 0 \text{ for } n \in \mathbb{Z}_1\}$, and let $\tau: V \to V$ denote the shift operator, $(\tau f)(t) = e_1(t)f(t)$, so that

$$\tau W_1 \subset W_1 \quad \text{and} \quad \tau^{-1}W_2 \subset W_2 . \tag{1.1}$$

A sesquilinear form $B: V \times V \to \mathbb{C}$ is said to be *positive* if $B(f,f) \geq 0$ for all f. The form is called *Toeplitz* if

$$B(\tau f,\tau g) = B(f,g) , \quad \forall(f,g) \in V \times V . \tag{1.2}$$

A sesquilinear form $B_0: W_1 \times W_2 \to \mathbb{C}$ is called *Hankel* if

$$B_0(\tau f,g) = B_0(f,\tau^{-1}g) , \quad \forall(f,g) \in W_1 \times W_2 , \tag{1.2a}$$

so that the restrictions of the Toeplitz forms to $W_1 \times W_2$ are Hankel.

For each kernel $K: \mathbb{Z} \times \mathbb{Z} \to \mathbb{C}$ there is a unique form $B_K: V \times V \to \mathbb{C}$ such that $K(m,n) = B_K(e_m,e_n)$ for all $m, n \in \mathbb{Z}$, and $K \mapsto B_K$ is a bijection, so that properties of kernels can be expressed in terms of forms. For instance, K is a *Toeplitz kernel*, i.e. $K(m,n) = K(m+1,n+1)$ iff B_K is a Toeplitz form, and K is a *positive definite* (p.d.) *kernel* iff B_K is a positive form.

To each $t \in \mathbb{T}$ corresponds a positive Toeplitz form B^t, defined by

$$B^t(f,g) = f(t)\overline{g(t)} = \int f\bar{g}\,d\delta_t \tag{1.3}$$

(δ_t the Dirac measure at t), and such forms are called *elementary*. Similarly the elementary kernels are

$$K^t(m,n) = e_m(t)\overline{e_n(t)} = e_{m-n}(t) . \tag{1.3a}$$

More generally, each positive measure μ in $\mathbb{T}$ determines a p.d. Toeplitz kernel K and a positive Toeplitz form B, given by

$$B(f,g) = \int f\bar{g}\,d\mu \quad \text{and} \quad K(m,n) = \int e_{m-n}(t)\,d\mu = \hat{\mu}(m-n) . \tag{1.4}$$

The Bochner-Herglotz theorem asserts that all p.d. Toeplitz kernels K and all positive Toeplitz forms B are given in this way or, equivalently, that all B's and K's have an expansion in elementary forms or kernels,

$$B = \int B^t d\mu(t) , \quad K = \int K^t d\mu(t) . \tag{1.4a}$$

Every positive sesquilinear form B satisfies the Schwarz inequality

$$(1.5) \qquad |B(f,g)| \le |B(f,f)|^{1/2}|B(g,g)|^{1/2}$$

which is related to the following notion of boundedness.

Given two seminorms σ_1 and σ_2 in V, a form $B: V \times V \to \mathbb{C}$ (respectively, $B_0: W_1 \times W_2 \to \mathbb{C}$) is said to be *bounded with respect to* σ_1, σ_2 if

$$(1.6) \qquad |B(f,g)| \le \sigma_1(f)\sigma_2(g) , \qquad \forall(f,g) \in V \times V$$

(respectively,

$$(1.6a) \qquad |B_0(f,g)| \le \sigma_1(f)\sigma_2(g) , \qquad \forall(f,g) \in W_1 \times W_2)$$

and in such cases we write $B \le (\sigma_1,\sigma_2)$ $(B_0 \prec (\sigma_1,\sigma_2))$ and say that (B,σ_1,σ_2) (or (B_0,σ_1,σ_2)) is a *bounded triplet*. If B_0 is the restriction of B to $W_1 \times W_2$ and only (1.6a) holds, then we say that B is *weakly bounded with respect to* σ_1, σ_2, or that (B,σ_1,σ_2) is a *weakly bounded triplet*, and write $B \prec (\sigma_1,\sigma_2)$.

If B_1 and B_2 are two positive sesquilinear forms in $V \times V$ they define two Hilbertian seminorms in V by $\sigma_1(f) = B_1(f,f)^{1/2}$, $\sigma_2(f) = B_2(f,f)^{1/2}$. Accordingly, for $B: V \times V \to \mathbb{C}$ and $B_0: W_1 \times W_2 \to \mathbb{C}$, $B \le (B_1,B_2)$ and $B_0 \prec (B_1,B_2)$ respectively mean that

$$(1.7) \qquad |B(f,g)|^2 \le B_1(f,f)B_2(g,g) , \quad \forall(f,g) \in V \times V$$

or

$$(1.7a) \qquad |B_0(f,g)|^2 \le B_1(f,f)B_2(g,g) , \quad \forall(f,g) \in W_1 \times W_2 ,$$

and we say that the triplets (B,B_1,B_2) and (B_0,B_1,B_2) are bounded. If (1.7) holds only for $(f,g) \in W_1 \times W_2$ we write $B \prec (B_1,B_2)$ and say that the triplet (B,B_1,B_2) is weakly bounded.

If B_0 and B_1 are Toeplitz forms and B_0 is a Hankel form (respectively, B is a Toeplitz form) (B_0,B_1,B_2) is a *Toeplitz-Hankel triplet* (respectively, (B,B_1,B_2) is a *Toeplitz triplet*). Thus, (B,B_1,B_2) is a bounded Toeplitz triplet if the three forms are Toeplitz and $B < (B_1,B_2)$.

As observed in (1.5), every positive Toeplitz form B gives rise to a special bounded Toeplitz triplet (B,B,B) as well as to a special bounded Toeplitz-Hankel triplet (B_0,B,B), where $B_0 = B|_{W_1 \times W_2}$, and the Bochner theorem provides an integral representation for such special bounded triplets. In this section, and in Section 3, the Bochner integral representation is extended to general bounded Toeplitz-Hankel triplets (B_0,B_1,B_2). A corresponding representation for bounded triplets (B_0,σ_1,σ_2), where B_0 is a Hankel form, is given in Section 5.

To each Toeplitz-Hankel triplet (B_0,B_1,B_2) is associated a quadruple of forms $(B_{\alpha\beta})$, $\alpha,\beta = 1,2$, where

(1.8) $\quad B_{11} = B_1,\ B_{22} = B_2,\ B_{12} = B_0$ and $B_{21} = B_0^*$

in the sense that $B_{21}: W_2 \times W_1 \to \mathbb{C}$ and $B_{21}(f,g) = \overline{B_0(g,f)}$ for $(f,g) \in W_2 \times W_1$. Since $W_1 \oplus W_2 = V$, there is also a unique associated form $B: V \times V \to \mathbb{C}$, satisfying, for each $\alpha,\beta = 1,2$,

(1.9) $\quad B(f,g) = B_{\alpha\beta}(f,g)$ if $(f,g) \in W_\alpha \times W_\beta$.

A form B as in (1.9) is called a *generalized Toeplitz form*, GTF, and it is easy to see that a Toeplitz-Hankel triplet (B_0,B_1,B_2) is bounded iff the associated GTF B is positive, and iff the associated quadruple $(B_{\alpha\beta})$ satisfies

(1.9a) $\quad \sum_{\alpha,\beta=1,2} B_{\alpha\beta}(f_\alpha,\bar{f}_\beta) \geq 0 \ , \quad \forall (f_1,f_2) \in W_1 \times W_2$.

Setting, for each $(f_1,f_2),\ (g_1,g_2) \in W_1 \times W_2$, $\langle (f_1,f_2),(g_1,g_2)\rangle = \sum_{\alpha,\beta} B_{\alpha\beta}(f_\alpha,g_\beta)$, (1.9a) expresses that $\langle\ ,\ \rangle$ defines a Hilbert metric in $W_1 \times W_2$, giving rise to an associated Hilbert space H.

A kernel $K: \mathbb{Z} \times \mathbb{Z} \to \mathbb{C}$ is a *generalized Toeplitz kernel*, GTK, if B_K is a GTF or, equivalently, if there exist four Toeplitz kernels $(K_{\alpha\beta})$, $\alpha,\beta = 1,2$, such that, for each $\alpha,\beta = 1,2$,

(1.10) $\quad K(m,n) = K_{\alpha\beta}(m,n)$ if $(m,n) \in \mathbb{Z}_\alpha \times \mathbb{Z}_\beta$.

Thus, *it is the same to give a p.d. GTK or a positive GTF or a bounded Toeplitz-Hankel triplet.*

Given three complex measures $\mu_0,\ \mu_1,\ \mu_2$ in $\mathbb{T}$, we write $\mu_0 \leq (\mu_1,\mu_2)$ if

(1.11) $\quad \mu_1 \geq 0,\ \mu_2 \geq 0$ and $|\mu_0(\Delta)|^2 \leq \mu_1(\Delta)\mu_2(\Delta)$

for every Borel set Δ in $\mathbb{T}$. Similarly, $(\mu_{\alpha\beta}) \geq 0$ stands for $\mu_0 \leq (\mu_1,\mu_2)$ if $\mu_{11} = \mu_1,\ \mu_{12} = \mu_0,\ \mu_{21} = \bar{\mu}_0$ and $\mu_{22} = \mu_2$.

If $\mu_0 \leq (\mu_1,\mu_2)$, setting $\rho = \mu_1 + \mu_2$, both μ_1 and μ_2 are absolutely continuous with respect to ρ and, therefore, so is μ_0, and we can write $d\mu_\alpha = \omega_\alpha(t)d\rho$ for $\alpha = 0,1,2$, with

(1.11a) $\quad \omega_1 \geq 0,\ \omega_2 \geq 0$ and $|\omega_0(t)|^2 \leq \omega_1(t)\omega_2(t)$ a.e. ρ

or, equivalently, $d\mu_{\alpha\beta} = \omega_{\alpha\beta}(t)d\rho$, $\alpha,\beta = 1,2$,

(1.11b) $\quad (\omega_{\alpha\beta}(t)) \geq 0$ a.e. ρ .

If $\mu_0 \leq (\mu_1,\mu_2)$ and, for $\alpha = 0,1,2$, the ω_α's are as in (1.11a), set

(1.12) $\quad B_\alpha^{t\omega}(f,g) = f(t)\overline{g(t)}\omega_\alpha(t)$, and call $(B_0^{t\omega}, B_1^{t\omega}, B_2^{t\omega})$ an *elementary triplet*. Note that (1.11a) insures the boundedness of the elementary triplets. Similarly, the *elementary kernels* $K^{t\omega}$ are defined by

(1.12a) $$K^{t\omega}(m,n) = \omega_{\alpha\beta}(t)e_{m-n}(t) \quad \text{if} \quad (m,n) \in \mathbf{Z}_\alpha \times \mathbf{Z}_\beta, \quad \alpha,\beta = 1,2 .$$

THEOREM I (*GBT in* $\mathbf{Z}$ [20]). *Given a bounded Toeplitz-Hankel triplet* (B_0,B_1,B_2), *there exist three complex measures* μ_0, μ_1, μ_2 *in* $\mathbf{T}$, $\mu_0 \leq (\mu_1,\mu_2)$ *as in* (1.11), *such that, for* $\alpha = 1,2$,

(1.13)
$$B_0(f,g) = \int f\bar{g}\, d\mu_0, \quad \forall(f,g) \in W_1 \times W_2 ,$$
$$B_\alpha(f,g) = \int f\bar{g}\, d\mu_\alpha, \quad \forall(f,g) \in V \times V .$$

Similarly, for every p.d. GTK K, *there exists a* 2×2 *measure matrix* $(\mu_{\alpha\beta}) \geq 0$, *such that, for* $\alpha,\beta = 1,2$,

(1.13a) $$K(m,n) = \hat{\mu}_{\alpha\beta}(n-m) = \int e_{m-n}(t)d\mu_{\alpha\beta} \quad \textit{if} \quad (m,n) \in \mathbf{Z}_\alpha \times \mathbf{Z}_\beta .$$

The integral representation (1.13) expresses that (B_0,B_1,B_2) has an expansion in elementary triplets $(B_0^{t\omega}, B_1^{t\omega}, B_2^{t\omega})$, i.e., that for $\alpha = 0,1,2$,

(1.14) $$B_\alpha = \int B_\alpha^{t\omega} d\rho(t) .$$

Similarly,

(1.14a) $$K = \int K^{t\omega} d\rho(t) .$$

Theorem I has been proved in several ways (cf. [12], [57]). One of them ([9]) is based on the fact that the shift τ induces an isometric operator in the associated Hilbert space H (see (1.9a)), and every generalized spectral measure E of this operator provides the desired measures through the formulae

(1.15) $$\mu_1 = \langle Ee_0,e_0\rangle, \quad \mu_2 = \langle Ee_{-1},e_{-1}\rangle, \quad \mu_0 = \langle Ee_0,e_{-1}\rangle .$$

In Section 5 the idea of another proof (cf. [21] and [28]) is given, to include the case of forms bounded with respect to non-Hilbertian seminorms. Theorem I extends to more general settings:

(1) Given a Hilbert space N we set $L(N)$ as the set of bounded linear operators in N, $V(N) = \{\sum_{\text{finite}} a_n e_n(t) \colon a_n \in N\}$, and consistently define $W_1(N)$, $W_2(N)$ and $\tau\colon V(N) \to V(N)$. Then, there is a 1-1 correspondence between the forms $B\colon V(N) \times V(N) \to \mathbf{C}$ and the operator-valued kernels $K\colon \mathbf{Z} \times \mathbf{Z} \to L(N)$. As shown in [10], [4], [27], Theorem I is still valid for GTFs in $V(N)$ or, equivalently, for $L(N)$-valued p.d. GTKs. The difference in such cases is that the measures $\mu_{\alpha\beta}$, $\alpha,\beta = 1,2$, in the representation (1.13) are $L(N)$-valued, and $(\mu_{\alpha\beta}) \geq 0$ has to be interpreted in the obvious Hilbert space sense.

(2) More generally, given two Hilbert spaces, N_1 and N_2, we set $V(N_1,N_2) = \{\sum_{finite} a_n e_n(t): a_n \in N_1 \text{ for } n \geq 0, a_n \in N_2 \text{ for } n < 0\}$. Now the notion of Toeplitz forms or kernels makes no sense, since the shift is not defined in the whole of $V(N_1,N_2)$, but we still can speak of GTFs B, given by kernels K such that $K(m,n) \in L(N_\alpha,N_\beta)$ if $(m,n) \in \mathbb{Z}_\alpha \times \mathbb{Z}_\beta$, $\alpha,\beta = 1,2$, and such K is called a GTK of (N_1,N_2)-type. Theorem I holds in this case, with $\mu_{\alpha\beta}(\Delta) \in L(N_\alpha,N_\beta)$, $\alpha,\beta = 1,2$ [4].

2. SOME APPLICATIONS TO HARMONIC ANALYSIS. In this section H stands for the Hilbert transform in $\mathbb{T}$, which can be defined as the operator $H: L^2(\mathbb{T}) \to L^2(\mathbb{T})$ satisfying

$$(2.1) \qquad Hf_1 = -if_1 \text{ for } f \in W_1, \qquad Hf_2 = if_2 \text{ for } f_2 \in W_2 .$$

We will consistently use the notation $h \in H^1(T)$ for $h \in L^1(\mathbb{T})$ and $\hat{h}(n) = 0$ if $n < 0$.

(a) *Two measures problem for the Hilbert transform in* $L^2(T)$. We want to characterize the pairs of measures $\mu \geq 0$, $\nu \geq 0$ in $\mathbb{T}$, such that, for some constant M,

$$(2.2) \qquad \int |Hf|^2 d\mu \leq M \int |f|^2 d\nu , \quad \forall f \in V .$$

Letting $f = f_1 + f_2$, where $(f_1,f_2) \in W_1 \times W_2$, we have $Hf = -i(f_1-f_2)$, and (2.2) can be rewritten as $\sum_{\alpha,\beta} \mu_{\alpha\beta}(f_\alpha \bar{f}_\beta) \geq 0$ for all $(f_1,f_2) \in W_1 \times W_2$, where $\mu_{11} = \mu_{22} = \mu_1 = \mu_2$, $\mu_{12} = \mu_{21} = \mu_0$, and

$$(2.3) \qquad \mu_1 = \mu_2 = M\nu - \mu , \quad \mu_0 = M\nu + \mu ,$$

or as $B_0 \prec (B_1,B_2)$, where

$$(2.4) \qquad B_1(f,g) = B_2(f,g) = \int f\bar{g} d\mu_1 , \quad B_0(f,g) = 2 \operatorname{Re} \int f\bar{g} d\mu_0 .$$

The application of Theorem I leads to the following result.

COROLLARY 1 [20]. (i) *Two positive measures* μ *and* ν *in* $\mathbb{T}$ *satisfy* (2.2) *iff* $\mu(\Delta) \leq M\nu(\Delta)$ *and there exists* $h \in H^1(T)$ *such that*

$$(2.5) \qquad |(M\nu+\mu-h dt)(\Delta)| \leq (M\nu-\mu)(\Delta) , \quad \forall\Delta \textit{ Borel set.}$$

In particular, $d\mu = \omega(t)dt$ *for* $\omega \in L^1$.

(ii) *If, in addition,* $d\nu = \rho(t)dt$, $\rho \in L^1$, *then* (2.2) *imply* $\omega(t) \leq M\rho(t)$ *a.e. and there exists* $h \in H^1(T)$ *such that*

$$(2.6) \qquad |\arg h(t)| < \frac{\pi}{2}, \qquad 2\omega(t) \leq \operatorname{Re} h(t) \leq |h(t)| \leq 2M\rho(t) \textit{ a.e.}$$

Conversely, (2.6) *implies* (2.2) *with a different constant, whenever* $\rho \leq C\omega$.

(iii) *In the case* $\mu = \nu$, $d\mu = \omega(t)dt$ *with*

$$(2.7) \qquad \omega = \exp(u+Hv) , \quad \|u\|_\infty \leq c_M , \quad \|v\|_\infty \leq \frac{\pi}{2} - \varepsilon_M ,$$

with control over c_M *and* ε_M. *Conversely,* (2.7) *implies* (2.2) *for* $\mu = \nu$ *with different constant.*

Proof. Inequality (2.5) follows immediately from (2.3) and (1.11), and taking $|\Delta| = 0$, it is $\mu(\Delta) = 0$, so $d\mu = \omega\, dt$ for $\omega \in L^1$.

Under the additional hypothesis $d\nu = \rho dt$, (2.5) can be rewritten as

(2.5a) $\quad |M\rho(t)+\omega(t)-h(t)| \le M\rho(t) - \omega(t)$ a.e.

which implies (2.6), and conversely with a change in constants, whenever $\rho \le C\omega$.

If furthermore $\mu = \nu$, for $u = \log \omega/|h|$ and $v = \arg h$, (2.6) says that $\|u\|_\infty \le c_M$ and $\|v\|_\infty \le \frac{\pi}{2} - \varepsilon_M$. Since for $h \in H^1$, $\log|h| = H(\arg h)$, (2.7) follows. From (2.7), taking $h = c\exp(Hv-iv)$, we get $|h| = ce^{-u}\omega$ and $\mathrm{Re}\, h = ce^{-u}\cos v$, so (2.6) follows for $\rho = \omega$, with different constants.

The equivalence of (2.7) and (2.2) for $\mu = \nu$ is the classical theorem of Helson-Szegö [37].

(b) *The Nehari theorem.* In the special case when $B_1(f,g) = B_2(f,g) = \int f\bar{g}dt$, Theorem I contains the classical theorem of Nehari [51]. It asserts that, given a sequence $s(n)$, $n > 0$, there exists a bounded function F such that $\|F\|_\infty \le 1$ and $\hat{F}(n) = s(n)$, $n > 0$, iff the Hankel form $B_0(f,g) = \sum_{n,k} s(n-k)\hat{f}(n)\overline{\hat{g}(k)}$ satisfies $|B_0(f,g)| \le \|f\|_2\|g\|_2 = B_1(f,f)^{1/2}B_2(g,g)^{1/2}$, for all $(f,g) \in W_1 \times W_2$. In fact, the measures μ_0, μ_1, μ_2, provided by Theorem I, satisfy, in this case, $d\mu_1 = d\mu_2 = dt$, so that $d\mu_0 = F(t)dt$ with $|F(t)| \le 1$.

Observe that defining $G(t) = \sum_{n>0} s(n)e^{int} + \sum_{n<0} \overline{s(-n)}e^{int}$, G is a real function, $h = F - G$ is conjugate analytic, and $G = \mathrm{Re}\, h - \mathrm{Re}\, F \in L^\infty + HL^\infty$, since $\mathrm{Re}\, h = -H(\mathrm{Im}\, h) = -H(\mathrm{Im}\, F)$ with $F \in L^\infty$.

(c) L^∞ *and BMO.* A function $G \in L^1(\mathbf{T})$ belongs to BMO, with norm $\|G\|_* \le 1$, iff $G = G_1 + HG_2$, with $\|G_1\|_\infty^2 + \|G_2\|_\infty^2 \le 1$. Thus, Corollary 1(iii) says that if $\mu = \nu$ satisfies (2.2) then $d\mu = \exp G dt$ with $G \in \mathrm{BMO}$, and both the theorems of Helson-Szegö and of Nehari are BMO results. The following consequence of Theorem I makes more clear the connection between the GBT and BMO, and shows that the relation between bounded and weakly bounded Toeplitz triplets corresponds to that of L^∞ and BMO.

COROLLARY 2 [12]. *For each real* $G \in L^1(\mathbf{T})$, *let* $B_G(f,g) = \int f\bar{g}G dt$, *and set* $B_1(f,g) = B_2(f,g) = \int f\bar{g}dt$. *Then* (i) $G \in L^\infty$, *with* $\|G\|_\infty \le 1$, *iff* (B_G, B_1, B_2) *is a bounded Toeplitz triplet, and* (ii) $G \in \mathrm{BMO}$, *with* $\|G\|_* \le 1$, *iff* (B_G, B_1, B_2) *is a weakly bounded Toeplitz triplet.*

Proof. (i) follows directly from (1.11a). If $B_G \prec (B_1, B_2)$ then G is as in the remark following Nehari's theorem.

(d) *Prediction problems.* The theorems of Helson-Szegö and Helson-Sarason mentioned in (a) lead to important developments in prediction theory. As shown in [12], the GBT gives refinements of such results, through the control of the norm of the operators. This was studied in detail in [33].

For applications of the GBT to the invertibility of Toeplitz operators (as in the Devinatz-Widom theorem), see [9]. Other applications to harmonic analysis are at the end of Sections 4 and 5.

3. EXTENSIONS TO SCATTERING SYSTEMS AND PARAMETRIZATION. Results analogous to those of Section 1 are valid in a more general setting. We say that $[V;W_1,W_2;\tau]$ is an *algebraic scattering system* if V is a vector space, W_1 and W_2 are two linear subspaces of V, and $\tau: V \to V$ is a linear bijection satisfying

$$(3.1) \qquad \tau W_1 \subset W_1 \quad \text{and} \quad \tau^{-1} W_2 \subset W_2 .$$

The trigonometric polynomials of Section 1 (with scalar- or vector-valued coefficients) are a first example of such scattering systems, that owe their name to the following observation. If V is a Hilbert space, W_1 and W_2, closed subspaces, and τ a unitary operator, and if, furthermore,

$$(3.1a) \qquad \bigcap_{n=1}^{\infty} \tau^n W_1 = \{0\} = \bigcap_{n=1}^{\infty} \tau^{-n} W_2 ,$$

then $[V;W_1,W_2;\tau]$ is an *Adamjan-Arov* (A-A) *scattering system*. If in addition $W_1 \perp W_2$, and

$$(3.1b) \qquad \bigvee_{n=\infty}^{\infty} \tau^n W_1 = \bigvee_{n=-\infty}^{\infty} \tau^n W_2 = V ,$$

$[V;W_1,W_2;\tau]$ is a classical *Lax-Phillips scattering system.*

Toeplitz and Hankel forms, as well as bounded Toeplitz and Toeplitz-Hankel triplets, can be defined in algebraic scattering systems, in a manner totally analogous to that of Section 1.

Another example of such systems, that will be called *function systems*, is as follows. Given an arbitrary set E, two subsets $E_1 \subset E$, $E_2 \subset E$, and a bijection $\tau: E \to E$ such that

$$(3.2) \qquad \tau E_1 \subset E_1 \quad \text{and} \quad \tau^{-1} E_2 \subset E_2 ,$$

let

$$(3.3) \qquad \begin{aligned} &V = V(E) = \{f: E \to \mathbb{C}, \text{ finitely supported}\} , \\ &W_\alpha = W(E_\alpha) = \{f \in V;\ \text{supp } f \subset E_\alpha\}, \quad \alpha = 1,2, \\ &(\tau f)(x) = f(\tau x), \quad \forall f \in V, \quad \forall x \in E. \end{aligned}$$

The case where $E = E_1 \cup E_2$ and $E_1 \cap E_2 = \{0\}$ is of special interest. In such a case, for every kernel $K: E \times E \to \mathbb{C}$ there exists a unique sesquilinear form $B = B_K: V \times V \to \mathbb{C}$ satisfying $B(1_x, 1_y) = K(x,y)$, and the correspondence

$K \to B_K$ is bijective. K is a *Toeplitz kernel*, $K(\tau x,\tau y) = K(x,y)$ $\forall x,y \in E$, iff B_K is a Toeplitz form, and K is a *positive definite* (p.d.) *kernel* iff B_K is a positive form. Thus, in this case all previous results can be reformulated in terms of kernels. If $E = \mathbb{Z}$, $E_1 = \mathbb{Z}_1$, $E_2 = \mathbb{Z}_2$, and $\tau n = n+1$, $n \in \mathbb{Z}$, we get back to the trigonometric case, by identifying the trigonometric polynomials with the finite sequences of their Fourier coefficients.

In [1] Adamjan and Arov gave a functional realization of the A-A systems in terms of two Hilbert spaces N_1 and N_2, and a function $S \in L^\infty(\mathbf{T};L(N_1,N_2))$ the *scattering function* of the system, providing in particular two canonical isomorphisms,

$$(3.4) \qquad j_1\colon W_1 \to H^2_+(\mathbf{T};N_1) \quad \text{and} \quad j_2\colon W_2 \to H^2_-(\mathbf{T};N_2)$$

(where $H^2_\pm$ are the analytic and antianalytic parts of the L^2 space of N_α-valued functions, $\alpha = 1,2$, defined in $\mathbf{T}$). In case of the Lax-Phillips systems, $S(t)$ coincides with the Heisenberg scattering function, as defined by those authors. Details of the Adamjan-Arov functional realization, as well as proofs of the following theorems, can be found in [31].

THEOREM II (*GBT in scattering systems* [31]). *If* $[V;W_1,W_2;\tau]$ *is an A-A scattering system, there exist two Hilbert spaces* N_1 *and* N_2 *and two isometries* j_1 *and* j_2 *as in* (3.4), *such that, for every bounded Toeplitz-Hankel triplet* (B_0,B_1,B_2), *there exist a p.d. 2×2-matrix measure* $(\mu_{\alpha\beta})$, *where for each* $\alpha,\beta = 1,2$, $\mu_{\alpha\beta}$ *is a* $L(N_\alpha,N_\beta)$*-valued measure in* $\mathbf{T}$, *such that*

$$(3.5) \qquad B_{\alpha\beta}(f_\alpha,f_\beta) = \int \langle d\mu_{\alpha\beta} j_\alpha f_\alpha, j_\beta f_\beta\rangle$$

for all $(f_1,f_2) \in W_1 \times W_2$, *for* $B_{11} = B_1$, $B_{22} = B_2$, $B_{12} = B_0$, $B_{21} = B_0^*$.

A pair of positive Toeplitz forms, B_1, B_2, is called *regular* if there exists no sequence (f_n), $n \geq 0$, such that $f_n \in \tau^n W_1$ for all $n \geq 0$ and $B_1(f_n-f_m,f_n-f_m) \to 0$, or such that $f_n \in \tau^{-n} W_2$ for all $n \geq 0$ and $B_2(f_n-f_m,f_n-f_m) \to 0$, $n, m \to \infty$. Since B_1 and B_2 define hilbertian seudometrics in W_1 and W_2, respectively, if B_1, B_2 is a regular pair, then $\cap_{n>0} \tau^n \bar{W}_1 = \{0\} = \cap_{n>0} \tau^{-n} \bar{W}_2$, where $\bar{W}_1$ and $\bar{W}_2$ are the closures of W_1 and W_2 taken in the corresponding seudometrics, thus giving rise to an associated A-A system, with scattering function $S(t)$.

THEOREM III (*GBT in algebraic scattering systems* [31]). *If* $[V;W_1,W_2;\tau]$ *is an algebraic scattering system and* B_1, B_2 *is a regular pair of positive Toeplitz forms, then for every bounded Toeplitz-Hankel triplet* (B_0,B_1,B_2) *there exist Hilbert spaces* N_1 *and* N_2, *and two isometries* j_1 *and* j_2 *as in* (3.4), *such that the integral representation* (3.5) *holds with*

(3.6) $d\mu_{11} = I_{N_1} dt$, $d\mu_{22} = I_{N_2} dt$ *and* $d\mu_{12} = S(t)dt$,

where $S(t)$ *is the scattering function of the associated A-A system.*

The matrix measures $(\mu_{\alpha\beta}) \geq 0$ that give the integral representations in Theorems I, II and III are not unique: in each case, μ_{11} and μ_{22} are uniquely determined, but not so $\mu_{12} = \mu_{21}^*$, which thus play a special role. Since in Theorem III, μ_{12} is given by $S(t)$, while in Corollary 2 of Theorem I, $\mu_{12} \in BMO$, we can say that the GBT unifies the notions of scattering and BMO functions.

As observed in Section 2, the Nehari theorem is a special case of Theorem I, and in this case all measures $\mu_0 = \mu_{12}$ where described by Adamjan, Arov and Krein [2]. In [5] the A-A-K parametrization was extended to general scalar GTKs, and further developed in [3] in the following constructive form which applies also to matrix-valued GTKs: for each GTK there is a sequence of polynomials (P_k) and two numerical sequences (c_k) and (d_k) (given by explicit formulae) such that all μ_0's are given by $\mu_0 = \mu_0' + h_\phi dt$,

$$h_\phi(t) = \sum_{n=1}^{\infty} t^n \sum_{k=1}^{n} (\phi(t)P_{k-1}(\phi)(t) - c_{k-1})d_{n-k} ,$$

where μ_0' is one of the measures μ_0 and ϕ runs over the unit ball of H^1.

4. LIFTINGS WITH MULTIPARAMETRIC SHIFTS AND THE NAGY-FOIAS THEOREM. It was already observed in the trigonometric example that if $B = B_1 = B_2$, where B is Toeplitz and $B(1,1) \geq 0$, then inequality (1.7) holds for all $(f,g) \in V \times V$ iff it holds for $(f,g) \in W_1 \times W_2$. But this is not true for general Toeplitz triplets (B,B_1,B_2), where $B \prec (B_1,B_2)$ does not imply $B \leq (B_1,B_2)$. However, there is a substitute property that holds:

THEOREM IV (*Lifting theorem* [27]). *Given* $[V;W_1,W_2;\tau]$, *an algebraic scattering system, then* (B_0,B_1,B_2) *is a bounded Toeplitz-Hankel triplet, i.e.* $B_0 \prec (B_1,B_2)$, *iff there exists a bounded Toeplitz triplet* (B,B_1,B_2), *i.e.* $B: V \times V \to \mathbb{C}$, $B \leq (B_1,B_2)$, *such that* $B = B_0$ *in* $W_1 \times W_2$.

B is a *lifting* of B_0 and such liftings are in 1-1 correspondence with those provided by the Nagy-Foias theorem for a certain associated intertwining contraction (see below).

Whenever the GBT is valid (as in the situations of Theorems I, II and III), it implies the lifting property, but Theorem IV by itself does not give integral representations.

In the trigonometric example, the liftings are in 1-1 correspondence with the (μ_0,μ_1,μ_2) of Theorem I.

In the case of forms corresponding to GTKs of (N_1,N_2) type (see end of Section 1), the lifting theorem, together with the Bochner theorem for vector-valued kernels, are equivalent to a dilation property, as follows:

COROLLARY 3 (*Dilation property* [10]). *If* K *is a p.d. GTK of* (N_1,N_2)*-type there exists a Hilbert space* H, *a unitary operator* $U \in L(H)$, *and two continuous linear applications,* $\phi_1: N_1 \to H$, $\phi_2: N_2 \to H$, *such that, for* $\alpha,\beta = 1,2$,

(4.1) $\langle K(m,n)\xi_\alpha,\xi_\beta\rangle = \langle U^{m-n}\phi_\alpha\xi_\alpha,\phi_\beta\xi_\beta\rangle$ *if* $(m,n) \in \mathbf{Z}_\alpha \times \mathbf{Z}_\beta$, $(\xi_\alpha,\xi_\beta) \in N_\alpha \times N_\beta$.

Analogous dilation properties hold for the systems in Theorems II and III; for details, see [27].

If K is an ordinary p.d. Toeplitz kernel and $N_1 = N_2 = N$, then Corollary 3 reduces to the classical NAIMARK DILATION THEOREM: there is a mapping $\phi: N \to H$ such that

$$\langle K(m,n)\xi,\eta\rangle = \langle U^{m-n}\phi\xi,\phi\eta\rangle, \quad \forall(m,n), \quad \forall(\xi,\eta) \in N \times N .$$

If $A: N \to N$ is a contraction, $\|A\| \leq 1$, then there is a p.d. Toeplitz kernel K such that $K(m,n) = A^{m-n}$ if $m-n \geq 0$, and this gives the

NAGY DILATION THEOREM. *For each contraction* $A \in L(N)$ *there is a Hilbert space* $H \supset N$ *and a unitary operator* $U \in L(H)$, *such that* A^n *is the compression of* U^n *to* N, *for* $n \geq 0$.

This theorem was followed by the

NAGY-FOIAS LIFTING THEOREM. *If* $U_1 \in L(H_1)$ *and* $U_2 \in L(H_2)$ *are the unitary dilations of the contractions* $A_1 \in L(N_1)$ *and* $A_2 \in L(N_2)$, *and if* $T: N_1 \to N_2$, $\|T\| \leq 1$, *satisfies*

(4.2) $A_1T = TA_2$,

then there is an intertwining contraction $Y: H_1 \to H_2$, $\|Y\| \leq 1$,

(4.2a) $U_1Y = YA_2$,

such that

(4.2b) $TP_1 = P_2Y$, P_α: *projection of* H_α *onto* N_α, $\alpha = 1,2$.

As shown in [29], the lifting theorem IV is equivalent to a special case of the Nagy-Foias lifting theorem, and the 1-1 correspondence of liftings holds (see also [27]).

On the other hand, as shown in [8], if A_1, A_2, T, U_1 and U_2 are as in (4.2), and if a GTK is defined by $K(m,n) = A_\alpha^{m-n}$ for $(m,n) \in \mathbf{Z}_\alpha \times \mathbf{Z}_\alpha$, $\alpha = 1,2$, and $K(m,n) = A_2^{m-n}T$ for $(m,n) \in \mathbf{Z}_1 \times \mathbf{Z}_2$, $m-n > 0$, then K is p.d. and

Theorem I for the operator-valued kernels of Section 1 leads to the Nagy-Foias theorem.

In fact, the Nagy-Foias theorem, together with the Bochner theorem for operator-valued kernels, is logically equivalent to the GBT for such kernels in $\mathbb{Z}\times\mathbb{Z}$, but the other GBT (Theorems II and III) are already outside the Nagy-Foias framework.

The construction of [4] was applied in [6] and [17] to linear systems or colligations. In [6] and [43] it was shown that Theorem I also provides a refinement of a result of Davis which extends the Nagy dilation theorem to Krein spaces. In [46], [44] the properties of GTKs were applied to colligations in Krein spaces.

The lifting theorem extends to the case of systems with multiparametric shifts, as follows.

Let $[V;W_1,W_2;\tau]$ be an algebraic scattering system, and assume that another linear bijection, $\sigma: V\to V$, is given, such that $\tau\sigma=\sigma\tau$. A sesquilinear form $B: V\times V\to\mathbb{C}$ is (τ,σ)-Toeplitz if it is invariant with respect to both τ and σ. A sesquilinear form $B_0: W_1\times W_2\to\mathbb{C}$ is (τ,σ)-Hankel if it is the restriction to $W_1\times W_2$ of a (τ,σ)-Toeplitz form.

THEOREM V (*two-parametric lifting* [31]). *Given an algebraic scattering system* $[V;W_1,W_2;\tau]$ *and a commuting bijection* $\sigma: V\to V$, $\sigma\tau=\tau\sigma$, *if* (B_0,B_1,B_2) *is a bounded* (τ,σ)*-Toeplitz-Hankel triplet, then there exists a bounded* (τ,σ)*-Toeplitz triplet* (B,B_1,B_2), *such that* $B=B_0$ *in* $W_1^\sigma\times W_2^\sigma$, *where, for* $\alpha=1,2$,

$$W_\alpha^\sigma=\{f\in W_\alpha:\ \sigma^p f\in W_\alpha,\ \forall p\in\mathbb{Z}\}\ . \tag{4.3}$$

This theorem provides integral representation for the restriction of (τ,σ)-Hankel forms to $W_1^\sigma\times W_2^\sigma$.

Note that when σ is the identity in V, Theorem V reduces to Theorem IV so that it can be considered as a two-parametric extension of the Nagy-Foias theorem.

While in Theorem IV the existence of the lifting is also a sufficient condition for $B_0\prec(B_1,B_2)$ in $W_1\times W_2$, this is not true in Theorem V, since $W_1^\sigma\times W_2^\sigma$ can be much smaller than $W_1\times W_2$. However, if σ is also a shift, i.e.,

$$\sigma W_1\subset W_1 \quad\text{and}\quad \sigma^{-1}W_2\subset W_2\ ,$$

and if additional hypotheses are satisfied, it is possible to give a two-parametric lifting theorem where the existence of the lifting is again both necessary and sufficient for $B_0\prec(B_1,B_2)$ in $W_1\times W_2$. This provides, in particular, extensions of the Nehari and the Helson-Szegö theorems to half-planes of

Helson-Lowdenslager [35]. We shall concentrate here on the simplest of such examples.

Let $S_0 = \{(m,n) \in \mathbb{Z}^2: n > 0 \text{ if } m < 0 \text{ and } n \geq 0 \text{ if } m \leq 0\}$, and consider the function system consisting of the trigonometric polynomials in $\mathbb{T}^2$, $V = V(\mathbb{T}^2) = \{f: \mathbb{T}^2 \to \mathbb{C};\ f(t,s) = \sum_{\text{finite}} \hat{f}(m,n) \exp i(mt+ns)\}$, $W_1 = W_1(S_0) = \{f \in V: \hat{f}(m,n) = 0 \text{ if } (m,n) \notin S_0\}$, $W_2 = W_2(S_0) = \{f \in V: \hat{f}(m,n) = 0 \text{ if } (m,n) \in S_0\}$, $\tau(m,n) = (m,n+1)$ and $\sigma(m,n) = (m+1,n)$, $\forall (m,n) \in \mathbb{Z}^2$.

THEOREM VI. *Given the trigonometric system in* $\mathbb{T}^2$, $[V; W_1(S_0), W_2(S_0); \tau]$ *and* σ, *where* τ *and* σ *respectively are the vertical and horizontal transitions in* $\mathbb{Z}^2$, *if* (B_0, B_1, B_2) *is a bounded Toeplitz-Hankel triplet, there exist two positive matrix measures* $(\mu_{\alpha\beta}) \geq 0$ *and* $(\nu_{\alpha\beta}) \geq 0$, *where for* $\alpha,\beta = 1,2$, *each* $\mu_{\alpha\beta}$ *is a measure defined in* $\mathbb{T}^2$ *and each* $\nu_{\alpha\beta}$ *is a measure defined in* $\mathbb{T}$, *such that, for* W_α^σ *as in* (4.3),

$$B_{\alpha\beta}(f,g) = \iint f\bar{g}\,d\mu_{\alpha\beta} \quad \textit{for} \quad (f_\alpha, f_\beta) \in W_\alpha^\sigma \times W_\beta^\sigma, \tag{4.4}$$

and

$$B_{\alpha\beta}(f,g) = \int f\bar{g}\,d\nu_{\alpha\beta} \quad \textit{for} \quad (f,g) \in W(\mathbb{Z}_1) \times W(\mathbb{Z}_2) \tag{4.4a}$$

where $W(\mathbb{Z}_1) = \{f \in V: \hat{f}(m,n) = 0 \text{ unless } m \geq 0,\ n = 0\}$ *and* $W(\mathbb{Z}_2) = \{f \in V: \hat{f}(m,n) = 0 \text{ unless } m < 0,\ n = 0\}$.

Note that the functions in $W(\mathbb{Z}_1)$ and $W(\mathbb{Z}_2)$ depend only on $t \in \mathbb{T}$.

COROLLARY 4 (*Nehari theorem in* S_0). *Given the system of Theorem VI, if* $B_1(f,g) = B_2(f,g) = \iint f\bar{g}\,dt\,ds$, *then a Toeplitz-Hankel triplet is bounded iff there exist two bounded functions* $F(t,s)$, $(t,s) \in \mathbb{T}^2$, *and* $G(t)$, $t \in \mathbb{T}$, *such that*

$$\begin{aligned} B_0(f,g) &= \iint f\bar{g}F\,dt\,ds \quad \textit{for} \quad (f,g) \in W_1^\sigma \times W_2^\sigma, \\ B_0(f,g) &= \int f\bar{g}G\,dt \quad \textit{for} \quad (f,g) \in W(\mathbb{Z}_1) \times W(\mathbb{Z}_2). \end{aligned} \tag{4.5}$$

COROLLARY 5 (*Helson-Szegö theorem in* S_0). *Two positive measures* μ *and* ν *defined in* $\mathbb{T}^2$ *satisfy*

$$\iint |H_{S_0} f|^2 d\mu \leq M \iint |f|^2 d\nu, \quad \forall f \in V(\mathbb{T}^2) \tag{4.6}$$

for $(H_{S_0} f)\hat{}\,(m,n) = -i\hat{f}(m,n)$ *if* $f \in W_1(S_0)$, *and* $= i\hat{f}(m,n)$ *if* $f \in W_2(S_0)$ *iff there exist two functions* H, *defined in* $\mathbb{T}^2$, *such that* $\hat{H}(m,n) = 0$ *for* $n < 0$, *and* h, *defined in* $\mathbb{T}$, *such that* $\hat{h}(m) = 0$ *for* $m < 0$, *satisfying, for all Borel sets,*

$$\begin{aligned} &|(M\nu+\mu)(\Delta) - \iint_\Delta H(t,s)\,dt\,ds| \leq (M\nu-\mu)(\Delta), \quad \forall \Delta \subset \mathbb{T}^2 \\ &\textit{and} \\ &|(M\nu+\mu)(B) - \int_B h(t)\,dt| \leq (M\nu-\mu)(B), \quad \forall B \subset \mathbb{T}. \end{aligned} \tag{4.7}$$

5. THE GBT IN FUNCTION ALGEBRAS WITH FORMS BOUNDED BY SEMINORMS. In the case of the trigonometric polynomials, the Toeplitz forms were defined as those invariant with respect to the shift, and this definition led to the general concepts of Section 3. However, in the trigonometric case, the Toeplitz forms can also be defined, without invoking the shift, as those sesquilinear applications, $B: V\times V \to \mathbb{C}$, satisfying $B(f,g) = I(f\bar{g})$ for all $(f,g) \in V\times V$, for some linear form $I: V \to \mathbb{C}$ (that can be taken as $I(f) = B(f,1)$). From (1.9a) it then follows that it is equivalent to give a bounded triplet (B_0,B_1,B_2), $B_0 \prec (B_1,B_2)$ or four linear functionals $(I_{11},I_{12},I_{21},I_{22}) \in V'\times W_1'\times W_2'\times V'$, i.e., $I_{11}: V\to\mathbb{C}$, $I_{12}: W_1\to\mathbb{C}$, $I_{21}: W_2\to\mathbb{C}$, $I_{22}: V\to\mathbb{C}$, such that

$$(5.1) \qquad \sum_{\alpha,\beta=1,2} I_{\alpha\beta}(f_\alpha\bar{f}_\beta) \geq 0\,, \quad \forall(f_1,f_2) \in W_1\times W_2\,.$$

Observe that in the trigonometric case $(f_1\bar{f}_1,f_1\bar{f}_2,\bar{f}_1f_2,f_2\bar{f}_2) \in V\times W_1\times W_2\times V$ if $(f_1,f_2) \in W_1\times W_2$.

This point of view leads to another extension of the GBT, as follows. Let now V be an algebra, whose elements are complex-valued functions defined in a certain set, and W_1 and W_2 are two linear subspaces of V satisfying (i) $f\in V$ implies $\operatorname{Re} f \in V$ and $|f| \in V$; (ii) $f \in W_2$ implies $\bar{f} \in W_1$, and (iii) $(f,g) \in W_1\times W_2$ implies $f\bar{g} \in W_1$ and $\bar{f}g \in W_2$.

Let $X = \{f = (f_{11},f_{12},f_{21},f_{22}) \in V\times W_1\times W_2\times V\colon f_{21} = \bar{f}_{12}\}$ and $X' = \{I = (I_{11}, I_{12},I_{21},I_{22}) \in V'\times W_1'\times W_2'\times V'\colon I_{21}(f_{21}) = \overline{I_{12}(f_{12})},\ \forall f_{12} \in W_1\}$, $V' = \{I: V\to\mathbb{C},$ I linear functional$\}$, and for each $I \in X'$, $f \in X$, set $I(f) = \sum_{\alpha\beta} I_{\alpha\beta}(f_{\alpha\beta})$. Since I_{21} is determined by I_{12}, the quadruple $(I_{\alpha\beta})$ is really a triplet of linear functionals. We write $f = (f_{\alpha\beta}) \geq 0$ if there exists $(\phi_1,\phi_2) \in W_1\times W_2$ such that $f_{\alpha\beta} = \phi_\alpha\bar{\phi}_\beta$, $\alpha,\beta = 1,2$, and $I = (I_{\alpha\beta}) \geq 0$ if $I(f) \geq 0$ for all $f \geq 0$.

Let us assume that there exists a set $E \subset X'$, called the *set of elementary quadruples* such that, for all $J = (J_{\alpha\beta}) \in E$,

(a) $J_{11}(\operatorname{Re} f) = \operatorname{Re} J_{11}(f)$ and $J_{22}(\operatorname{Re} f) = \operatorname{Re} J_{22}(f)$, $\forall f \in V$

(b) If the quadruple $f \in X$ satisfies $J(f) \geq 0$, $\forall J \in E$, then $f = g+h$ with $g \geq 0$ and $h \geq 0$.

EXAMPLE. Let V, W_1 and W_2 be the closures in $C(\mathbf{T})$ of the V, W_1 and W_2 of the trigonometric example in Section 1, so that $V = C(\mathbf{T})$, and write $J = (J_{\alpha\beta}) \in E$ if there exist $t_0 \in \mathbf{T}$ and c_1, $c_2 \in \mathbb{C}$ such that $J(f) = \sum_{\alpha\beta} c_\alpha\bar{c}_\beta f_{\alpha\beta}(t_0)$, for each $f = (f_{\alpha\beta}) \in X$. Then condition (a) is evidently fulfilled and using the classical Fejér-Riesz theorem, condition (b) is easily seen to be satisfied, since the elements of W_1 are analytic functions. For details see Theorem 1 of [28]. Note that these $J \in E$ are the elementary quadruples (or triplets) defined in Section 1.

By the GBT of Section 1 (Theorem I), every $I = (I_{\alpha\beta}) \geq 0$ has an integral representation, which is equivalent to an expansion in elementary J's. We are going to extend this result for the general function algebra setting, where the problem splits into two non-equivalent ones: that of elementary expansions and that of integral representations, which are obtained from the first under additional hypotheses.

Let now σ_1 and σ_2 be two seminorms in V such that $f \geq 0$ implies $\sigma_1(f) \geq 0$ and $\sigma_2(f) \geq 0$ and such that there exist $\mu_1 \in V'$, $\mu_2 \in V'$ satisfying, for all $f \in V$ and $\alpha = 1,2$,

$$(5.3) \qquad \begin{aligned} &\sigma_\alpha(f) = \mu_\alpha(|f|), \quad |\mu_\alpha(f)| \leq \sigma_\alpha(f) , \\ &\text{Re } \mu_\alpha(f) = \mu_\alpha(\text{Re } f) . \end{aligned}$$

With these notations, we have the following theorem proved in [28] in the case of the example above.

THEOREM VII (*Expansion into elementary triplets*). *If* $I_{12} \in W_1'$ *is bounded by* σ_1, σ_2, $I_{12} \prec (\sigma_1,\sigma_2)$, *in the sense that for each* $f = (f_{\alpha\beta}) \geq 0$,

$$\sigma(f_{11}) + I_{12}(f_{12}) + \overline{I_{12}(\overline{f_{21}})} + \sigma_2(f_{22}) \geq 0 ,$$

then there exists a net $J^p = (J_{\alpha\beta}{}^p)$ *of convex combinations of elementary quadruples, such that* $J_{12}{}^p(f_{12}) \to J_{12}(f_{12})$ *and* $|J_{11}{}^p(f_{11})| \leq \sigma_1(f_{11})$, $|J_{22}{}^p(f_{22})| \leq \sigma_2(f_{22})$, $\forall f \in X$.

If $\sigma_1(f) = \mu_1(|f|)$, $\sigma_2(f) = \mu_2(|f|)$ *for some* $\mu_1, \mu_2 \in V'$ *then it is also* $J_{\alpha\alpha}{}^p(f_{\alpha\alpha}) \to J_{\alpha\alpha}(f_{\alpha\alpha})$ *for* $\alpha = 1,2$, *so that* $(\mu_1,I_{12},\mu_2) = \lim_p (J_{11}{}^p,J_{12}{}^p,J_{22}{}^p)$.

In the case of the example above Theorem VII leads to

THEOREM VIII (*GBT for forms bounded by seminorms in* $C(\mathbb{T})$ [28]). *If* V, W_1 *and* W_2 *are as in Section 1,* σ_1 *and* σ_2 *are two seminorms in* $C(\mathbb{T})$ *as described in* (5.3), *and* B_0 *is a Hankel form such that* $B_0 \prec (\sigma_1,\sigma_2)$, *then there exists a measure* μ *in* $\mathbb{T}$ *such that* $B_0(f,g) = \int f\bar{g}d\mu$, $\forall (f,g) \in W_1 \times W_2$ *and* $\mu \leq (\sigma_1,\sigma_2)$ *in the sense that*

$$(5.4) \qquad \left|\int f\bar{g}d\mu\right| \leq \sigma_1(f)\sigma_2(g) , \qquad \forall (f,g) \in V \times V.$$

For $\sigma_\alpha(f) = B_\alpha(f,f)^{1/2}$, $\alpha = 1,2$, Theorem VIII reduces to Theorem I. If $\sigma_1 = \sigma_2 = \sigma$ and $B_0(f,g) = \int f\bar{g}\omega\, dt$, then the measure μ in Theorem VIII satisfies $d\mu = \theta(t)dt$ with

$$(5.5) \qquad \left|\int f(t)\theta(t)dt\right| \leq \sigma(f) , \qquad \forall f \in C(\mathbb{T}) ,$$

which gives an extension of the theorem of Nehari, that for $\sigma(f) = \|f\|_{p'}$, $1 < p \leq \infty$, reduces to that of [34].

If in addition σ is *non-deterministic*, i.e.,

$$(5.6)\qquad \sigma(|f_1|^2)^{1/2}+\sigma(|f_2|)^2)^{1/2} \le c\sigma(|f_1+f_2|^2)^{1/2}, \quad \forall(f_1,f_2)\in W_1\times W_2$$

(for instance, if $\sigma(f) = \|f\|_p$, $1<p<\infty$), then

COROLLARY 6 [23]. *Let* V, W_1, W_2 *and* B_0 *be as in Theorem VIII and* $\sigma_1 = \sigma_2 = \sigma$ *be a non-deterministic seminorm in* V. *If*

$$(5.7)\qquad B_0(f_1,f_2) \le \sigma(|f_1+f_2|^2)^{1/2}, \quad \forall(f_1,f_2)\in W_1\times W_2,$$

then there exists a measure μ *in* $\mathbf{T}$ *such that* $B_0(f_1,f_2) = \int f_1\bar{f}_2 d\mu$ *for* $(f_1,f_2)\in W_1\times W_2$ *and* $\mu\le\sigma$ *in the sense that*

$$(5.8)\qquad \left|\int f\bar{g}d\mu\right| \le \sigma(|f+g|^2)^{1/2}, \quad \forall(f,g)\in V\times V.$$

If, in addition, $B_0(f,g) = \int f\bar{g}\omega\, dt$ *and thus* $d\mu = \theta\, dt$, *then* (5.8) *is strengthened to*

$$(5.8a)\qquad \int |f(t)|(|\theta(t)|^2/\mathrm{Re}\,\theta(t))dt \le c\sigma(f), \quad \forall f\in C(\mathbf{T}).$$

Observe now that, if $B_0 = B_K$ for a kernel K, then the condition (5.7) can be rewritten as

$$(5.7a)\qquad \left|\sum_{m,n\in\mathbf{Z}} K(m,n)\lambda(m)\overline{\lambda(n)}\right| \le \sigma\left(\left|\sum_{m\in\mathbf{Z}}\lambda(m)e_m(t)\right|^2\right),$$

for every sequence $(\lambda(n))$ of finite support.

In [24] inequality (5.8a) was used to prove that the following Grothendieck type inequality holds for a large class of seminorms σ.

COROLLARY 7 [24]. *Given a GTK by* $K(m,n) = \hat{\omega}_{\alpha\beta}(m-n)$ *for* $(m,n)\in\mathbf{Z}_\alpha\times\mathbf{Z}_\beta$, $\alpha,\beta = 1,2$, *where* $\omega_{\alpha\beta}\in L^1(\mathbf{T})$ *and* H *a Hilbert space, if* (5.7a) *holds for every sequence* $\lambda\colon \mathbf{Z}\to\mathbf{C}$ *of finite support, then it holds for every* $\xi\colon\mathbf{Z}\to H$, *i.e.,*

$$(5.7b)\qquad \left|\sum_{m,n} K(m,n)\langle\xi(m),\xi(n)\rangle_H\right| \le c\sigma\left(\left\|\sum_m \xi(m)e_m(t)\right\|_H\right)^2$$

with c *a universal constant.*

Theorem VIII also led to a generalization of the Helson-Szegö theorem to L^p, $p\ne 2$ [22]. A linear transformation A is *u-bounded* in L^p if there exists $c>0$ such that for every $f\in L^p$, $\|f\|_p\le 1$, there is a $g\in L^p$ with $|f|\le g$ and $\|Ag\|_p\le c$. With this notation, a weight $\omega\ge 0$ satisfies

$$(5.9)\qquad \int |Hf|^p\omega dt \le M\int |f|^p\omega dt, \quad \forall f\in C(\mathbf{T})$$

iff T_ω is u-bounded in $L^{p/(p-2)}$, where, for $2\le p<\infty$,

$$(5.10)\qquad T_\omega f = \omega^{-2/p}C(\omega^{2/p}f), \quad C(f) = f+iHf,$$

or iff for every $0\le\phi\in L^{p/(p-2)}$, $\|\phi\|\le 1$, there exist two functions $h = h_\phi\in H^1$, $0\le a\in L^{p/(p-2)}$, $\|a\|\le c_M$, such that

(5.11) $2\phi\omega^{2/p} \leq \text{Re}\, h \leq |h| \leq a\omega^{2/p}$.

For $p=2$, (5.11) reduces to condition (2.6) in Corollary 1 for the case of one weight. Corresponding statements hold for a pair of positive measures μ and ν satisfying

$$\int |Hf|^p d\mu \leq M \int |f|^p d\nu , \quad \forall f \in C(\mathbf{T}) .$$

Details of their characterization can be found in [24] and [58].

Rubio de Francia in [56] extended the equivalence (5.9)-(5.10) from L^p to 2-convex spaces, using his powerful extrapolation theory.

Condition (5.11), being the extension for $2 \leq p < \infty$ of the Helson-Szegö condition (2.7), is equivalent to $\omega \in A_p$, the Muckenhoupt condition [38].

Theorem VIII provides also a generalization of the Paley lacunary inequality for weighted seminorms [26]. Since Grothendieck's inequality can be deduced from Paley's inequality and, on the other hand, the above results of Rubio, as those on harmonizable processes in Section 7, are based on Grothendieck's theorem, it follows that there is a close connection between it and the lifting theorems IV and VII.

There are other variants of the definition of the elementary family E and of the corresponding Theorem VII, which apply to the case where

$$W_1(N) = W_2(N) = \{f: \mathbb{R} \to \mathbb{C};\ \hat{f}(x) = 0 \text{ for } |x| > N\} , \quad N \text{ fixed.}$$

The corresponding Theorem VII then gives generalizations of the results of Section 2 for Toeplitz forms defined in Paley-Wiener spaces. This leads, in particular, to refinements of the Nehari theorem in Paley-Wiener spaces due to R. Rochberg [55]. These, as well as results for entire functions of exponential type, will be treated in a forthcoming paper.

6. CONNECTIONS WITH THE KREIN METHOD AND LOCAL SEMIGROUPS. The definition of GTK extends in an obvious way to continuous kernels $K: \mathbb{R}\times\mathbb{R} \to \mathbb{C}$ or to operator-valued kernels defined in $\mathbb{R}\times\mathbb{R}$, and such kernels give rise to GTF $B: V\times V \to \mathbb{C}$, where V can be taken as the set of all continuous functions in $\mathbb{R}$ with compact support. More generally these notions extend to algebraic scattering systems with a continuous 1-parameter group of linear bijections, i.e., when instead of one bijection $\tau: V \to V$ we have a group of bijections $\{\tau^t: V \to V,\ t \in \mathbb{R}\}$. As shown in [29], Theorem IV extends to such groups. In particular, in the case of $\mathbb{R}$, this gives the GBT for distribution-valued GTKs or GTFs in $\mathbb{R}$, which is a generalized Bochner-Schwartz theorem (see [30]). Theorems II and III also extend to such $\{\tau^t\}$ and Theorem V extends to continuous Toeplitz forms in $\mathbb{R}^2$.

Different proofs of Theorem I in $\mathbb{R}$ were given in [8] and [12]. The notion of GTK in $\mathbb{R}$ leads to that of *local group of contractions*, as follows. If $K: \mathbb{R}\times\mathbb{R} \to \mathbb{C}$ is a p.d. GTK, then, by definition, there exist four ordinary Toeplitz kernels, $K_{\alpha\beta}$, $\alpha,\beta = 1,2$, such that $K(m,n) = K_{\alpha\beta}(m,n)$ if $(m,n) \in \mathbf{Z}_\alpha \times \mathbf{Z}_\beta$, and the relation $K(t,s) = K(t+r,s+r)$, $r \geq 0$, holds only if t and

s belong to the set $\mathbb{R}(r) = \{t \in \mathbb{R}: t \notin (-r,0)\}$. Therefore, if H is the Hilbert space associated to K (see (1.9a)), then $\{\tau^t\}$ does not act in H as a unitary group but as a local semigroup of isometries (cf. [14] for precise definitions). As shown in [14], each local semigroup of contractions can be dilated to a unitary group and in the case of local semigroups associated with GTKs, the spectral measures of the unitary groups provide the measures $(\mu_{\alpha\beta})$, $\alpha,\beta = 1,2$, of Theorem I. This method allows one not only to prove Theorem I for GTKs in $\mathbb{R}$, but also for GTKs in finite intervals of $\mathbb{R}$, which, in the case of ordinary Toeplitz kernels reduces to the Krein generalization of the Bochner theorem. In [14] the method of local semigroups was combined with that of Berezanskii [13] leading to extensions of Theorem I to distribution-valued GTKs.

The theory of M. G. Krein, mentioned in the Introduction, perfected by H. Langer, extends Bochner's expansion in eigenfunctionals to general symmetric operators D having a family of so-called *directing eigenfunctionals* $e_t(f)$, with operator-valued measures μ in the integral representation [41]. On the other hand, for a large class of self-adjoint operators D, the classical Gelfand-Kostuchenko spectral theorem provides an expansion of Bochner's type into eigenfunctionals $\phi_t(f)$, with a scalar measure ν. In [17] the $e_t(f)$, μ in Krein-Langer theorem were related to the $\phi_t(f)$, ν of the Gelfand-Kostuchenko theorem for the same D. Moreover, a refinement of the Krein-Langer method was given there, which allows one to apply the method to GTKs and to more general kernels that are symmetric with respect to general differential operators D. When $D = i\, d/dx$, this reduces to Theorem I for GTKs in $\mathbb{R}$.

The generalized Bochner-Krein theorem in finite intervals proved in [15] extends also to continuous algebraic scattering structures $[V;W_1,W_2;\tau^t, t \in \mathbb{R}]$, in the function case when $V = V(E)$, $W_1 = W_1(E_1)$, $W = W(E_2)$, etc. This leads to applications to *reduced moment problems*, as for instance, the one that follows.

THEOREM IX. *Let* $E_1 = \{(m,n) \in \mathbb{Z}^2: n \geq 0\} \cap [-A,A]^2$, $E_2 = \{(m,n) \in \mathbb{Z}^2: n < 0\} \cap [-A,A]^2$ *and, for* $\alpha = 0,1,2$, *let* $K_\alpha: \mathbb{Z}^2 \times \mathbb{Z}^2 \to \mathbb{C}$ *be Toeplitz kernels. If the associated Toeplitz forms* B_α *satisfy* $B_0 \prec (B_1,B_2)$ *in* $W(E_1) \times W(E_2)$ *and if the restrictions of* K_1 *and* K_2 *to* $(E_1 \cup E_2) \times (E_1 \cup E_2)$ *have unique p.d. Toeplitz extensions to* $\mathbb{Z}$, *in each variable, then there exists a matrix measure* $(\mu_{\alpha\beta}) \geq 0$ *in* $\mathbb{T}^2$ *such that*

$$\hat{\mu}_{12}(m-m',n-n') = K_0((m,n),(m',n')), \quad (m,n) \in E_1, \quad (m',n') \in E_2$$

(6.1) *and*

$$\hat{\mu}_{\alpha\alpha}(m-m',n-n') = K_\alpha((m,n),(m',n')), \quad (m,n), (m',n') \in E_1 \cup E_2, \ \alpha = 1,2.$$

In the case when $K_0 = K_1 = K_2$, Theorem IX reduces to a result of Livshitz and Devinatz [32]. The details concerning this and other reduced bidimensional moment problems will be given elsewhere.

7. HILBERT SPACE STOCHASTIC PROCESSES. A function $X\colon \mathbf{Z} \to H$, H a Hilbert space, is called a discrete *H-valued stochastic process* and the p.d. kernel $K(m,n) = \langle X(m), X(n)\rangle$ is called the *covariance kernel* of X.

A bounded H-valued measure ν defined in $\mathbf{T}$, i.e., a continuous linear mapping $\nu\colon C(\mathbf{T}) \to H$, is said to be *orthogonally scattered*, o.s., if $\langle \nu(f), \nu(g)\rangle = 0$ whenever $f\cdot\bar{g} = 0$. Two such measures, ν_1 and ν_2, are said to be *mutually orthogonally scattered*, m.o.s., if, for $\alpha,\beta = 1,2$, $\langle \nu_\alpha(f), \nu_\beta(g)\rangle = 0$ whenever $f\cdot\bar{g} = 0$. By Masani and Niemi (cf. [25]), if ν_1 and ν_2 are m.o.s., there exists a 2×2 matrix measure $(\mu_{\alpha\beta}) \geq 0$ such that

$$\langle \nu_\alpha(f), \nu_\beta(g)\rangle = \mu_{\alpha\beta}(f\bar{g}), \quad \alpha,\beta = 1,2 . \tag{7.1}$$

The process X is called *stationary* if its covariance kernel is Toeplitz and the classical Bochner-Khinchine theorem asserts that X is stationary iff $K(m,n) = \hat{\mu}(m-n)$, for all $m, n \in \mathbf{Z}$ and for some positive measure μ defined in $\mathbf{T}$, and iff $X(n) = \hat{\nu}(n)$ for all $n \in \mathbf{Z}$ and for some H-valued o.s. measure ν defined in $\mathbf{T}$.

Similarly, the process X is called *generalized stationary* if its covariance kernel is a GTK. Combining the characterization (7.1) with Theorem I we get

COROLLARY 8 [25]. *A process* $X\colon \mathbf{Z} \to H$ *is generalized stationary iff its covariance kernel* K *is given by*

$$K(m,n) = \hat{\mu}_{\alpha\beta}(m-n)\ , \quad (m,n) \in \mathbf{Z}_\alpha \times \mathbf{Z}_\beta\ , \quad \alpha,\beta = 1,2\ , \tag{7.2}$$

for some matrix measure $(\mu_{\alpha\beta}) \geq 0$, *and iff there exists a Hilbert space* $N \supset H$ *and two N-valued m.o.s. measures* ν_1 *and* ν_2 *defined in* $\mathbf{T}$, *such that*

$$X(n) = \hat{\nu}_1(n) \text{ for } n \geq 0 \text{ and } X(n) = \hat{\nu}_2(n) \text{ for } n < 0\ . \tag{7.2a}$$

The process X is called *V-bounded* in the sense of Bochner if its kernel K satisfies

$$\Big|\sum_{m,n} K(m,n)\lambda(m)\overline{\lambda(n)}\Big| \leq c\Big\|\sum_m \lambda(m)e_m(t)\Big\|_\infty^2 \tag{7.3}$$

for some $c > 0$ and every finitely supported sequence $(\lambda(m))$. According to (7.3) we say that the process X is *generalized V-bounded* if there exist two measures μ_1 and μ_2 such that for $\alpha = 1,2$,

$$\Big|\sum_{m,n} K(m,n)\lambda(m)\overline{\lambda(n)}\Big| \leq \mu_\alpha\Big(\Big|\sum_m \lambda(m)e_m(t)\Big|^2\Big) \tag{7.3a}$$

for $\sum\lambda(m)e_m(t) \in W_\alpha$.

As shown in [25], X is generalized V-bounded iff its covariance kernel is majorized by a GTK, i.e., by the kernel of a generalized stationary process. It is known that a process is V-bounded iff it is *harmonizable* in the sense of Rozanov, and there is an extensive theory of harmonizable processes (cf. [52] and its bibliography), generalizing that of stationary processes. The notion of generalized V-boundedness was shown in [25] as equivalent to a notion of generalized harmonizability, and the basic properties of harmonizable processes have corresponding ones for generalized harmonizable processes.

We saw in Section 4 that if $A \in L(N)$ is a contraction in the Hilbert space N, then there exists a Toeplitz kernel K such that $K(m,n) = A^{m-n}$ for $m-n \geq 0$, and A has a unitary dilation $U \in L(H)$, $H \supset N$. If in addition $A^n \to 0$ and $A^{*n} \to 0$ strongly, then H can be decomposed as $H = D_1 \oplus N \oplus D_2$, so that $[H;D_1,D_2;U]$ is a Lax-Phillips scattering system, and A^n is the compression of U^n to N, for all $n \geq 0$. Moreover by a theorem of Lewis and Thomas, there exists a *white noise process* $\xi: \mathbf{Z} \to L(N,H)$ such that the process

$$X_n = U^n|_N \in L(N,H)$$

satisfies the Langevin equation

$$X_{n+1} = X_n A + \xi(n)(1-A^*A)^{1/2} .$$

If we consider $L(N)$-valued GTKs, the analogue to $K(m,n) = A^{m-n}$ is a GTK of the form

$$K(m,n) = C_\beta^* A^{m-n} C_\alpha \quad \text{for} \quad (m,n) \in \mathbf{Z}_\alpha \times \mathbf{Z}_\beta, \quad \alpha,\beta = 1,2,$$

and certain operators C_1 and C_2. As shown in [17], such kernels have properties similar to those of the scalar-valued ones, and they are given by solutions of a "generalized Langevin equation".

The theory of operator-valued GTKs led to a notion of generalized Hilbert space processes, which in turn provides a natural motivation to the notion of GTK, as follows. After Kolmogorov, some authors (cf. [17] and its bibliography) define a *Hilbert space process of type* $(H;N)$ as a function X assigning to each $n \in \mathbf{Z}$ an operator $X_n \in L(N,H)$, where H and N are two Hilbert spaces. The covariance kernel of X is the function

$$(7.4) \qquad K: \mathbf{Z} \times \mathbf{Z} \to L(N) \quad \text{defined by} \quad K(m,n) = X_n^* X_m ,$$

and X is called stationary if this $L(N)$-valued kernel is Toeplitz.

Fix now those Hilbert spaces H, N_1 and N_2 and say that X is a process of type $(H;N_1,N_2)$ if $X_n \in L(N_1,H)$ for $n \geq 0$ and $X_n \in L(N_2,H)$ for $n < 0$. The covariance kernel of such X is then a kernel of type (N_1,N_2) (in the sense of Section 1), given by

(7.4a) $K(m,n) = X_n^* X_m \in L(N_\alpha, N_\beta)$ if $(m,n) \in \mathbb{Z}_\alpha \times \mathbb{Z}_\beta$, $\alpha,\beta = 1,2$.

If $N_1 \neq N_2$, this covariance kernel cannot be Toeplitz, since the Toeplitz condition $K(m+1,n+1) = K(m,n)$, $\forall m,n$, does not make sense for it if either $m = -1$ or $n = -1$. However, K can be a GTK, if the condition holds for all $m \neq -1$, $n \neq -1$. Theorem I provides then the analogue of the Bochner-Khinchine theorem for generalized stationary processes of type $(H;N_1,N_2)$. The theory of such processes is still to be explored, especially in the case of $\mathbb{R}$ where local semigroups of isometries are associated to them.

8. OTHER EXTENSIONS. Let us mention briefly some of the developments that we are currently considering.

(A) A kernel $K: \mathbb{Z}\times\mathbb{Z} \to \mathbb{C}$ is *conditionally p.d.* if $B_K(f,f) \geq 0$ only for f such that $\sum\hat{f}(n) = 0$ (compare this condition with the fact that B is a GTF if $B(\tau f,\tau g) = B(f,g)$ only for $\hat{f}(-1) = \hat{g}(-1) = 0$).

The analogue of the Bochner theorem for conditionally p.d. Toeplitz kernels is the classical Levy-Khinchine theorem, which has applications in convolution semigroup theory, Markov processes and physics. In [25] the Levy-Khinchine and convolution semigroup property were extended to GTKs. This leads to the problem of lifting generalized stationary Markov processes to ordinary ones.

(B) The results of Section 5, and in particular Theorem VII, apply to interpolation problems for entire functions of exponential type and related questions concerning Levitan translations. On the other hand, Theorem VIII extends Toeplitz triplets satisfying a weaker boundedness condition, relating to the notion of Osterwalder-Schrader (OS) positivity and providing a solution of the two-weighted problem for the Poisson operator in the circle [23].

More precisely, let E, $E_1 \subset E$, $E_2 \subset E$ and τ be as in Section 3, and assume that $E_2 = E_1^*$, where * is an involution in E ("time reflection"). As before, set $W_1 = W(E_1) = \{f \in V(E): \text{supp } f \subset E_1\}$, but, instead of W_2, consider $\tilde{W}_2 = \{g \in V(E): \text{supp } g \subset E_2 \text{ and } g(x^*) = \overline{f(x)} \text{ for } x \in E_1,\ f \in W_1\}$. Given a triplet of kernels (K_0,K_1,K_2), write $K_0 \prec\!\prec (K_1,K_2)$ if

$$|B_{K_0}(f,g)|^2 \leq B_{K_1}(f,f)B_{K_2}(g,g) \quad \text{only for} \quad (f,g) \in W_1 \times \tilde{W}_2 ,$$

so that a Toeplitz kernel K is OS positive if $-K \prec\!\prec (K_1,K_2)$ for $K_1 = K_2 = 0$. When $E = \mathbb{Z}$, $E_1 = \mathbb{Z}_+$, $E_2 = \mathbb{Z}_-$, $x^* = -x$, the GBT extends for the weaker boundedness relation $\prec\!\prec$, providing a characterization of the weights u and v for which the Poisson operator is continuous from $L^p(\mathbb{T},u\,dt)$ to $L^p(\mathbb{T},v\,dt)$ (cf. [23] and [57]).

The operator-valued versions of these results and their extensions to general algebraic scattering systems are still to be developed.

(C) As discussed in Section 4 the representation and lifting properties of Toeplitz triplets can be viewed as a development of the Naimark-Nagy dilation theory. This theory was developed in another direction for completely positive maps, irreversible or dissipative semigroups and dynamical systems, of quantum and quasi-free quantum processes, etc. (cf. the papers in Quantum Probability and Applications to Quantum Theory of Irreversible Processes, Lecture Notes in Math. 1055, Springer-Verlag, New York, 1984, and their bibliographies), and the notion of bounded Toeplitz triplets can also be developed in these contexts.

BIBLIOGRAPHY

1. V. M. Adamjan and D. Z. Arov, "On unitary couplings of semiunitary operators," Matem. Issledovanya, 1:2 (1966), 3-64 (in Russian).

2. V. M. Adamjan, D. Z. Arov and M. G. Krein, "Infinite Hankel matrices and generalized Carathéodory-Fejér and I. Schur problems," Func. Anal. Appl., 2 (1968), 1-17 (in Russian).

3. P. Alegría, "Parametrization of the representing measures of a GTK," Master Thesis, U.C.V., Caracas, 1987.

4. R. Arocena, "Toeplitz kernels and dilations of intertwining operators," Integral Eqs. and Operator Theory, 6 (1983), 759-778.

5. ______, "On generalized Toeplitz kernels and their relation with a paper of Adamjan, Arov and Krein," North Holland Math. Studies, 86 (1984), 1-22.

6. ______, "Scattering functions, linear systems, Fourier transforms of measures and unitary dilations to Krein spaces: a unified approach," Publ. Math. d'Orsay, 85-02 (1985), 1-57.

7. ______, "Une classe de formes invariantes par translation et le théorème de Helson et Szegö," C. R. Acad. Sci. Paris I, 302 (1986), 107-109.

8. ______, "Generalized Toeplitz kernels and dilations of intertwining operators," to appear in Acta Sci. Math. (Szeged).

9. R. Arocena and M. Cotlar, "Generalized Toeplitz kernels, Hankel forms and Sarason's commutation theorem," Acta Cient. Venez., 33 (1982), 89-98.

10. ______, "Dilation of generalized Toeplitz kernels and L^2-weighted problems," Lecture Notes in Math., Springer-Verlag, 908 (1982), 169-188.

11. R. Arocena, M. Cotlar and J. León, "Toeplitz kernels, scattering structures and covariant systems," North-Holland Math. Lib., 34 (1985), 1-19.

12. R. Arocena, M. Cotlar and C. Sadosky, "Weighted inequalities in L^2 and lifting properties," Adv. Math. Suppl. Studies, 7A (1981), 95-128.

13. Iu. M. Berezanskii, Expansions in eigenfunctions of selfadjoint operators, Transl. Math. Monographs, Amer. Math. Soc., Providence, R.I., 1968.

14. R. Bruzual, "Local semigroups of contractions and some applications to Fourier representation theorems," to appear in Integral Eqs. and Operator Theory.

15. ______, "Integral representations of kernels invariant with respect to local semigroups of isometries and generalization of the Krein-Schwartz theorem," to appear in Acta Cient. Venez.

16. M. Cotlar, "Equipación en espacios de Hilbert," Cursos y Sem. de Matem., 15, FCEN, UBA, Buenos Aires, 1970.

17. M. Cotlar, J. León and M. C. Pereyra, "Eigenfunction expansions of covariance kernels," to appear in Acta Cient. Venez.

18. M. Cotlar and C. Sadosky, "A moment theory approach to the Riesz theorem on the conjugate function for general measures," Studia Math., 53 (1975), 75-101.

19. ______, "Transformée de Hilbert, théorème de Bochner et le problème des moments," I, II, C. R. Acad. Sci., Paris, A 285 (1977), 433-435, 661-665.

20. ______, "On the Helson-Szegö theorem and a related class of modified Toeplitz kernels," Proc. Symp. Pure Math. AMS, 34 (1979), 383-407.

21. ______, "Characterization of two measures satisfying the Riesz inequality for the Hilbert transforms in L^2," Acta Cient. Venez., 30 (1979), 346-348.

22. ______, "On some L^p versions of the Helson-Szegö theorem," in Conf. Harmonic Analysis in honor of A. Zygmund (Eds.: W. Beckner, A. P. Calderón, R. Fefferman, P. W. Jones), Wadsworth Int. Math. Series (1982), 306-317.

23. ______, "Majorized Toeplitz forms and weighted inequalities with general norms," Lecture Notes in Math., Springer-Verlag, 908 (1982), 132-168.

24. ______, "Vectorial inequalities of Marcinkiewicz-Zygmund and Grothendieck type for generalized Toeplitz kernels," Lectures Notes in Math., Springer-Verlag, 992 (1983), 278-308.

25. ______, "Generalized Toeplitz kernels, stationarity and harmonizability," J. Anal. Math., 44 (1985), 117-133.

26. ______, "Inégalités à poid pour les coefficients lacunaires de certaines fonctions analytiques," C. R. Acad. Sci. Paris I, 299 (1984), 591-594.

27. ______, "A lifting theorem for subordinated invariant kernels," J. Func. Anal., 67 (1986), 345-359.

28. ______, "Lifting properties, Nehari theorem and Paley lacunary inequality," Rev. Mat. Iberoamericana, 2 (1986), 55-71.

29. ______, "Prolongements des formes de Hankel généralisées en formes de Toeplitz," C. R. Acad. Sci., Paris, I 305 (1987), 167-170.

30. ______, "Toeplitz liftings of Hankel forms," to appear in Function Spaces and Applications, Lund 1986 (Eds.: J. Peetre, Y. Sagher and H. Wallin), Lecture Notes in Math., Springer-Verlag, New York.

31. ______, "Integral representations of bounded Hankel forms defined in scattering systems with a multiparametric evolution group," to appear in Operator Theory: Advances and Applications, Birkhäuser Verlag, Basle-Boston.

32. A. Devinatz, "On extensions of positive definite functions," Acta Math., 102 (1959), 109-134.

33. M. Domínguez, "Procesos contínuos completamente regulares," Master Thesis, U.C.V., Caracas, 1984.

34. P. Duren and D. L. Williams, "Interpolation problems in function spaces," J. Func. Anal., 9 (1972), 75-86.

35. H. Helson and D. Lowdenslager, "Prediction theory and Fourier series in several variables," Acta Math., 99 (1958), 165-202.

36. H. Helson and D. Sarason, "Past and future," Math. Scand., 21 (1967), 5-16.

37. H. Helson and G. Szegö, "A problem in prediction theory," Ann. Math. Pure Appl., 51 (1960), 107-138.

38. R. Hunt, B. Muckenhoupt and R. L. Wheeden, "Weighted norm inequalities for the conjugate function and the Hilbert transform," Trans. Amer. Math. Soc., 176 (1973), 227-252.

39. M. G. Krein, "Sur le problème du prolongement des fonctions hermitiennes positives et continues," Dokl. Akad. Nauk SSSR, 26 (1940), 17-22 (in Russian).

40. M. G. Krein, "On hermitian operators with directed functionals," Akad. Nauk Ukrain. RSR Zbirnik Proc. Inst. Mat. (1947), 104-129 (in Russian).

41. H. Langer, "Über die Methode der Richtenden Funktionale von M. G. Krein," Acta Math. Acad. Sci. Hung., 21 (1970), 207-224.

42. P. Lax and R. Phillips, Scattering theory, Academic Press, New York, 1967.

43. S. Marcantognini, "Dilataciones unitarias a espacios de Krein," Master Thesis, U.C.V., Caracas, 1985.

44. ______, "Colligations in Krein spaces," preprint, U.C.V.-Conicit, Caracas, 1987.

45. K. Maurin, General eigenfunction expansions and unitary representations of topological groups, Monografie Matemat., 48, PWN, Warsaw, 1968.

46. M. D. Morán, "Parametrización de Foias-Ceausescu y núcleos de Toeplitz generalizados," Ph.D. Thesis, U.C.V., Caracas, 1987.

47. B. Sz. Nagy and C. Foias, Analyse harmonique des opérateurs de l'espace de Hilbert, Masson-Akad. Kiado, Paris and Budapest, 1967; Engl. trans. North-Holland-Akad. Kiado, Amsterdam and Budapest, 1970.

48. ______, "Dilatation des commutants d'opérateurs," C. R. Acad. Sci. Paris, A 266 (1968), 493-495.

49. M. A. Naimark, "Extremal spectral functions of a symmetric operator," I. V. Akad. Nauk SSSR Ser. Mat., 11 (1947), 327-344 (in Russian).

50. T. Nakazi, "Weighted norm inequalities and uniform algebras," to appear in Proc. Amer. Math. Soc.

51. Z. Nehari, "On bilinear forms," Ann. Math., 68 (1957), 153-162.

52. H. Niemi and A. Weron, "Dilation theorems for positive definite operator kernels having majorants," J. Func. Anal., 40 (1981), 54-65.

53. N. K. Nikolskii, Lectures on the shift operator, Mir, Moscow, 1980; Engl. trans. Springer-Verlag, New York, 1985.

54. V. V. Peller and S. V. Khrushchev, "Hankel operators, best approximations and stationary Gaussian processes," Uspehi Mat. Nauk, 37 (1982), 53-124 (in Russian).

55. R. Rochberg, "Toeplitz and Hankel operators on a Paley-Wiener space," to appear in Integral Eqs. and Operator Theory.

56. J. L. Rubio de Francia, "Linear operators in Banach lattices and weighted L^2 inequalities," Math. Nachr. 133 (1987), 197-209.

57. C. Sadosky, "Some applications of majorized Toeplitz kernels," Topics in Modern Harm. Anal., II, Ist. Alta Mat. Roma (1983), 581-626.

58. L. Schwartz, Mathemática y Física Cuántica, Cursos y Sem. de Matem., 1, Fac. de Ciencias, U.B.A., Buenos Aires (1959), 1-266.

59. ______, "Sous-espaces hilbertiens d'espaces vectoriels topologiques et noyaux associés," J. Anal. Math., 13 (1964), 115-256.

60. T. Yamamoto, "On the generalization of the theorem of Helson and Szegö," Hokkaido Math. J., 14 (1985), 1-11.

MATHEMATICAL SCIENCES RESEARCH INSTITUTE
Berkeley, CA 94720

ADDENDUM. After this paper was written, R. Rochberg kindly called to our attention that the Helson-Szegö theorem for Helson-Lowdenslager half planes follows immediately from the corresponding result for weak-* Dirichlet algebras appeared in his joint paper with I. I. Hirschman, "Conjugate Function Theory in Weak-* Dirichlet Algebras," J. Funct. Anal. 16 (1974), 359-371.

DIFFERENTIAL ESTIMATES AND COMMUTATORS IN INTERPOLATION THEORY

by

M. Cwikel, B. Jawerth[1], M. Milman and R. Rochberg[2]

Abstract: Associated with the construction of intermediate Banach spaces in interpolation theory are unbounded non-linear operators, Ω, which are, roughly, the differentials of the fundamental mappings used in the construction of the intermediate spaces. We look at the theory of these operators and their applications to interpolation theory and to other topics in analysis. We see that the mapping properties of Ω are strongly related to the structure of the interpolation scale. We also see that computations using Ω are closely related to certain computations in the theory of hypercontractive semigroups.

I. Introduction and Summary: Several constructions in interpolation theory start with a pair of Banach spaces (for example, L^1 and L^∞) and construct a parameterized family of spaces (in the example, L^p, $1 < p < \infty$) which are, in an appropriate technical sense, intermediate to the original pair. In particular it will be true that if a linear operator T is bounded on both spaces of the starting pair, then T will also be bounded on the intermediate spaces.

Associated to the construction of intermediate spaces are mappings Ω, generally unbounded and non-linear, which can be obtained by differentiation with respect to certain parameters used in the construction of the intermediate spaces. In the case of L^p spaces, one instance of Ω is given by

$$\Omega f = f \log(|f|/\|f\|_{L^p}). \tag{1.1}$$

It is shown in [RW] and [JRW] that, for T as above, the commutator $[T,\Omega] = T\Omega - \Omega T$ is bounded on the intermediate L^p spaces and that T is bounded on the domain of definition of Ω in

[1]Partially supported by NSF grant DMS 8404528.

[2]Partially supported by NSF grant DMS 8701271.

L^p. That is, given p, $1 < p < \infty$, there are constants c_p so that for all f in L^p

(1.2) $$\|[T,\Omega]f\|_p \leq c_p \|f\|_p,$$

and

(1.3) $$\|Tf\|_p + \|\Omega Tf\|_p \leq c_p (\|f\|_p + \|\Omega f\|_p).$$

In this survey we present (without full proofs but with complete references and some proofs) most of what is known about the operators such as Ω. In Sections 2 through 5 we relate these operators to the general theories of real and complex interpolation. In Sections 6 and 7 we return to the relation between these results and other areas in analysis. In particular, readers less interested in interpolation theory may want to skip to those later sections.

We have tried to be brief and have concentrated on versions of the results which can be presented efficiently. More general and more precise statements are in the papers mentioned in the references.

Greatest attention is devoted to the most familiar interpolation methods, the K and J methods of real interpolation theory. We are more brief with analogous results for the E and F methods and the complex method.

Here are the contents in more detail.

In Section 2 we give the basic constructions of operators such as Ω, and we recall the main results of [RW] and [JRW].

In Section 3 we give relations between the Ω's and the duality and reiteration theorems of interpolation theory. Some of the results are expressed in terms of a twisted direct sum construction.

In Section 4 we give explicit descriptions of the spaces Dom(Ω), the domain of definition of Ω, and Ran(Ω), the range of Ω. We will see that there is a duality pairing between those spaces. Also, although the operator Ω generally depends on the interpolation method considered, we will see that the space

Dom(Ω) does not; nor, when appropriately defined, does Ran(Ω). Some of this material is from [CJM], other parts are new. The main tool in this section is the use of local retracts to reduce consideration to weighted sequence spaces.

In Section 5 we see that the norm of Ωa is closely related to the derivative of the norm of a with respect to one of the parameters which index the interpolation spaces. Note, as a motivating instance, that for a function f which satisfies $|f| \geq 1$, $\|f\|_1 = 1$, and with Ω given by (1.1),

$$\frac{d}{dp} \|f\|_p^p \Big|_{p=1} = \|\Omega f\|_1.$$

By relating the norm of Ωa to the derivative of the norm of a we give criteria which insure that a linear operator which is bounded on one of the intermediate spaces will be bounded on a family of spaces. Such extrapolation results are partial converses to (1.2) and (1.3). We also show that if Ω is bounded on one of the intermediate spaces then the intermediate spaces are all the same. This is related to the fact that the classical interpolation spaces $A_{\theta,q}$ (often denoted $\bar{A}_{\theta,q}$) must depend effectively on the parameters. We also give a Wolff type theorem on "welding" of interpolation scales.

In Section 6 we describe the recent work of Kalton on estimates such as (1.2) for lattices of functions and for operator ideals.

In Section 7 we use the operators Ω to develop estimates in classical contexts. We see that logarithmic Sovolev estimates for generators of hypercontractive semigroups are closely related to our abstract results. We also obtain boundedness results for the commutator between Ω and sublinear operators such as the maximal operator associated with fractional order integration.

In the final section we pose some problems.

<u>**II. The Basic Constructions:**</u> In this section we describe constructions of operators Ω which satisfy the basic commutation result, Theorem 2.4 below. Most of these results are from [RW]

and [JRW].

A. The K and J method: Let $\bar{A} = (A_0, A_1)$ be a Banach couple, that is, a pair of Banach spaces which are both continuously embedded in a topological Hausdorff vector space. We will also call $\bar{A}$ a compatible couple. The reader not familiar with this general viewpoint should think of (L^1, L^∞).

A basic theme in the analysis here is that real interpolation spaces are constructed by finding efficient ways of decomposing elements of $A_0 + A_1$ into pieces which can then be studied seperately.

First consider the interpolation method based on the K and J functionals. For a in $A_0 + A_1$, we define the K functional for each positive t by

$$K(t,a;\bar{A}) = \inf_{\substack{a=a_0+a_1 \\ a_i \in A_i}} (\|a_0\|_{A_0} + t\|a_1\|_{A_1}).$$

For each t let us say that a decomposition $a = a_0(t) + a_1(t)$ is almost optimal (for the K method) if

$$\|a_0(t)\|_{A_0} + t\|a_1(t)\|_{A_1} \leq cK(t,a;\bar{A}) \tag{2.1}$$

(where c is a constant whose value is fixed during our discussion). We write $D_K(t,\bar{A})a = D_K(t)a = a_0(t)$. Recall that for θ and q which satisfy

$$0 < \theta < 1, \quad 1 \leq q \leq \infty, \tag{2.2}$$

the interpolation space $A_{\theta,q,K}$ is defined by

$$A_{\theta,q,K} = \left\{a: \left[\int_0^\infty (t^{-\theta}K(t,a,\bar{A}))^q \frac{dt}{t}\right]^{1/q} < \infty\right\},$$

with the obvious norm.

The functional J, which is dual to K, is defined on $A_0 \cap A_1$ by, for each positive t,

$$J(t,a;\bar{A}) = \max\{\|a\|_{A_0}, t\|a\|_{A_1}\}.$$

For θ and q which satisfy (2.1) the intermediate spaces $A_{\theta,q,J}$ obtained by the J method of interpolation are defined by the norming functionals

$$\inf\left\{\left[\int_0^\infty (t^{-\theta}J(t,u(t);\bar{A}))^q \frac{dt}{t}\right]^{1/q}\right\},$$

where the infimum is over all strongly measurable $A_0 \cap A_1$ valued functions u(t) for which $\int_0^\infty u(t)\,\frac{dt}{t} = a$. We call a choice of u(t) for which the infimum is almost attained an almost optimal J decomposition and denote it $D_J(t) = D_J(t)a = D_J(t,\bar{A})a$. Thus, $\int_0^\infty D_J(t)\,\frac{dt}{t} = a$ and

$$\text{(2.3)} \qquad \left[\int_0^\infty (t^{-\theta}J(t,D_J(t)a,\bar{A}))^q \frac{dt}{t}\right]^{1/q} \le c\,\|a\|_{A_{\theta,q,J}}$$

Roughly, $D_K(t) = \int_0^t D_J(s)\,\frac{ds}{s}$. (Precisely, there are choices of almost optimal decompositions for which this relation holds. It is also possible to choose a single decomposition $D_J(t)$ which is simultaneously almost optimal for all values of θ and q.)

It is well known that $A_{\theta,q,K}$ and $A_{\theta,q,J}$ coincide to within equivalence of norm. For this reason we will often omit those letters in the notation.

Suppose T is a bounded linear operator between the Banach couples $\bar{A}$ and $\bar{B}$. Given a in $A_{\theta,q,K}$, we can apply the almost optimal decompositions both before and after applying T. This leads to the study of the "local commutator"

$$[T,D_K(t)]a = TD_K(t,\bar{A})a - D_K(t,\bar{B})Ta.$$

It follows readily from the definitions that

(2.4) $$J(t,[T,D_K(t)]a;\overline{B}) \le cK(t,a;\overline{A})$$

where c is a constant depending only on the norm of T on the couples. The estimate (2.4) can be used to study the nonlinear operator $[T,\Omega]$ defined by

$$[T,\Omega]a = \int_0^\infty [T,D_K(t)]a \frac{dt}{t}.$$

Moreover, the operator $[T,\Omega]$ can be represented

$$[T,\Omega]a = T\Omega_{K,\overline{A}}a - \Omega_{K,\overline{B}}Ta$$

where, for a Banach pair $\overline{H}$, $\Omega_{K,\overline{H}}$ is defined by

(2.5) $$\Omega_{K,\overline{H}}a = \int_0^1 D_K(t)a \frac{dt}{t} - \int_1^\infty (I - D_K(t))a \frac{dt}{t}.$$

Thus, if $\overline{A} = \overline{B}$, then $[T,\Omega]$ is, in fact, a commutator.

We will often omit one or both subscripts from $\Omega_{K,\overline{H}}$.

Observe that Ω is not a linear operator, nor is it bounded on the pair $\overline{H}$. In fact, we will see below that Ω is generally not bounded on any of the intermediate $H_{\theta,q}$.

Whether or not Ω is positive homogenous depends on the choice of D_K. It is quite natural to suppose that D_K is chosen to be positive homogenous and we make a standing assumption that this is the case. In that case Ω_K will also be positive homogenous.

In addition to commutators with Ω, we will also be interested in the domain of definition of Ω. For fixed $A = A_{\theta,q}$ we define the domain of Ω, $\mathrm{Dom}(\Omega) = \mathrm{Dom}_A(\Omega)$ to be $\{a\in A : \Omega a\in A\}$. We norm $\mathrm{Dom}(\Omega)$ by

$$\|a\|_{\mathrm{Dom}} = \|a\|_A + \|\Omega a\|_A.$$

The basic result concerning Ω is

<u>**Theorem 2.6:**</u> ([JRW]) Suppose $T:\overline{A} \to \overline{B}$ is a bounded linear operator

and θ, q satisfy (2.2). Then

(i) $[T,\Omega]$ is a bounded operator from $A_{\theta,q}$ to $B_{\theta,q}$ and

(ii) T is a bounded operator from $\mathrm{Dom}_{A_{\theta,q}}(\Omega)$ to $\mathrm{Dom}_{B_{\theta,q}}(\Omega)$

Part (i) of the theorem is a straightforward consequence of (2.4) and the equivalence of the K and the J methods of interpolation. Part (ii) then follows quickly from part (i).

The definition of Ω, and hence apparently also of the space $\mathrm{Dom}(\Omega)$, depends on the particular choice of almost optimal decomposition. Such decompositions are far from unique and different choices give rise to different operators Ω. However, if Ω and $\tilde{\Omega}$ are operators defined as above using different almost optimal decompositions, then, using part (i) of the theorem with $T = I$ (= identity operator), we see that $[I,\Omega] = \Omega - \tilde{\Omega}$ is bounded. Thus Ω is "well defined up to a bounded error." In particular, the space $\mathrm{Dom}(\Omega)$ is independent of the particular choice of Ω.

Because the operator Ω is not generally linear, it is not clear that $\mathrm{Dom}(\Omega)$ is a linear space. However it is. This can be seen by applying the theorem to the couples $\bar{A} \oplus \bar{A}$ and $\bar{A}$ and the map $T((a_1,a_2)) = a_1 + a_2$. It is straightforward to check that one can take $\Omega_{A\oplus A} = \Omega_A \oplus \Omega_A$. Hence

(2.7) $$\|\Omega(a+b) - \Omega(a) - \Omega(b)\| \leq c(\|a\| + \|b\|).$$

As is always true in interpolation theory, it is important to make the abstract theory explicit for the standard examples:

Consider the couple $\bar{A} = (L^1([0,1]), L^\infty([0,1]))$. Recall (see [BL]) that an almost optimal decomposition for the computation of $K(t,f,L^1,L^\infty)$ can be described using $f^*(t)$, the non-increasing rearrangement of f. We have $f = f_0(t) + f_1(t) = f\chi_{\{|f|>f^*(t)\}} + f\chi_{\{|f|\leq f^*(t)\}}$. Direct computation now gives

(2.8) $$\Omega f(x) = f(x)\,|\log r_f(x)|.$$

Here $r(x) = r_f(x)$ is the so-called rank function of f defined by

$$r_f(x) = |\{y : |f(y)| > |f(x)| \text{ or } |f(y)| = |f(x)| \text{ any } y \leq x\}|.$$

(The terminology can be motivated by considering a positive f defined on a probablity space or on Z^+.)

(To extend this result to general measure space we need to introduce a measurable ordering. However, the ordering plays no role in the theory except in this explicit formula for r_f. We note here that some of the computations in [JRW] relating to r_f are wrong due to ignoring case in which $\{y : |f(y)| = |f(x)|\}$ has positive measure.)

One can define another operator $\Omega = \Omega_J$ using the almost optimal decomposition associated with the J method of interpolation. The natural definition turns out to be

$$\Omega_J a = \int_0^\infty D_J(t) a \log t \frac{dt}{t}. \tag{2.9}$$

In [CJM] it was shown that for any Banach couple $\bar{A}$ we can choose decompositions so that the corresponding K and J operators satisfy

$$\Omega_K a = -\Omega_J a. \tag{2.10}$$

Thus the previous results also apply to Ω_J.

B. The E and F methods of interpolation: There are results parallel to those we have been discussing in the context of the E method of interpolation. We replace the K functional by the E functional defined by

$$E(t,a;\bar{A}) = \inf_{\substack{\|a_1\|_{A_1} \le t \\ a_0 + a_1 = a}} \|a_0\|_{A_0} .$$

Let $D_E(t,\bar{A})a = D_E(t)a = a_0(t)$ be the first term of the corresponding almost optimal decomposition. We then define the associated $\Omega = \Omega_E$ by

$$\Omega a = \int_1^\infty D_E(t) a \frac{dt}{t} - \int_0^1 (I - D_E(t)) a \frac{dt}{t}.$$

The relationship of this approach to the K and J functionals, the introduction of the F functional which is "dual" to the E functional, and the introduction of more general families of functionals E_α, F_α is given in [JRW]. (E here is tE_1 there.) For the moment we want to note two things. First, the exact analog of Theorem 2.6 holds for Ω_E. Second, the couple (L^1, L^∞) gives an example in which $\Omega_E \neq \Omega_K$. (More precisely, since each of the operators is only well-defined up to a bounded error, the difference $\Omega_E - \Omega_K$ is not bounded.) To see this we note that an allowable choice for D_E is given by $D_E(t)f = f\chi_{\{|f|>t\}}$ and then, by direct computation, $\Omega_E f = f \log|f|$, which is closely related to the Ω of (1.1). Note that here however Ω is not homogenous. That complicates the functional analysis associated with Ω_E.

Although Ω_K and Ω_E differ, we will see later that the domain spaces, $\mathrm{Dom}(\Omega_K)$ and $\mathrm{Dom}(\Omega_E)$ coincide.

C. The Complex Method: For a couple $\bar{A}$ the complex interpolation spaces A_θ, $0 < \theta < 1$, are defined using the norm

$$\|a\|_\theta = \inf \{\max \{\sup_y \|F(iy)\|_0, \sup_y \|F(1+iy)\|_1\}\}.$$

Here F is in $\mathcal{F}(A_0, A_1)$, the class of $A_0 + A_1$ valued function which are analytic in $\{0 < \mathrm{Re}\, z < 1\}$ (and satisfy certain technical growth conditions, see [BL]), and F satisfies $F(\theta) = a$. If some particular F, which we write F_a, satisfies

$$\max \{\sup_y \|F_a(iy)\|_0, \sup_y \|F_a(1+iy)\|_1\} \le c\|a\|_\theta$$

then we call F_a almost optimal for the complex method and define the associated Ω by $\Omega_{\mathbb{C}} a = \Omega a = c_\theta F_a'(\theta)$. c_θ is an numerical factor which is unimportant here. Again we have the same pattern of results. The analog of Theorem 2.6 holds, Ω is well defined up to a bounded error, the domain of Ω is a linear space, etc. All of this is in [RW] where it is also shown that for the L^p

spaces Ω is a constant multiple of the operator given in (1.1).

III. The Abstract Theory: In this section we describe some of the properties of Ω_K. As noted in (2.10), $\Omega_K = -\Omega_J$ and thus analogous results hold for Ω_J.

A. Iteration: Suppose $\bar{A}$ is a couple of Banach spaces and $0 < \theta_0 < \theta_1 < 1 \leq q_0, q_1 \leq \infty$. We can then form the new couple $\bar{B} = (A_{\theta_0,q_0}, A_{\theta_1,q_1})$. The Iteration Theorem in interpolation theory gives the relation between the interpolation spaces for $\bar{A}$ and for $\bar{B}$. If θ, q satisfy (2.2) then $B_{\eta,q} = A_{\theta,q}$ with $\theta = (1-\eta)\theta_0+\eta\theta_1$. It is then possible to compare the corresponding operators $\Omega_{K,\bar{A}}$ and $\Omega_{K,\bar{B}}$. The result is that

$$\Omega_{K,\bar{B}} = \Delta\ \Omega_{K,\bar{A}}. \tag{3.1}$$

with $\Delta = \theta_1 - \theta_0$. (That is, there are choices of D_K so that...) This precise relationship will be important in a computation in Section 7.

B. The Lemma of Sneiberg and Zafran: Suppose that θ, q are fixed and satisfy (2.2). Given an a in $A_{\theta,q}$, let $D_J(t)a$ be an almost optimal J decomposition. For small ϵ define R_ϵ by

$$R_\epsilon a = \int_0^\infty t^\epsilon\ D_J(t)a\ \frac{dt}{t}.$$

Certainly $R_0 a = a$ and it follows at once from the definition of the interpolation spaces using the J method that

$$\|R_\epsilon a\|_{A_{\theta+\epsilon,q,J}} \leq c\|a\|_{A_{\theta,q,J}}. \tag{3.2}$$

Also,

$$\frac{d}{d\epsilon} R_\epsilon a\bigg|_{\epsilon=0} = \Omega_J a. \tag{3.3}$$

(Again we are putting aside some technical issues; the domain of R_ϵ should be restricted to, say, $A_0 \cap A_1$, and the operator R_ϵ actually depends on the choice of almost optimal decomposition.)

R_ε was considered in Section 5 of [JRW] (using different notation) and was also used by Zafran [Z] in comparing the real and complex methods of interpolation and in extending results of Sneiberg. We will use the result of Zafran which shows that, to some extent, (3.2) can be reversed.

First let us recall (cf. [BL] p. 54) that if $A_0 \cap A_1$ is dense in A_i for $i = 0, 1$, then the dual space of $A_{\theta,q}$ is $A^*_{\theta,q'}$ where $A^* = (A_0^*, A_1^*)$ is the dual couple of A, $1/q + 1/q' = 1$, and $1 < q < \infty$. For simplicity let us suppose throughout this subsection that q is finite, that the couple A^* satisfies the analogous density conditions, and that A_0 and A_1, and therefore also $A_{\theta,q'}$, are reflexive.

We denote the bilinear pairing between a in $A_{\theta,q}$ and a^* in $A^*_{\theta,q'}$ by $\langle a,a^* \rangle$.

__Lemma 3.4__ (Zafran [Z]): There are constants $c > 0$, $\varepsilon_0 > 0$ such that for all ε, $|\varepsilon| \le \varepsilon_0$, for all a in $A_{\theta,q}$

(3.5) $$\|a\|_{A_{\theta,q,J}} \le c\,\|R_\varepsilon a\|_{A_{\theta+\varepsilon,q,J}} .$$

We give some of the details of the argument in [Z]. In a moment we will take it further.

It follows from (3.2) that, for small ε

(3.6) $$|\langle R_\varepsilon a,\ R^*_\varepsilon a^* \rangle| \le c\,\|a\|_{A_{\theta,q}}\,\|a^*\|_{A^*_{\theta,q'}} .$$

(Here R^*_ε is the analog of R_ε for the couple $\bar{A}^*$.)

The crucial observation now is that the definition of R_ε continues to make sense for complex values of ε. Furthermore, if $\mathrm{Re}(\varepsilon)$ is sufficiently small then (3.2), and thus also (3.6), holds in a strip in the complex plane: $\{\varepsilon : |\mathrm{Re}(\varepsilon)| \le \varepsilon_0\}$. By (3.6) the function $f(\varepsilon) = (R_\varepsilon a,\ R^*_\varepsilon a^*)$ is a bounded analytic function in the strip. The basic estimate for the modulus of continuity of a bounded analytic function, the Schwarz-Pick

Lemma, now yields the basic analytic estimate for Zafran's proof.

The Schwarz-Pick Lemma also insures that

$$|f'(0)| \le c\, c'\, \|a\|_{A_{\theta,q}} \|a^*\|_{A^*_{\theta,q'}}$$

Here c is the same constant as in (3.6) and c' depends only on ϵ_0. When we explicitly compute the derivative using (3.3) we obtain

Proposition 3.7: The operator $\Omega = \Omega_{J,\bar{A}}$ for the couple $\bar{A}$ and the operator $\Omega_* = \Omega_{J,\bar{A}^*}$ for the dual couple $\bar{A}^*$ are related by

(3.8) $$|\langle \Omega a, a^* \rangle + \langle a, \Omega_* a^* \rangle| \le c \|a\|_{A_{\theta,q}} \|a^*\|_{A^*_{\theta,q'}} .$$

In fact the Ω's are only well defined up to bounded errors (that is, a change in the choice of decomposition used in the definition will produce a bounded change in Ω). Thus, (3.8) should be seen as saying $\langle \Omega a, a^* \rangle = - \langle a, \Omega_* a^* \rangle$ modulo the standard errors in this theory.

If it is possible to make a choice of $D_J(t)a$ which is a linear function of a, then the mappings R_ϵ and Ω will be linear maps. Couples for which this is possible are called quasilinearizable. The case in which A_0 isan L^p space and A_1 is an L^p space with a weight is an example of a quasilinearizable couple. The couple (L^1, L^∞) is not quasilinearizable. For quasilinearizable couples we can consider the adjoints of Ω and of R_ϵ. In particular, by (3.8), Ω^*, the adjoint of Ω, satisfies:

(3.9) $$\Omega^* \equiv -\Omega_* \text{ (modulo bounded operators).}$$

Proposition 3.10: Suppose the couple $\bar{A}$ is quasilinearizable. Suppose θ, q are given and satisfy (2.1). There is an $\epsilon_0 > 0$ and a $c > 0$ such that for all ϵ, $|\epsilon| \le \epsilon_0$, R_ϵ is an invertible map of $A_{\theta,q}$ onto $A_{\theta+\epsilon,q}$ which satisfies $\|R_\epsilon\| \|R_\epsilon^{-1}\| \le c$.

Proof: Pick and fix a small ϵ. We have just obtained upper and lower bounds on R_ϵ, (3.2) and (3.5). These show that R_ϵ is a bounded invertible map onto its range and that the range is a closed subspace of $A_{\theta+\epsilon,q}$. We need to show this subspace is the entire space. Suppose not; then there is a linear functional, b, in $A^*_{\theta+\epsilon,q'}$ which has norm one and is zero on all of $R_\epsilon(A_{\theta,q})$. Let S_δ be a map which is analogous to R_ϵ but defined on the dual couple $\bar{A}^*$. Thus S_δ maps $A^*_{\theta+\epsilon,q'}$ to $A^*_{\theta+\epsilon+\delta,q'}$. Pick a in $A_{\theta,q}$ of norm one and consider the analytic function $F(\delta) = \langle R_\delta a, S_{-\epsilon+\delta} b\rangle$. By construction $F(\epsilon) = 0$. As in Zafran's proof we have a uniform bound on $|F(\delta)|$ for $|\delta| \le K$, where K can be choosen to be independent of the small number ϵ. Thus, by the Schwarz-Pick lemma, $F(0) = O(\epsilon)$. However, the supremum of all such $|F(0)|$ is the norm of the linear functional $S_{-\epsilon}b$ acting on $A_{\theta,q}$ which, by Lemma 3.4 applied to the family S_ϵ, can't be too small. This contradiction completes the proof.

<u>Corollary 3.11:</u> Suppose $\bar{A}$ is quasilinearizable. Given $0 < \theta_0 < \theta_1 < 1 \le q \le \infty$, the spaces $A_{\theta_0,q}$ and $A_{\theta_1,q}$ are isomorphic Banach spaces.

<u>Proof:</u> By the previous proposition, given θ in (0,1) there is a neighborhood N of θ so that all of the spaces $A_{\varphi,q}$, φ in N, are isomorphic. Pick a finite set of such neighborhoods to cover the compact interval $[\theta_0,\theta_1]$. Since isomorphism is transitive, we are done.

Note: In general, for fixed θ and $q_1 \ne q_2$, A_{θ,q_1} will not be isomorphic to A_{θ,q_2}. See, for instance, the discussion in Section 3 of [JNP].

<u>Corollary 3.12:</u> The couple (L^1,L^∞) is not quasilinearizable.

<u>Proof:</u> If (L^1,L^∞) were quasilinearizable then, by the previous result, all of the Lorentz spaces $L^{p,2}$ would be isomorphic to each other, $1 < p < \infty$. However this is impossible since they do

not all have the same Boyd indices. (This follows e.g. from a theorem of Krivine ([LT] pg. 141).

C. The Twisted Direct Sum and the Duality Pairing: Given $A = A_{\theta,q}$ and given Ω we define the Ω-twisted direct sum of A with itself to be the set of pairs (a,b) in $(A_0 + A_1, A_0 + A_1)$ for which

$$\|(a,b)\| = \|b - \Omega a\|_A + \|a\|_A \tag{3.13}$$

is finite. It follows from (2.7) that $\|\cdot\|$ satisfies the triangle inequality but with a constant factor on the right. Hence, if Ω is also positive homogenous, then the twisted direct sum is a linear space and (3.13) defines a quasinorm. We will see in a moment that, for Ω_K, this space has an equivalent norm.

We denote this quasinormed space by $A\ {}_\Omega\!\oplus A$. The same terminology, and the notation $A \oplus_\Omega A$, is used in [K] for the analogous construction with the role of the two coordinates reversed. We use this formulation because of the analogy with the spaces $A^{(2)}$ of [RW]. In fact, by Lemma 2.9 on p. 325 of [RW], $A^{(2)}$ is equivalent to the space $A\ {}_\Omega\!\oplus A$ defined by (3.13) with $\Omega = \Omega_{\mathbb{C}}$.

It is an easy consequence of Theorem 2.4 that the operator T induces a bounded operator on $A\ {}_\Omega\!\oplus A$, namely $T(a,b) = (Ta,Tb)$.

We now establish a duality result for $A\ {}_\Omega\!\oplus A$ and then show how that yields a description of the dual of the domain of Ω.

We continue to write A^* for the couple dual to A and we abuse notation and also write A^* for $A^*_{\theta,q'}$, the space dual to a particular $A = A_{\theta,q}$. We assume for the rest of this subsection that the couples A and A^* satisfy the same density and reflexivity conditions as required in the previous subsection. This insures that $A^{**} = A$. Hence, among other things, that the duality theorem holds.

Using the notation Ω_* of Proposition 3.7 we form the twisted direct sum $A^* {}_{\Omega_*}\oplus A^*$.

We now introduce a bilinear pairing of $A {}_{\Omega}\oplus A$ and $A^* {}_{\Omega_*}\oplus A^*$. Given (a_1,a_2) in $A {}_{\Omega}\oplus A$ and (b_1,b_2) in $A^* {}_{\Omega_*}\oplus A^*$ we set

$$\langle (a_1,a_2),(b_1,b_2)\rangle = \langle a_1,b_2\rangle + \langle a_2,b_1\rangle \tag{3.14}$$

Theorem 3.15: There are constants c_1, c_2 such that

(i) for (a_1,a_2) in $A {}_{\Omega}\oplus A$ and (b_1,b_2) in $A^* {}_{\Omega_*}\oplus A^*$ we have

$$|\langle (a_1,a_2),(b_1,b_2)\rangle| \le c_1 \|(a_1,a_2)\| \, \|(b_1,b_2)\|;$$

(ii) for (a_1,a_2) in $A {}_{\Omega}\oplus A$

$$\|(a_1,a_2)\| \le c_2 \sup \left\{ |\langle (a_1,a_2),(b_1,b_2)\rangle| / \|(b_1,b_2)\| \right\}.$$

Here the supremum is over all nonzero (b_1,b_2) in $A^* {}_{\Omega_*}\oplus A^*$.

(iii) A statement similar to (ii) holds for (b_1,b_2) in $A^* {}_{\Omega_*}\oplus A^*$.

Proof: Suppose we are given (a_1,a_2) in $A {}_{\Omega}\oplus A$. We write

$$\langle a_1,a_2\rangle = \langle a_1,\Omega a_1\rangle + \langle 0,a_2 - \Omega a_1\rangle. \tag{3.16}$$

To establish (i) we may consider the two terms on the right separately. For the first term we have

$$\begin{aligned}\langle (a_1,\Omega a_1),(b_1,b_2)\rangle &= \langle a_1,b_2\rangle + \langle \Omega a_1,b_1\rangle \\ &= \langle a_1,b_2 - \Omega_* b_1\rangle + \langle \Omega a_1,b_1\rangle + \langle a_1,\Omega_* b_1\rangle.\end{aligned}$$

By the definitions, the first term on the right is bounded by

$$\|a_1\| \, \|b_2 - \Omega_* b_1\| \le c \, \|(a_1,a_2)\| \, \|(b_1,b_2)\|.$$

By Proposition 3.7, the second and third terms together are dominated by

$$c \, \|a_1\| \, \|b_1\| \le c \, \|(a_1,a_2)\| \, \|(b_1,b_2)\|.$$

For the second term in (3.16) we have the estimates

$$|\langle(0,a_2 - \Omega a_1),(b_1,b_2)\rangle| = |\langle a_2 - \Omega a_1,b_1\rangle|$$
$$\leq c\,\|a_2 - \Omega a_1\|\,\|b_1\|$$
$$\leq c\,\|(a_1,a_2)\|\,\|(b_1,b_2)\|.$$

We now go to (ii). For any a in A let Ja be an element of A^* for which $\|Ja\| = 1$ and $\langle a,Ja\rangle = \|a\|$. Set

$$S = \sup\left\{|\langle(a_1,a_2),(b_1,b_2)\rangle|/\|(b_1,b_2)\|\right\}.$$

By making the selection $(b_1,b_2) = (0,Ja_1)$ we get

$$(3.17)\qquad S \geq |\langle(a_1,a_2),(0,Ja_1)\rangle|/\|(0,Ja_1)\|$$
$$\geq |\langle a_1,Ja_1\rangle|/\|Ja_1\| \geq \|a_1\|.$$

Now set $b = J(a_2-\Omega a_1)$ and estimate S using the choice $(b_1,b_2) = (b,\Omega_* b)$. Thus

$$S \geq |\langle(a_1,a_2),(b,\Omega_* b)\rangle|/\|(b,\Omega_* b)\|.$$

Note that $\|(b,\Omega_* b)\| = \|b\| = 1$. Thus we continue with

$$S \geq |\langle a_2,b\rangle + \langle a_1,\Omega_* b\rangle|$$
$$= |\langle a_2,b\rangle - \langle\Omega a_1,b\rangle + \langle\Omega a_1,b\rangle + \langle a_1,\Omega_* b\rangle|$$
$$= |\langle a_2- \Omega a_1,b\rangle + \langle\Omega a_1,b\rangle + \langle a_1,\Omega_* b\rangle|$$
$$\geq |\langle a_2- \Omega a_1,b\rangle| - |\langle\Omega a_1,b\rangle + \langle a_1,\Omega_* b\rangle|.$$

The first term is evaluated using the choice of b and the second is estimated using Proposition 3.7. This gives

$$(3.18)\qquad S \geq \|a_2 - \Omega a_1\| - c\,\|a_1\|\|b\|.$$

To finish the proof, note that $\|b\| = 1$ and add $(c + 1)$ times (3.17) to (3.18).

Part (iii) is the same as part (ii).

We now put norms on $A\ {}_\Omega\!\oplus A$ and $A^*\ {}_{\Omega_*}\!\oplus A^*$. For (a,b) in $A\ {}_\Omega\!\oplus A$ define

$$\|(a,b)\|_* = \inf\left\{\sum_{n=1}^{N}\|(a_n,b_n)\| :\right.$$
$$\left.(a_n,b_n)\in A\ {}_\Omega\!\oplus A \text{ and } \sum(a_n,b_n) = (a,b)\right\}.$$

Define $\|\cdot\|^*$ on $A^* {}_{\Omega_*}\oplus A^*$ similarly.

Theorem 3.19: $\|\cdot\|_*$ is a norm on $A {}_\Omega\oplus A$ which is equivalent to the quasinorm $\|\cdot\|$; similarly for $\|\cdot\|^*$ on the space $A^* {}_{\Omega_*}\oplus A^*$. Also, we have the duality relation

$$(A {}_\Omega\oplus A)^* \cong A^* {}_{\Omega_*}\oplus A^*.$$

Proof: Clearly $\|\cdot\|_*$ is positive homogenous, satisfies the triangle inequality, and is dominated by $\|\cdot\|$. From (i) of the previous theorem it follows that

$$|\langle(a_1,a_2),(b_1,b_2)\rangle| \le c_1 \|(a_1,a_2)\|_* \|(b_1,b_2)\|.$$

Dividing by $\|(b_1,b_2)\|$, taking the supremum over all (b_1,b_2), and invoking (ii) of the previous theorem gives the other inequality. The argument for $\|\cdot\|^*$ is similar.

These equivalences and the previous theorem are enough to establish that every (b_1,b_2) in $A^* {}_{\Omega_*}\oplus A^*$ gives a continuous linear functional on $A {}_\Omega\oplus A$ and that we get enough functionals this way to determine norms in $A {}_\Omega\oplus A$. It remains to show that there are no other elements in $(A {}_\Omega\oplus A)^*$. Pick L, a linear functional on $A {}_\Omega\oplus A$. For any a_2 in A,

$$|L(0,a_2)| \le \|L\| \|(0,a_2)\| \le c \|L\| \|a_2\|_A.$$

Thus $L(0,a_2)$ defines a continuous linear functional on A and hence there is a b_1 in A^* so that $L(0,a_2) = \langle a_2,b_1\rangle$. Define K on $A {}_\Omega\oplus A$ by $K(a_1,a_2) = \langle a_2,b_1\rangle$. Since $\langle a_2,b_1\rangle = \langle(a_1,a_2),(b_1,0)\rangle$ it follows immediately from part (i) of the previous theorem that K is a continuous functional on $A {}_\Omega\oplus A$. Now consider the functional $J = L - K$. $J(a_1,a_2)$ is independent of a_2. Thus $J(a_1,0) = J(a_1,-\Omega a_1)$. Now using the fact that J is continuous on $A {}_\Omega\oplus A$ we conclude that $J(a_1,0)$ is a continuous functional on A.

Thus, for some b_2 in A^*, $J(a_1,a_2) = \langle a_1,b_2\rangle$. Thus $L(a_1,a_2) = \langle(a_1,a_2),(b_1,b_2)\rangle$ and (b_1,b_2) is in $A^* {}_{\Omega_*}\oplus A^*$ by part (iii) of Theorem (3.15).

We had previously described $\mathrm{Dom}(\Omega)$, the domain of definition of the operator Ω, as the set of those a for which $\|a\| + \|\Omega a\|$ is finite. Now note that $\mathrm{Dom}(\Omega)$ can be realized as $A {}_{\Omega}\oplus \{0\}$ contained isometrically in $A {}_{\Omega}\oplus A$. Using the previous theorem we can identify the dual space of $A {}_{\Omega}\oplus \{0\}$ as the quotient space

$$(A^* {}_{\Omega_*}\oplus A^*)/(A^* {}_{\Omega_*}\oplus \{0\}).$$

That is, the quotient space consists of cosets $[(0,b)]$ with norm given by

$$\|[(0,b)]\| = \inf_{c\in A^*} \|(0,b) + (c,0)\| = \inf_{c} (\|b - \Omega_* c\| + \|c\|) \tag{3.20}$$

<u>Corollary 3.21:</u> The space $\mathrm{Dom}(\Omega)$ and the space

$$\{(0,b) : b \text{ in } A^*\}$$

normed by (3.20) are dual to each other with respect to the pairing which establishes the duality of A and A^*.

In the next section we will give an alternative and more explicit description of the space $\mathrm{Dom}(\Omega)$ and its norm and we will relate the dual of $\mathrm{Dom}(\Omega)$ to the range of Ω_*.

Although the previous discussion was for the Ω's associated with the K and J methods, the proof only uses Theorem 2.6 and Poroposition 3.7. Since analogs of those results also hold for the Ω's associated with the complex method (by the same proof; see, for instance the proof of Proposition 2.11 of [RW]) we also have analogs of the previous results in that context. The situation for the Ω's associated with the E and F method is complicated by the lack of homogeneity of those Ω's and we have not investigated it.

IV. Explicit Descriptions of the Domains and Ranges:

A. The K and J Methods: By Theorem 2.6 the space Dom(Ω) is an interpolation space for the couple $\bar{A}$. Hence it is plausible that the norm on Dom(Ω) could be described directly in terms of the classical interpolation constructions using, for instance, the K or J functionals. In fact, that is done in [CJM]. In the next theorem we recall that result and give the analogous result for the range of Ω. The situation with the range is a bit more complicated because the set $\{\Omega_K a: a \in A\}$ is not independent of the choices of almost optimal decomposition. (Recall that Dom(Ω_K) is independent of those choices.)

We begin with some definitions. Associated with each $\Omega = \Omega_K, \Omega_J$ there is a constant c_Ω measuring how close the various decompositions associated with Ω are to being optimal. c_Ω is the smallest c for which (2.1) or (2.3) holds for all a in $A_{\theta,q}$ for all θ and q. For each integer N and for $\Omega = \Omega_J$ or Ω_K we now define

$$\begin{aligned} \mathrm{Ran}(\Omega) &= \mathrm{Ran}_{\theta q,N}(\Omega) \\ &= \{a \in \Sigma(\bar{A}) : a = \Omega b \text{ for some } b \in A_{\theta,q} \\ &\qquad \text{and some } \Omega \text{ with } c_\Omega \leq N\} \end{aligned}$$

and equip this set with a topology using

$$\|a\|_{\mathrm{Ran}} = \inf_{\Omega} \inf_{b} \{\|b\|_{A_{\theta,q}} : a = \Omega b \text{ and } c_\Omega \leq N\}.$$

We will show that for large N Ran(Ω) is independent of N and furthermore that it is normable.

Theorem 4.1:

(i) The topology on Dom(Ω) given by the quasinorm $\|a\| + \|\Omega a\|$ is also given by either of the following two quantities:

$$\|a\|_{\log,\theta,q,K} = \left[\int_0^\infty (t^{-\theta}(1 + |\log t|)\, K(t,a,\bar{A}))^q \frac{dt}{t}\right]^{1/q}$$

and

$$\|a\|_{\log,\theta,q,J} = \inf \left[\int_0^\infty (t^{-\theta}(1 + |\log t|)\ J(t,u(t),\bar{A}))^q \frac{dt}{t}\right]^{1/q}$$

where, as usual, the infimum is over all strongly measurable

$A_0 \cap A_1$ valued functions $u(t)$ for which $\int_0^\infty u(t)\ \frac{dt}{t} = a$.

(ii) Given θ, q there is an $N_0 = N_0(\theta,q)$ so that for all $N \geq N_0$ the topology just described for $\mathrm{Ran}(\Omega)$ is the same as that given by either of the following two quantities:

$$\|a\|_{\log^{-1},\theta,q,K} = \left[\int_0^\infty (t^{-\theta}(1 + |\log t|)^{-1}\ K(t,a,\bar{A}))^q \frac{dt}{t}\right]^{1/q}$$

and

$$\|a\|_{\log^{-1},\theta,q,J} = \inf \left[\int_0^\infty (t^{-\theta}(1 + |\log t|)^{-1}\ J(t,u(t),\bar{A}))^q \frac{dt}{t}\right]^{1/q}$$

with the infimum over the usual set of $u(t)$.

(iii) If the couples A and A^* satisfy the density and reflexivity conditons imposed in the previous subsection then $\mathrm{Dom}(\Omega)$ and $\mathrm{Ran}(\Omega_*)$ are dual to each other under the duality pairing of A and A^*.

Notes: (1) In (i) we are not making the (much stronger) claim that $K(t,\Omega a,\bar{A}))$ is comparable to $(1 + |\log t|)\ K(t,a,\bar{A}))$.

(2) The analog of (iii) for complex interpolation is in [RW].

(3) Norms of the sort just described have been considered extensively (for other reasons). See [DRS].

Proof: Part (i) is proved in [CJM]. The general theory of the K and J functionals insures that $\|a\|_{\log^{-1},\theta,q,K} \cong \|a\|_{\log^{-1},\theta,q,J}$. Thus to establish (ii) we need only consider the case of the J functional.

We wish to show

$$\|a\|_{\mathrm{Ran}} \cong \|a\|_{\log^{-1},\theta,q,J}.$$

If $a \in \mathrm{Ran}(\Omega)$ then there exists $b \in A_{\theta,q}$ with $a = \Omega b$. As we noted in (2.10) we may assume that $\Omega = \Omega_J$. Since $|\log t|/(1+|\log t|) \le 1$

$$\|a\|_{\log^{-1},\theta,q,J} \le \left(\int_0^\infty \left[t^{-\theta}(1+|\log t|)^{-1}J(t,D_J(t)b\ \log t)\right]^q \frac{dt}{t}\right)^{1/q}$$

$$\le \left(\int_0^\infty \left[t^{-\theta}J(t,D_J(t)b)\right]^q \frac{dt}{t}\right)^{1/q} \le C_\Omega \|b\|_{A_{\theta,q,J}}.$$

By taking infima, we find

$$\|a\|_{\log^{-1},\theta,q,J} \le C\|a\|_{Ran}.$$

To prove the converse we first notice that, as in [CJM], by reiteration and a local retract argument it suffices to consider the case of weighted ℓ^1-spaces, $\bar{A} = (\ell^1(2^{-n\theta_0}),\ell^1(2^{-n\theta_1}))$ with $0 < \theta_0 < \theta < \theta_1 < 1$. It is now easy to see that an almost optimal J-decomposition (for all $0 < \theta < 1$, say, and $1 \le q \le \infty$) of the sequence $\lambda = \{\lambda_n\}$ is given by

$$(D_J(t)\lambda)_n = \begin{cases} \lambda_n/\log 2^\Delta & \text{if } \mu = n \\ 0 & \text{otherwise} \end{cases}$$

if $2^{\mu\Delta} \le t < 2^{(\mu+1)\Delta}$. Here $\Delta = \theta_1 - \theta_0$. Calculating Ω_J we find that

$$(\Omega_J\lambda)_n = c\ \lambda_n(2n+1)$$

for some (irrelevant) constant c. Given $\beta = \{\beta_n\} \in (\ell^1(2^{-n\theta_0}),\ell^1(2^{-n\theta_1}))_{\eta q} = \ell^q(2^{-n\theta})$, $\theta = (1-\eta)\theta_0 + \eta\theta_1$, we let $\lambda = \{\lambda_n\} = \{\beta_n/(2n+1)\}$. Then $\beta = \Omega_J\lambda$ and

$$\|\lambda\|_{A_{\eta,q;J}} \le C\ \|\lambda\|_{\ell^q(2^{-n\theta})} \le C\ \|\{\beta_n/(2n+1)\}\|_{\ell^q(2^{-n\theta})}$$

$$\le \|\beta\|_{\log^{-1},\eta,q,J}.$$

A similar argument can also be used to give a proof of part (i) which is a little shorter than that appearing in of [CJM].

Part (iii) follows from the first two parts and the standard duality results of interpolation theory.

B. The E and F Methods: Analogs of the previous results hold for the E and F interpolation methods. This is implicit in [CJM] and can be made explicit using those results and the descriptions of domain and range norms developed in Theorem 4.2 below.

For instance, for the E functional we obtain

$$\|a\|_{Dom(\Omega_E)} \cong \left[\int_0^\infty ((1 + |\log s|)\ s^\theta\ E(s,a)^{1-\theta})^q \frac{ds}{s}\right]^{1/q}.$$

C. The Complex Method: As we noted before, in the context of complex interpolation the space $A\ _\Omega\oplus\ A$ can be given an explicit description. $A\ _\Omega\oplus\ A$ is the space

$$\{(a,b): a = F(\theta),\ b = F'(\theta)\ F \in \mathcal{F}\}$$

with the natural norm. From this we get an explicit description of the domain of Ω, namely as

$$\{(a,0): a = F(\theta),\ F \in \mathcal{F}\ F'(\theta) = 0\}$$

with the natural norm. Similarly we can describe the range of Ω as

$$\mathcal{F}'(\theta) = \{a: a = F'(\theta): F\epsilon\mathcal{F}\}.$$

These spaces have been studied before; explicitly by Schechter in [S] and implicitly in the work of Feissner which we discuss in Section 7.

D. The Relation of the Various Domain and Range Spaces: We have described a number of different operators Ω. It is easy to see that the difference between the operator given in (1.1) and the one given in (2.8) is an unbounded operator. Thus it is not generally true that that the various operators differ from each other by a bounded error. Hence there is no reason to suppose that there is an especially close relation between their domains of definition. However, we have the following (which extends a result in [CJM]):

Theorem 4.2:

(i) Suppose the couple $\bar{A}$ and the interpolation spaces $A_{\theta,q}$ are given. The spaces $Dom(\Omega)$ are the same for the Ω based on the various interpolation methods; Ω_J, Ω_K, Ω_E, and Ω_F. More precisely, the vector spaces are the same and the various (quasi)norms are equivalent on vectors of approximately unit length.

(ii) Given $0 < \theta_0, \theta_1, \eta < 1 \leq q_0, q_1 \leq \infty$ and θ, q defined by

$$\theta = (1-\eta)\theta_0 + \eta\theta_1, \; q^{-1} = (1-\eta)q_0^{-1} + \eta q_1^{-1}.$$

Suppose the couple $\bar{A}$ is given and let $\bar{B}$ be the couple $(A_{\theta_0,q_0}, A_{\theta_1,q_1})$. Then (as is well known) the complex interpolation space $\bar{B}_{[\eta]} = A_{\theta,q}$. In this case the space $\mathrm{Dom}(\Omega_{\mathbb{C}})$ agrees with the space described in (i).

Note: We mentioned earlier that the functional which we call E here is part of a larger family. The functional E is, in the notation of [JRW], tE_1. In [JRW] a family of functionals E_α, $1 \leq \alpha$, is described. To each E_α can be associated an operator $\Omega_{E_\alpha} = \Omega_\alpha$ and a domain space $\mathrm{Dom}_\alpha = \mathrm{Dom}(\Omega_\alpha)$. For simplicity we had avoided discussion of these more general constructions. However they arise in this proof. In fact the proof shows that the spaces Dom_α are all the same. For brevity we refer to [JRW] for information about the E_α and Ω_α.

We will not give a complete proof.

Proof outline: We start with part (i). θ and q are given. Suppose $\alpha \geq 1$ is also given. We want to show that

$$\|a\|_{\mathrm{Dom}_\alpha(A_{\theta,q})} \cong \|a\|_{\mathrm{Dom}_K(A_{\theta,q})} \text{ if } \|a\|_{A_{\theta,q}} \cong 1.$$

(The normalization is necessary because the various domain norms have different homogenieties.) Pick $0 < \theta_0 < \theta < \theta_1 < 1$ and set $\bar{B} = (A_{\theta_0,q_0}, A_{\theta_1,q_1})$ where q_0, q_1 will be selected later. Define η by $\theta = (1 - \eta)\theta_0 + \eta\theta_1$. By the reiteration theorem $B_{\eta,q} = A_{\theta,q}$ (independently of the choice of q_0, q_1). Also, from [JRW], defining β by $\beta = (\alpha - \theta_0)/(\theta_1 - \theta_0)$, we have

$$\Omega_{E_\beta,\bar{B}} = \Omega_{E_\alpha,\bar{A}},$$

and

$$\Omega_{K,\overline{B}} = c\,\Omega_{K,\overline{A}}.$$

Combining these facts gives

Step 1: It is enough to prove that the domain of Ω_{E_β} for the space $B_{\eta,q}$ equals the domain of Ω_K on the same space. Furthermore q_0 and q_1 are at our disposal.

Step 2: It is enough to work with the case $B_i = \ell^{q_i}(2^{-n\theta_i})$, $i = 0, 1$. This follows from a local retract argument as in [CJM].

Step 3: Having selected a convenient couple for $\overline{B}$ we now select the q's to make computations easier. We choose q_0, q_1 so that $\beta = q_1/(q_1 - q_0)$. For this choice, an optimal decomposition of a sequence can be computed one coordinate at a time. The result is, writing $A = (\theta_1 q_1 - \theta_0 q_0)/(q_1 - q_0)$, and letting λ be a function on $\mathbb{Z}$,

$$E_\beta(t,\lambda,\overline{B})^{q_0\beta} \cong \sum_{|\lambda_n|>t2^{nA}} \left|\frac{\lambda_n 2^{-n\theta_0}}{t}\right|^{q_0} + \sum_{|\lambda_n|\le t2^{nA}} \left|\frac{\lambda_n 2^{-n\theta_1}}{t}\right|^{q_1}.$$

Step 4: Using the definition of Ω_β and doing the explicit integration gives

$$(\Omega_\beta\lambda)_n = \lambda_n \log 2^{-nA} |\lambda_n|.$$

Thus the domain norm corrersponding to Ω_β is

$$\|\lambda\|^q \cong \sum \left[(1 + |\log 2^{-nA}|\lambda_n||)\ 2^{-n\theta}\ |\lambda_n|\right]^q. \tag{4.3}$$

Step 5: Using the results of [JRW] or [CJM] the corresponding expression for the domain of Ω_K is

$$\|\lambda\|^q \cong \sum \left[(1 + |\log 2^n|)\ 2^{-n\theta}\ |\lambda_n|\right]^q. \tag{4.4}$$

Step 6: To finish we must show that the expressions given

by (4.3) and (4.4) are comparable. Using the fact that $\delta = A - \theta$ is positive we are reduced to the following

Lemma: If μ_n are positive, $\sum \mu_n \cong 1$, $\delta > 0$, then

$$\sum \left[1 + |\log 2^{-n\delta}\mu_n|\right]^q \mu_n \cong \sum (1 + |n|)^q \mu_n.$$

Proof: We split the index set into three pieces,

$$A = \{n: n < 0,\ 2^{-n\delta}\mu_n < 1\},$$

$$B = \{n: n < 0,\ 2^{-n\delta}\mu_n \geq 1\},$$

$$C = \{n: n \geq 0,\ 2^{-n\delta}\mu_n < 1\}.$$

(There are no terms in the sum corresponding to the potential index set D.) This splits the left hand side as $\sum_A$, $\sum_B$, and $\sum_C$. For A we write

$$\sum_A \left[1 + |\log 2^{-n\delta}\mu_n|\right]^q \mu_n = \sum_A \left[1 + |\log 2^{-n\delta}\mu_n|\right]^q 2^{-n\delta}\mu_n\, 2^{n\delta}.$$

Using the fact that $x(1 + \log(1/x))^q$ is bounded for $0 \leq x \leq 1$ we can control this by $\sum_{n<0} 2^{n\delta}$. For B, we use the fact that $\mu_n \leq 1$ and hence

$$\left[1 + |\log 2^{-n\delta}\mu_n|\right]^q \leq \left[1 + |\log 2^{-n\delta}|\right]^q .$$

To estimate the third sum we split C into two;

$$C_1 = \{n \in C: 1 > 2^{-n\delta}\mu_n \geq 2^{-10n\delta}\}$$

and C_2 is the rest. In C_1

$$\left[1 + |\log 2^{-n\delta}\mu_n|\right]^q \leq \left[1 + |\log 2^{-10n\delta}|\right]^q$$

and there is no problem. For n in C_2 we write

$$\left[1 + |\log 2^{-n\delta}\mu_n|\right]^q \mu_n = \left[1 + |\log 2^{-n\delta}\mu_n|\right]^q 2^{-n\delta}\mu_n\, 2^{n\delta}$$
$$\leq C\, 2^{.9n\delta}$$

using the fact that $x(1 + \log(1/x))^q \leq C\, x^{.9}$ for x between 0 and 1.

We now need to show that the right hand side is dominated by the left hand side. The arguments are similar but this time the index set B must be subdivided instead of C. We omit the details.

We now go on to part (ii). Again, by a local retract argument it is sufficient to consider the spaces B_i in Step 2 of the first part of the proof. The results of [RW] then give an explicit formula for the norm on $\mathrm{Dom}(\Omega_{\mathbf{C}})$ at the interpolation space $\ell^q(2^{-n\theta})$. Write $N(\cdot)$ for the norm on $\ell^q(2^{-n\theta})$. Then

$$\|\lambda\|_{\mathrm{Dom}(\Omega_{\mathbf{C}})} \cong N(\{\lambda_n\}) + N(\{\lambda_n \log (2^{-nA} |\lambda_n/N(\{\lambda_n\}|)\})$$

with $A = (\theta_1 q_1 - \theta_0 q_0)/(q_1 - q_0)$. If $N(\lambda_n) \cong 1$ then this is the same as the domain norm for Ω_{E_β}, $\beta = q_1/(q_1 - q_0)$, described in the first part of the proof. We are done.

V. Differentiating the Norm: We assume throughout this section that the couple (A_0, A_1) satisfies $A_1 \subset A_0$. This insures that $\|a\|_{\theta,q}$ is an increasing function of θ. It also allows us to replace the interval of integration, $0 < t < \infty$, in the computation of $\|\cdot\|_{\theta,q,K}$ with the interval $0 < t < 1/2$. We will use the shorter interval without further explanation. The same comment applies to the norms with the logarithmic terms introduced in Theorem 4.1.

Throughout this section we develop a number of estimates which are actually only a priori estimates valid for vectors in a dense class. We will be informal about such issues.

A. Formulas for the Derivative of the Norm: In this section we look at the relation between the norm of Ωa. and the derivative of $\|a\|_{\theta,q} = \|a\|_{A_{\theta,q}}$ with respect to the parameters. The relation is most simple when q is fixed at $q = 1$ and we differentiate $\|a\|_{\theta,1} = \|a\|_{\theta,1,K}$ with respect to θ. Using

$$\|a\|_{\theta,q,K} = \left[\int_0^{1/2} (t^{-\theta} K(t,a))^q \frac{dt}{t}\right]^{1/q}, \tag{5.1}$$

we see that

(5.2) $$\frac{d}{d\theta} \|a\|_{\theta,1} = \int_0^{1/2} t^{-\theta} \log \frac{1}{t} K(t,a) \frac{dt}{t}$$

For $0 < t < 1/2$, $|\log t| \cong 1 + |\log t|$. Hence

$$\frac{d}{d\theta} \|a\|_{\theta,1} \approx \|a\|_{\log,\theta,1,K}.$$

We now invoke Theorem 4.1 and obtain

Theorem 5.2: If $A_1 \subset A_0$ then

(5.3) $$\frac{d}{d\theta} \|a\|_{\theta,1} \approx \|a\|_{\theta,1} + \|\Omega a\|_{\theta,1} .$$

(By Theorem 4.1 we may use Ω_K or Ω_J on the right.)

A similar analysis leads to similar, but more complicated, results involving the derivative of $\|a\|_{\theta,q}$ with respect to θ and with respect to q. The general case involves Ω_E and Ω_F as well. The one other case that is particularly natural and particularly simple is the one which is modeled on the computation of $\frac{d}{dp} \|a\|^p_{L^p}$. The result is

Theorem 5.4: Suppose $A_1 \subset A_0$. Let

$$|a|_\theta = \|a\|^{1/(1-\theta)}_{\theta,1/(1-\theta)}.$$

Then

(5.5) $$\frac{d}{d\theta} |a|_\theta \approx |a|_\theta + |\Omega a|_\theta.$$

Here Ω denotes Ω_E or Ω_F.

B. Extrapolation Theorems: In this section we give results based on the idea that information about the Ω's is differential information that can, in some sense, be integrated. The conclusions will be that an operator is bounded on a larger family of spaces than had been hypothesized. Such extrapolation theorems are partial converses to Theorem 2.6.

For the moment we have no applications of these results to operators in classical analysis. Rather we see the results as steps toward understanding the role that Ω plays in determining the structure of the interpolation scale. In some ways the role is similar to the role of the infinitesimal generator in semigroup theory.

For the first two results we will work with the spaces $A_{\theta,q}$ with $q = 1$ and we supress the index q: $A_\theta = A_{\theta,1}$.

Theorem 5.6: Given $0 < \theta < \theta+\varepsilon < 1$. Suppose that a linear operator T is bounded on A_θ and that for $\theta \leq \varphi \leq \theta+\varepsilon$, T is uniformly bounded on the domain spaces associated to A_φ; that is

$$\|Ta\|_\varphi + \|\Omega Ta\|_\varphi \leq c(\|a\|_\varphi + \|\Omega a\|_\varphi). \tag{5.7}$$

Then T is bounded on $A_{\theta+\varepsilon}$.

Proof: Clearly,

$$\|Ta\|_{\theta+\varepsilon} = \|Ta\|_\theta + \int_0^\varepsilon \frac{d}{ds}\|Ta\|_{\theta+s}\, ds.$$

By Theorem 5.2 we continue with

$$\|Ta\|_{\theta+\varepsilon} \leq \|Ta\|_\theta + c\int_0^\varepsilon \|Ta\|_{\theta+s} + \|\Omega Ta\|_{\theta+s}\, ds.$$

By (5.7)

$$\|Ta\|_{\theta+\varepsilon} \leq \|Ta\|_\theta + c\int_0^\varepsilon \|a\|_{\theta+s} + \|\Omega a\|_{\theta+s}\, ds.$$

We now use Theorem 5.2 again.

$$\begin{aligned}\|Ta\|_{\theta+\varepsilon} &\leq \|Ta\|_\theta + c\int_0^\varepsilon \frac{d}{ds}\|a\|_{\theta+s}\, ds \\ &\leq \|Ta\|_\theta + c\,(\|a\|_{\theta+\varepsilon} - \|a\|_\theta).\end{aligned}$$

The hypothesis that T is bounded on A_θ insures that

$$\|Ta\|_\theta + c\,(\|a\|_{\theta+\varepsilon} - \|a\|_\theta) \leq c\,\|a\|_{\theta+\varepsilon}$$

for sufficiently large c. Hence we have the required conclusion.

Notes: 1. The hypothesis that the bound of T on the domain spaces be uniform is more than is needed. Similar comments apply in some of the other results in this and the next section.

2. To go in the other direction and obtain a result for $A_{\theta-\varepsilon}$ would require an inequality in the opposite direction to that in (5.7). We could, for instance, assume the uniform boundedness of T^{-1}.

3. The theorem and proof extend directly to the case of a map T

from a space $A_{\theta,1}$ obtained from a couple $\bar{A}$ to a space $B_{\theta,1}$ obtained from a couple $\bar{B}$.

Theorem 5.8: Given $0 < \theta < \theta+\epsilon < 1$. Suppose that a linear operator T is bounded on A_θ and that for $\theta \leq \varphi \leq \theta+\epsilon$, T satisfies

$$\|\Omega Ta\|_\varphi \leq c(\|a\|_\varphi + \|\Omega a\|_\varphi). \tag{5.9}$$

Then T is bounded on $A_{\theta+\epsilon}$.

Note: This is a slight extension of the previous theorem. Our main interest is in the technique of proof which introduces ideas which we use again in the next section.

Proof: Again we start with

$$\|Ta\|_{\theta+\epsilon} = \|Ta\|_\theta + \int_0^\epsilon \frac{d}{ds} \|Ta\|_{\theta+s}\, ds.$$

$$\leq \|Ta\|_\theta + c\int_0^\epsilon \|Ta\|_{\theta+s} + \|\Omega Ta\|_{\theta+s}\, ds.$$

By (5.9)

$$\|Ta\|_{\theta+\epsilon} \leq \|Ta\|_\theta + c\int_0^\epsilon \|Ta\|_{\theta+s}\, ds + c\int_0^\epsilon \|a\|_{\theta+s} + \|\Omega a\|_{\theta+s}\, ds.$$

We analyze the second integral as in the previous proof; we use Theorem 5.2 and select a large c to obtain

$$\|Ta\|_{\theta+\epsilon} \leq c\,\|a\|_{\theta+\epsilon} + c\int_0^\epsilon \|Ta\|_{\theta+s}\, ds. \tag{5.10}$$

We now iterate this integral inequality. That is, by the same argument which led to (5.10) we also have, for $0 < s_2 < s < \epsilon$,

$$\|Ta\|_{\theta+s} \leq c\,\|a\|_{\theta+s} + c\int_0^s \|Ta\|_{\theta+s_2}\, ds_2.$$

Recalling that the norm of a is monotone in the parameter θ we have

$$\|Ta\|_{\theta+s} \leq c\,\|a\|_{\theta+\epsilon} + c\int_0^s \|Ta\|_{\theta+s_2}\, ds_2 \tag{5.11}$$

with the same constant c as in (5.10).

We now substitute (5.11) into (5.10) and evaluate the simple integral to get

(5.12) $\|Ta\|_{\theta+\epsilon} \leq (c + c^2\epsilon) \|a\|_{\theta+\epsilon} + c \int_0^\epsilon \int_0^s \|Ta\|_{\theta+s_2} \, ds_2 \, ds.$

Proceeding inductively gives

(5.13) $$\|Ta\|_{\theta+\epsilon} \leq c \left[\sum_1^N c^n\epsilon^n/n!\right] \|a\|_{\theta+\epsilon}$$

$$+ c \int_0^\epsilon \int_0^s \cdots \int_0^{s_{N-1}} \|Ta\|_{\theta+s_N} \, ds_N \cdots ds.$$

Recall that $\|Ta\|_{\theta+s_N} \leq \|Ta\|_{\theta+\epsilon}$. Hence if we start with any a for which $\|Ta\|_{\theta+\epsilon}$ is finite then the multiple integral tends to zero as N tends to infinity. Passing to the limit we have

$$\|Ta\|_{\theta+\epsilon} \leq c \left[\sum_1^\infty c^n\epsilon^n/n!\right] \|a\|_{\theta+\epsilon} = c' \|a\|_{\theta+\epsilon}$$

as required.

Suppose that (A_0,A_1) is quasilinearizable and that $A_{\theta,q}$ is fixed. Pick and fix a linear choice of $D_J(t)a$ which gives the almost optimal decompositions of elements of $A_{\theta,q}$. By Proposition 3.10, the function $t^\epsilon D_J(t)a$ is an allowable choice of almost optimal decomposition of the element $b = R_\epsilon a$ in the space $A_{\theta+\epsilon,q}$. Using this decomposition we can define Ω_ϵ by

(5.13) $$\Omega_\epsilon b = \int \log t \; t^\epsilon D_J(t)a \frac{dt}{t} .$$

This choice does give an allowable set of choices of Ω for the J method and also insures that $\frac{d}{d\epsilon} R_\epsilon a = \Omega_\epsilon R_\epsilon a$.

Theorem 5.14: Suppose that T is bounded on $A_{\theta,q}$ and that for ϵ, $|\epsilon| < \epsilon_0$, the commutator $[T,\Omega_\epsilon]$ is uniformly bounded on $A_{\theta+\epsilon,q}$

(5.15) $$\|[T,\Omega_\epsilon]\|_{op} \leq c.$$

Then there is an $\epsilon_1 > 0$ so that for ϵ, $|\epsilon| < \epsilon_1$, T is bounded on $A_{\theta+\epsilon,q}$.

Remark: We should emphasize that the hypothesis involves a

particular choice of the Ω's. That is in contrast to most of the other results we have discussed. The hypothesis (5.15) is not stable under replacement of Ω by a bounded perturbation of itself.

Proof: By Proposition 3.10 we know that for small ϵ the R_ϵ are invertible. Hence the required conclusion is equivalent to showing that the operators $S_\epsilon = R_\epsilon^{-1}TR_\epsilon$ are uniformly bounded on the fixed space $A_{\theta,q}$. $S_0 = T$ is bounded by hypotheses. Hence, by the Fundamental Theorem of Calculus, it suffices to estimate

$$\begin{aligned} S_\epsilon' &= (R_\epsilon^{-1})'TR_\epsilon + R_\epsilon^{-1}T(R_\epsilon'). \\ &= -(R_\epsilon^{-1})R_\epsilon'(R_\epsilon^{-1})TR_\epsilon + R_\epsilon^{-1}T(R_\epsilon') \\ &= R_\epsilon^{-1}\left\{-R_\epsilon'R_\epsilon^{-1}T + TR_\epsilon'R_\epsilon^{-1}\right\}R_\epsilon \\ &= R_\epsilon^{-1}\left\{-\Omega_\epsilon T + T\Omega_\epsilon\right\}R_\epsilon \\ &= R_\epsilon^{-1}[T,\Omega_\epsilon]R_\epsilon. \end{aligned}$$

However, again by Proposition 3.10, the boundedness of $R_\epsilon^{-1}[T,\Omega_\epsilon]R_\epsilon$ on $A_{\theta,q}$ is equivalent to the boundedness of $[T,\Omega_\epsilon]$ on $A_{\theta+\epsilon,q}$ which was hypothesized.

Note: The proof extends to give results for families of operators $T(\epsilon)$ which depend differentiably on ϵ. The required hypotheses would be that $[T(\epsilon),\Omega_\epsilon]$ and $T'(\epsilon)$ both be bounded on $A_{\theta+\epsilon,q}$.

C. Constancy and Splicing Theorems: The results in this section show how strongly the specification of one $A_{\theta,q}$ and of Ω determines the structure of the entire scale of spaces. First here is a very elementary result of that type.

Theorem 5.16: Suppose $A_1 \subset A_0$ and that for some $0 < \theta_0 < 1$, $\epsilon > 0$, Ω is a uniformly bounded map of $A_{\theta,1}$ to itself for all θ with

$|\theta_0 - \theta| \leq \varepsilon$. Then the couple $\bar{A}$ is trivial; that is, $A_0 = A_1$.

Proof: By (5.3)

$$\frac{d}{d\theta} \|a\|_{\theta,1} \leq c \, (\|a\|_{\theta,1} + \|\Omega a\|_{\theta,1}).$$

Thus, for the range of θ covered by the hypotheses we get

$$\frac{d}{d\theta} \|a\|_{\theta,1} \leq c \, \|a\|_{\theta,1}.$$

Integrating the differential inequality from $\theta_0 - \varepsilon/2$ to $\theta_0 + \varepsilon/2$ gives

$$\|a\|_{\theta_0-\varepsilon/2} \leq c \, \|a\|_{\theta_0+\varepsilon/2} .$$

Since we also know that $\|\cdot\|_\theta$ is an increasing function of θ we conclude

$$\|a\|_{\theta_0-\varepsilon/2} \cong \|a\|_{\theta_0+\varepsilon/2}.$$

Thus, for θ near θ_0 the interpolation spaces do not depend effectively on θ. The only way that can happen is if the couple is trivial. (See [BL] and, for more information, [JNP].)

Actually a much stronger result is true.

<u>**Theorem**</u> <u>**5.17:**</u> Suppose that for some θ,q, Ω is a bounded map of $A_{\theta,q}$ to itself. Then the couple $\bar{A}$ is trivial; that is, $A_0 = A_1$.

(It is interesting to note that the operator

$$\Omega f = f \log|f|$$

which is unbounded on the spaces L^p for $0 < p < \infty$, is <u>bounded</u> on the two end points of the scale, L^0 and L^∞. Perhaps this is an instance of a general fact.)

<u>**Proof:**</u> As in the previous proof we will show that the family of $A_{\theta,q}$ does not depend effectively on θ.

Pick and fix a choice of $D_J(t)a$, an almost optimal J decomposition for computing the norms on $A_{\theta,q}$. Suppose for now that the maps R_ε constructed using this choice of D_J are invertible for small ε. (For instance, by Proposition 3.10, this will be true if $\bar{A}$ is quasilinearizable.) By (3.2) and (3.5) we have, for any b,

$$\|R_{-\varepsilon}b\|_{\theta-\varepsilon,q} \cong \|b\|_{\theta,q}.$$

Suppose we want to establish

(5.18) $$\|a\|_{\theta,q} \lesssim c\ \|a\|_{\theta-\varepsilon,q}.$$

If we know $R_{-\varepsilon}$ is invertible then we can write $a = R_{-\varepsilon}b$ and we have

$$\|a\|_{\theta-\varepsilon,q} = \|R_{-\varepsilon}b\|_{\theta-\varepsilon,q} \cong \|b\|_{\theta,q}.$$

Thus we need to show

(5.19) $$\|R_{-\varepsilon}b\|_{\theta,q} \lesssim c\ \|b\|_{\theta,q}\ .$$

Similarly, to establish the inequality opposite to (5.18) we need the inequality opposite to (5.19). We will establish both by obtaining the estimate that, for any d and for all small ε,

(5.20) $$R_{\varepsilon}d = d + \varepsilon b,\ \|b\| = O(\|d\|).$$

The difference between any two allowable choices of Ω_J is a bounded operator. Hence our hypothesis insures that the Ω given by (2.7), with the choice that we have made of D_J, is a bounded operator. Since Ω is bounded, (5.20) would follow if we could establish that $R_{\varepsilon} = e^{\varepsilon\Omega}$. That is true in some cases. We now show that something closely related is true in general.

Pick and fix a in $A_{\theta,q}$. For $k = 0,1,2,\ldots$ define a_k by

$$a_k = \int_0^\infty D_J(t)a\ (\log t)^k \frac{dt}{t}$$

and let $u_k(t) = D_J(t)(a_k)$. For positive functions H defined on $(0,\infty)$ let

$$\Psi(H) = \left[\int_0^\infty (t^{-\theta}H(t))^q \frac{dt}{t}\right]^{1/q}.$$

Thus $\|a_k\| \cong \Psi(J(u_k))$. Set

$$c_k = \Psi(J(D_J(t)a\ (\log t)^k)).$$

We want to show that for some constants, A and B, which depend only on the operator norm of Ω,

(5.21) $$c_k \lesssim A\, B^k \,\|a\| \qquad k = 0,1,2,\ldots$$

For $k = 0$ this holds because D_J is an almost optimal decomposition. For $k = 1$,

$$c_1 = \Psi(J(D_J(t)a \log t)) \lesssim \Psi((1 + |\log t|)J(D_J(t)a\,)).$$

As we noted earlier, D_J can be chosen to be an almost optimal decomposition for all the $A_{\theta,q}$. In fact it can also be simultaneously almost optimal for the spaces with norms $\|\cdot\|_{\log,\theta,q,J}$ described in Theorem 4.1. Thus

$$c_1 \lesssim c\, \|a\|_{\log,\theta,q,J}.$$

By Theorem 4.1 we continue with $c_1 \lesssim c\,(\|a\| + \|\Omega a\|)$. By the hypothesis, we obtain (5.21) for $k = 1$.

We continue by induction. First note that the argument used for $k = 1$ shows that for any b in $A_{\theta,q}$

(5.22) $$\Psi(J(D_J(t)b \log t)) \lesssim c\, \|b\|.$$

Now

$$c_{k+1} = \Psi(J(D_J(t)a\,(\log t)^{k+1}))$$
$$\lesssim \Psi(J(D_J(t)a\,(\log t)^{k+1} - u_k(t) \log t\,))$$
$$+ \Psi(J(u_k(t) \log t\,)).$$

By (5.22) the second term is at most $c\,\|a_k\|$. Since c_k is an upper bound on $\|a_k\|$, the second term is at most $c\, c_k$. To estimate the first term we rewrite it as,

$$\Psi(\,(\,J(D_J(t)a\,(\log t)^k - u_k(t))\,)\log t\,)$$

and apply Proposition 5.2 of [JRW]. That proposition (or rather it's proof) shows that we may use the fact that

$$\int_0^\infty D_J(t)a\,(\log t)^k\,\frac{dt}{t} = \int_0^\infty u_k(t)\,\frac{dt}{t}$$

to conclude that

$$\Psi(\,(\,J(D_J(t)a\,(\log t)^k - u_k(t))\,)\log t\,)$$

$$\leq c\,\Psi((J(D_J(t)a\,(\log t)^k) + c\,\Psi(J(u_k(t))).$$

The first of these terms is $c\,c_k$ and the second is dominated by $c\,\|a_k\| = c\,c_k$. Combining the estimates we have

$$c_{k+1} \leq c\,c_k$$

which allows the induction to continue.

Since $R_\epsilon(a) = \sum \epsilon^k a_k / k!$, (5.20) now follows.

We now need to consider our assumption that R_ϵ is invertible. What is at issue is knowing that R_ϵ maps $A_{\theta,q}$ onto itself. However, it is sufficient to show that for given a in $A_{\theta,q}$ there is a choice of $D_J(t)a$ (which determines the construction of R_ϵ) so that, for that choice, a is in the range of R_ϵ (and, of course, the choice must be compatible with the proof just given). Suppose a is given and $D_J(t)a = u(t)$. Let

$$b = R_{-\epsilon}a = \int_0^\infty t^{-\epsilon}u(t)\,\frac{dt}{t}\;.$$

If we had $D_J(t)b = t^{-\epsilon}u(t)$ then we would also have $R_\epsilon b = a$ and we would be done. Hence we redefine D_J by setting $D_J(t)b = t^{-\epsilon}u(t)$. For this to be allowable we need to know that $\Psi(J(t^{-\epsilon}u(t))) \cong \|b\|$. By (5.20) $\|b\| \cong \|a\|$. Also, the proof of (5.20) actually shows that $\Psi(J(t^{-\epsilon}D_J(t)a)) \cong \|a\| + O(\epsilon\|a\|)$. We also need to know that this new choice of $D_J b$ is almost optimal for computing the norm of Ωb. Again, that follows directly using (5.21).

The previous proof is based on the idea of exponentiating Ω to obtain R_ϵ. The details differ because it is not true that $\Omega a_k = a_{k+1}$. Instead we had to estimate the difference between those two. We now give other results which share this general idea; results in which quantities similar to exponentials of Ω govern the change of the $\|\cdot\|_{\theta,q}$ with θ.

Suppose $A_1 \subset A_0$. Fix A_θ and fix a small positive ϵ_0. For $0 < \epsilon < \epsilon_0$,

$$\|a\|_{\theta+\epsilon} = \|a\|_\theta + \int_0^\epsilon \frac{d}{ds} \|a\|_{\theta+s}\, ds.$$

By Theorem 5.2 we continue with

$$\|a\|_{\theta+\epsilon} \lesssim \|a\|_\theta + c \int_0^\epsilon \|a\|_{\theta+s} + \|\Omega a\|_{\theta+s}\, ds. \tag{5.23}$$

As in the proof of Theorem (5.8) we now use (5.23) with $\epsilon = s$ to estimate both of the terms inside the integral sign in (5.23):

$$\begin{aligned}\|a\|_{\theta+\epsilon} \lesssim\ & \|a\|_\theta \\ & + c \int_0^\epsilon \left[\|a\|_\theta + c\int_0^s \|a\|_{\theta+s_1} + \|\Omega a\|_{\theta+s_1} ds_1\right] \\ & \qquad + \left[\|\Omega a\|_\theta + c\int_0^s \|\Omega a\|_{\theta+s_1} + \|\Omega^2 a\|_{\theta+s_1} ds_1\right] ds \\ & = (1 + c\epsilon)\ \|a\|_\theta + c\epsilon\ \|\Omega a\|_\theta + \\ & \quad + c^2 \int_0^\epsilon \int_0^s \|a\|_{\theta+s_1} + 2\ \|\Omega a\|_{\theta+s_1} + \|\Omega^2 a\|_{\theta+s_1}\ \ ds_1\, ds\end{aligned}$$

Continuing in this way we get

$$\|a\|_{\theta+\epsilon} \lesssim \sum_{0 \lesssim n \lesssim N} \frac{c^n \epsilon^n}{n!} \left[\sum_{0 \lesssim j \lesssim n} \binom{n}{j} \|\Omega^j a\|_\theta \right] + R_N \tag{5.24}$$

The remainder term, R_N, is given by

$$R_N = c^N \int_0^\epsilon \int_0^s \cdots \int_0^{s_N} \sum_{0 \lesssim j \lesssim N} \binom{N}{j} \|\Omega^j a\|_{\theta+s_N}\, ds_N \ldots ds.$$

If we knew that for some $c(a)$, and all η with $\theta \lesssim \eta \lesssim \theta + \epsilon$, and all j

$$\|\Omega^j a\|_{\theta+\eta} \lesssim (c(a) j)^j, \tag{5.25}$$

then we would have the estimate

$$R_N \lesssim c^N \int_0^\epsilon \int_0^s \cdots \int_0^{s_N} \sum_{0 \lesssim j \lesssim N} \binom{N}{j} (c(a) j)^j ds_N \ldots ds$$

$$\le c^N \int_0^\epsilon \int_0^s \cdots \int_0^{s_N} \sum_{0\le j\le N} \binom{N}{j} (c(a)N)^N ds_N \ldots ds$$

$$\le c^N \epsilon^N (N!)^{-1} \sum_{0\le j\le N} \binom{N}{j} (c(a)N)^N.$$

$\sum_{0\le j\le N} \binom{N}{j} = 2^N$ and, by Stirling's formula, $N! \ge (cN)^N$. Thus, if (5.25) holds then

$$R_N \le (\epsilon c\ c(a))^N.$$

In this case we can pass to the limit in (5.24) and for $\epsilon < (c\ c(a))^{-1}$ we have

$$\|a\|_{\theta+\epsilon} \le \sum_n \frac{c^n \epsilon^n}{n!} \left[\sum_{0\le j\le n} \binom{n}{j} \|\Omega^j a\|_\theta \right]. \tag{5.26}$$

We use new summation indices j and $k = n-j$. The sum in k can be evaluated explicitly and we have

$$\|a\|_{\theta+\epsilon} \le e^{C\epsilon} \sum \frac{c^j \epsilon^j}{j!} \|\Omega^j a\|_\theta. \tag{5.26}$$

If it were possible to move the summation inside the norm sign (and in some cases it is) we would write

$$e^{c(I+\Omega)} = e^c \sum \frac{c^j}{j!} \Omega^j$$

and would have the exponential of Ω controlling the growth of the norm. To study (5.26) further we introduce the notation

$$F(\alpha,a) = e^\alpha \sum \frac{\alpha^j}{j!} \|\Omega^j a\|_\theta.$$

Theorem 5.27: Suppose $A_1 \subset A_0$. Fix A_θ. There are constants c_1, c_2 and c_3 so that for all a in A_θ with $\|a\|_\theta \le 1$, for all ϵ with $0 \le \epsilon \le c_3$,

$$F(c_1\epsilon,a) \le \|a\|_{\theta+\epsilon} \le F(c_2\epsilon,a).$$

Proof: To obtain the upper estimate it suffices to show that for the a of interest we have (5.25). Since we have $A_1 \subset A_0$ the norm of a and of Ωa can both be computed using the J method and using integrals from $t = 0$ to $t = 1$ (rather than an upper limit of $t =$

∞). It follows immediately that

$$\|\Omega a\|_{\theta+\delta} \le \frac{c}{\delta} \|a\|_\theta$$

where c is a universal constant and the factor of δ^{-1} arises in estimating the maximum of $t^\delta \log(1/t)$ on (0,1). We can use this estimate n times and obtain

$$\|\Omega^n a\|_{\theta+c_3} \le \left[\frac{c}{c_3/n}\right]^n \|a\|_\theta$$

which is the required estimate.

To complete the proof it is enough to note that Theorem 5.2 also provides a lower estimate

$$\frac{d}{ds} \|a\|_{\theta+s} \ge c \left[\|a\|_{\theta+s} + \|\Omega a\|_{\theta+s} \right]$$

as in the proof of Theorem (5.8).

Notes: 1. These arguments can be extended to give (a slightly more awkward) result for negative ϵ. Also, some related results still hold without the assumption that $A_1 \subset A_0$. (The reason is that the upper estimate on the derivative of the norm hold even without that assumption.)

2. This result shows that if Ω is bounded on A_θ then the norm does not depend effectively on θ. This gives another route to Theorem 5.17, but under the additional hypothesis that $A_1 \subset A_0$.

We now use this theorem to show that two interpolation scales that agree at a single point and have the same Ω at that point can be spliced together. (This is analogous to the fact that two geodesics on a Riemannian manifold which share a point and a tangent vector must agree locally. To see why the analogy with Riemannian geometry is natural, see Section 8 of [CS].) We suppose that we have two couples $\bar{A}$ and $\bar{B}$ with $A_1 \subset A_0$ and $B_1 \subset B_0$. We also suppose that both couples are contained in a fixed large topological vector space. Suppose that for some θ and φ we have that for $A_\theta = B_\varphi$ (isometrically) and that the corresponding

Ω's agree; $\Omega_{A,K} = \Omega_{B,K}$.

Theorem 5.28:

(i) There are constants c_1, c_2, c_3, c_4 so that for all small positive ϵ, for all a,

$$\|a\|_{A_{\theta+c_1\epsilon,1}} \lesssim \|a\|_{B_{\varphi+c_2\epsilon,1}} \lesssim \|a\|_{A_{\theta+c_3\epsilon,1}} \lesssim \|a\|_{B_{\varphi+c_4\epsilon,1}}.$$

(ii) For all small positive ϵ, $A_{\theta+\epsilon}$ is an interpolation space for the couple (A_0,B_1). For all small positive ϵ, $B_{\varphi+\epsilon}$ is an interpolation space for the couple (B_0,A_1).

Proof: The first result follows directly on applying the previous theorem to the A's and B's and noting that the hypotheses insure that the comparison function F is the same in both cases. The second part follows from the first and Wolff's iteration theorem as generalized in [JNP].

VI. The work of Kalton: In this section we describe very briefly and informally some of the results in a pair of papers by Kalton, [K1] and [K2].

Suppose X is a rearrangement invariant function space on the nonatomic measure space (E,μ). Following Kalton we call a map Ω from X into functions on E a symmetric centralizer if there is a function δ: $\mathbb{R}_+ \longrightarrow \mathbb{R}_+$ such that

$$\|[T,\Omega]f\|_X \lesssim \delta(\|f\|_X) \tag{6.1}$$

for two particular classes of operators T. First, (6.1) must hold with Tf = uf for u in L^∞ with $\|u\|_\infty \lesssim 1$. Second, it must hold for $Tf(x) = f(\sigma(x))$ with σ a measure preserving map of E to itself. Kalton shows directly that a large class of functions are symmetric centralizers including, for instance,

$$\Omega f = |f| \left| \log |f| \right|^a \left| \log r_f \right|^b$$

for $0 \lesssim a,\ b \lesssim a+b \lesssim 1$ (r_f is the rank function introduced in Section II A). He then proves

Theorem 6.2: Suppose T is an operator on X which is of strong types (p_0,p_0) and (p_1,p_1), and the Boyd indices of X satisfy $p_0 < p_X \le q_X < p_1$. Then, (6.1) holds for this T and any symmetric centralizer (although δ may need to be changed).

One of the tools which is very important in Kalton's work is the pointwise product of functions which gives a bilinear map from $X \oplus X^*$ into $L^1(E,\mu)$. Using, among other things, the Boyd interpolation theorem, Kalton shows that if T satisfies the hypotheses then, for f in X, and g in X^*, $Tf\cdot g - f\cdot T^*g$ is in a space that is a rearrangement invariant analog of the Hardy space H^1. He then obtains Theorem 6.1 by studying the action of certain centralizers on that subspace of L^1.

Kalton also develops some of the systematic relationship between symmetric centralizers and the twisted direct sum $A \oplus_\Omega A$.

In [K2] Kalton uses related ideas to study certain bilinear maps acting on the Schatten ideals C_p. (These are often viewed as analogs of the Lebesgue spaces but with a noncommutative multiplication.) He obtains a striking identification of the linear span of commmutators of Hilbert-Schmidt operators with the Schatten analog of H^1.

VII. Applications in Analysis: Much of the work described in the earlier sections was in the framework of abstract interpolation theory. In this section we show ways these ideas interact with operators and function spaces which arise naturally in analysis.

A. Hypercontractive semigroups. Let $(M,d\mu)$ be a probability space. Throughout this section we write L^p for $L^p(M,\mu)$. Let H be a nonnegative self adjoint operator on L^2. (H will generally be unbounded above.) We say that the semigroup of operators $\{e^{-tH}\}_{t\ge0}$ is a <u>hypercontractive semigroup</u> if

(7.1) e^{-tH} is a contraction on L^p, $1 \le p \le \infty$, $t \ge 0$,

and

(7.2) $\exists T > 0$, e^{-TH} is a bounded map from L^2 to L^4.

To obtain the cleanest example we assume more; we suppose that e^{-TH} has operator norm at most one as a map from L^2 to L^4. For the same reason we assume that e^{-tH} maps positive functions to positive fuctions.

Finally we make the additional assumptions that

(7.1)′ the operators $e^{-t(I+H)}$ are uniformly bounded on L^p, $1 \leq p \leq \infty$, $t \geq 0$,

and

(7.2)′ $\exists T > 0$, $e^{-T(I+H)}$ is a bounded map of L^2 to L^4.

(7.1)′ and (7.2)′ are actually consequences of (7.1) and (7.2). In fact it will be true in this case that

(7.1)′′ the operator norm of $e^{-t(I+H)}$ decays exponentially for large t.

For these facts and more about such semigroups see [SH-K]. More recent references can be found in [S].

A good example of a hypercontractive semigroup is obtained by taking H to be the Laplace operator acting on functions on $\mathbb{R}^n$ and letting the underlying measure space be $\mathbb{R}^n$ with the normalized Gaussian measure $c_n e^{-|x|^2/2}dx$.

Many interesting consequences follow from (7.1) and (7.2). For instance we have the so called logarithmic Sobolev inequalities:

Theorem 7.3 (Gross, [G]): If f is in L^2 and $\|f\|_2 = 1$ then

$$\int |f|^2 \log |f| \, d\mu \leq \langle Hf,f\rangle.$$

One indication that this result is related to those of the previous section is that the inequality in Theorem 7.3 can be summarized by $\langle \Omega f,f\rangle \leq \langle Hf,f\rangle$ with $\Omega f = f \log |f|$, or, even more simply, by the operator inequality $\Omega \ll H$.

There are also the so called higher order logarithmic Sobolev inequalities such as

Theorem 7.4 (Feissner [F]): For m in $\mathbb{Z}$ and $p \geq 2$, $(I + H)^{-1}$ is a bounded map from the Orlicz space $L^p(\mathrm{Log}\ L)^m$ to the Orlicz space $L^p(\mathrm{Log}\ L)^{m+2}$.

Our goal here is to indicate briefly that the ideas of the previous section are closely related to the theory of hypercontractive semigroups.

If (7.1) and (7.2) hold then the analytic family of operators $\{e^{-zH}\}$ defined on the strip $0 \le \mathrm{Re}(z) \le T$, acting on the family of L^p spaces, can be studied using analytic interpolation. By interpolation we find that there is a δ so that for $p(t) = (2^{-1} - t\delta)^{-1}$

(7.5) $\qquad e^{-tH}$ is a contraction from L^2 to $L^{p(t)}$.

Define

$$r(t) = (2^{-1} + t\delta)^{-1}.$$

Since we are working on a probability space, the inclusion of $L^{p(t)}$ into $L^{r(t)}$ is norm decreasing. Hence, from (7.5) it follows that

(7.6) $\qquad e^{-tH}$ is a contraction from L^2 to $L^{r(t)}$.

Similarly $e^{-t(I+H)}$ is a bounded map of L^2 into both $L^{p(t)}$ and $L^{r(t)}$.

We now apply the basic commutator result, Theorem 2.4. In this case the couple $\bar{A}$ is the constant couple (L^2, L^2) and hence we may make the choice $\Omega_A = 0$. The couple $\bar{B} = \bar{B}_t$ is $(L^{p(t)}, L^{r(t)})$. Thus the choice of Ω_{B_t} depends on t. However we can regard all of the couples B_t as obtained from the couple (L^∞, L^1) by iteration. Thus by the iteration result for the Ω's, described in Section III A, we may make the single selection

$$\Omega_{B_t} = c\, t\, \Omega$$

with a fixed Ω obtained from the couple (L^∞, L^1).

We want to look at the mapping properties of $(I + H)^{-\alpha}$ for positive α. To do this we use the integral representation

$$(I + H)^{-\alpha} = c_\alpha \int_0^\infty e^{-t(I+H)}\, t^\alpha \frac{dt}{t}. \tag{7.7}$$

The question of interest here is the convergence (or failure of convergence) of such integrals for small t. The shift from H to I+H is to insure that integrals such as (7.6) have good convergence properties at infinity; by (7.2)'' this is so. We won't mention that end point again.

By Theorem 2.4 we have a uniform bound on

$$e^{-t(I+H)}\Omega_A - \Omega_{B_t} e^{-t(I+H)} = - \Omega_{B_t} e^{-t(I+H)}$$

$$= c\ t\ \Omega\ e^{-t(I+H)}.$$

Thus, for f in L^2 we have

$$\|\Omega\ e^{-t(I+H)} f\|_2 \leq c\ t^{-1}\ c(\|f\|_2).$$

Hence, if $\alpha > 1$ then

$$\int_0^\infty \Omega\ e^{-t(I+H)}\ t^\alpha \frac{dt}{t} \tag{7.8}$$

will be a convergent integral. Next we note that

$$\int_0^\infty \Omega\ e^{-t(I+H)}\ t^\alpha \frac{dt}{t} - \Omega \int_0^\infty e^{-t(I+H)}\ t^\alpha \frac{dt}{t} \tag{7.9}$$

is bounded. To see this we use Theorem 2.4 again, but now to a different pair of couples. The first couple will be with $A_0 = (L^1((0,\infty),dt);\ L^1(M,d\mu))$ and $A_1 = (L^1((0,\infty),dt);\ L^\infty(M,d\mu))$ Here $L^1((0,\infty),dt);\ X)$ denotes the space of X valued functions on $(0,\infty)$ normed by $\int\|f_t\|_X dt$. The second couple is $(L^1(M,d\mu),L^\infty(M,d\mu))$. Straightforward computation shows that a choice of Ω for the first couple is obtained by selecting an Ω for the second couple and applying it pointwise (in t). The map of $f_t(x)$ to $\int f_t(x)dt$ is clearly a continuous map between the two couples. Hence Theorem 2.4, applied at the point L^2, shows that (7.8) is bounded.

Combining the fact that (7.8) converges in L^2 with the bound on the expression in (7.9) we conclude that

$$\Omega \int_0^\infty e^{-t(I+H)}\ t^\alpha \frac{dt}{t}$$

is in L^2. We evaluate the integral using (7.7) and get

Theorem 7.10: If $\alpha > 1$ then $\Omega(I + H)^{-\alpha}f$ is in L^2 for f in L^2. That is, $(I + H)^{-\alpha}$ maps L^2 to the Orlicz space $L^2 \, (\mathrm{Log}\, L)^2$.

This is slightly weaker than the case p = 2, m = 0 of Feissner's result in that we don't get the end point $\alpha = 1$. That appears to be a bit more delicate and we return to it in a moment.

Direct estimates using (7.7) show that $(I + H)^{-\alpha}$ maps the couple (L^1, L^∞) to itself. Hence, by Theorem 2.4 $[\Omega, (I + H)^{-\alpha}]$ is bounded on all of the intermediate L^p spaces. In fact, since that boundedness is shown by an inequality between the J and K functionals, $[\Omega, (I + H)^{-\alpha}]$ will be bounded on any intermediate spaces for for which the K and J methods coincide. In particular, for all integers m,

(7.11) $[\Omega, (I + H)^{-\alpha}]$ is bounded on the Orlicz spaces $L^2(\mathrm{Log}\, L)^m$.

This, together with the previous theorem, is enough to get

Theorem 7.12: If $\alpha > 1$ then, for all real m, $(I + H)^{-\alpha}$ maps the Orlicz space $L^2 \, (\mathrm{Log}\, L)^m$ to the Orlicz space $L^2 \, (\mathrm{Log}\, L)^{m+2}$.

Proof: We have the result for m = 0.

First we move up by induction. Suppose we have the result for a given m. Pick f in $L^2 \, (\mathrm{Log}\, L)^{m+2}$. By direct estimates Ωf will be in $L^2 \, (\mathrm{Log}\, L)^m$. Hence by the induction hypothesis $T\Omega f$ will be in $L^2 \, (\mathrm{Log}\, L)^{m+2}$. By (7.11) we get ΩTf in $L^2 \, (\mathrm{Log}\, L)^{m+2}$. By direct estimates that implies that Tf is in $L^2 \, (\mathrm{Log}\, L)^{m+4}$.

Now we go down by induction. If f is in $L^2 \, (\mathrm{Log}\, L)^{m-2}$ then $f = \Omega g$ for some g in $L^2 \, (\mathrm{Log}\, L)^m$. By the induction hypothesis Tg is in $L^2 \, (\mathrm{Log}\, L)^{m+2}$. By direct estimates ΩTg is in $L^2 \, (\mathrm{Log}\, L)^m$. By (7.11) $T\Omega g = Tf$ is in $L^2 \, (\mathrm{Log}\, L)^m$.

These two inductions take care of all even integers m. The other cases follow by interpolation.

Similar arguments and further interpolations give results involving L^p for $p \neq 2$.

We now outline a proof of Theorem 7.3. Pick f in L^2 of norm 1. Since H is self adjoint, it follows from (7.6) that

$$e^{-tH} \text{ is a contraction from } L^{r(t)} \text{ to } L^2$$

Hence, for z in the right half plane

(7.13) e^{-zH} is a contraction from $L^{r(Re(z))}$ to L^2.

Let $f_z = \frac{f}{|f|}|f|^{1+2\delta z}$. Thus f_z is in $L^{r(Re(z))}$ and has norm at most one. Hence by (7.13)

(7.14) $$\|e^{-zH}(f_z)\|_2 \leq 1.$$

The derivative of $e^{-zH}(f_z)$ at 0 is $-Hf + \Omega f$ for the choice

$$\Omega f = c\, f \log|f|.$$

If we could differentiate (7.15) and evaluate at 0 we would obtain $\langle -Hf + \Omega f, f\rangle \leq 0$, or, more compactly, $\Omega \ll H$. To justify this computation we argue as follows. Consider the analytic function $F(z) = \langle e^{-zH}(f_z), e^{-zH}(\overline{(\bar{f})_z})\rangle$. By (7.15) we know $|F(z)| \leq 1$ in the right half plane and $|F(0)| = 1$. In this case the theory of the angular derivative of analytic functions insures that $F'(0)$ exists (as $\lim_{x\to 0^+}(F(x) - F(0))/x$) and satisfies $F'(0) \leq 0$ (with $F'(0) = -\infty$ allowed as a possibility). That is all that is needed.

The use of function theory here is not really necessary. In [G] Gross gives a direct, real variable, argument for differentiating (7.14). We described this route because it seemed natural in this context. It is slightly more delicate than some of the earlier arguments because we are estimating the derivative of an analytic function at the boundary of the domain of definition.

A similar argument can also be used to give Theorem 7.4. In this case the function to be differentiated is $\|e^{-zH}(e^{zW}f)\|_2$. Here $W = W(x)$ are various weight functions on M with the property that, by Holder's inequality, $e^{zW}f$ is in $L^{r(Re(z))}$ with norm at most one. The differentiation argument then produces the estimate $M_W \ll H$, where M_W represents multiplication by W. Given such an estimate for all suitably normalized W, the crucial case of Feissner's theorem ($\alpha = 1$, $m = 0$) follows by Orlicz space considerations (pgs 54 and 55 of [F]).

In short, a substantial part of the arguments of [G] and especially of [F] can be reformulated using Ω's.

It is interesting to note that Feissner's route from the case $m = 0$ to the general case has some similarities to the proof of the main result of [RW] and was done much earlier.

B. Commutator estimates for nonlinear operators. Theorem 2.4 states that if T is a bounded linear operator, $T:\bar{A}\longrightarrow\bar{B}$, then $[T,\Omega]:A_{\theta,q} \to B_{\theta,q}$. This result is a direct consequence of the estimate (2.2) and the equivalence of the K and J interpolation methods. For certain nonlinear choices of T we get a partial result.

Theorem 7.15: Let T be an operator mapping $A_0 + A_1$ to $B_0 + B_1$ which satisfies, for constants c_0 and c_1,

(i) if $f \in A_0$ then $\|Tf\|_{B_0} \leq c_0 \|f\|_{A_0}$, and

(ii) for $f, g \in A_0 + A_1$, $f - g \in A_1$,

$$\|Tf - Tg\|_{B_1} \leq c_1 \|f - g\|_{A_1}.$$

Then,

$$[T,\Omega]: A_{\theta,q} \longrightarrow B_{\theta,q}.$$

Notes: 1. The hypothesis (i) and (ii) are standard ones for interpolation results for nonlinear operators. In particular they imply that $T: A_{\theta,q} \longrightarrow B_{\theta,q}$.

2. The reason this is only a partial result is that $[T,\Omega]$ is, for us, defined by

$$[T,\Omega] = \int_0^\infty [T,D_K(t)]\,\frac{dt}{t}.$$

Thus, formally,

$$[T,\Omega] = \int_0^\infty TD_K(t)\,\frac{dt}{t} - \int_0^\infty D_K(t)T\,\frac{dt}{t}$$

$$= \int_0^\infty TD_K(t)\,\frac{dt}{t} - \Omega T.$$

In order to continue this equation with

$$= T\Omega - \Omega T$$

we would need to know that

$$\int_0^\infty TD_K(t)\,\frac{dt}{t} = T\int_0^\infty D_K(t)\,\frac{dt}{t}.$$

However, T is not assumed linear so we don't have this. In short, we have shown that $[T,\Omega]$ is bounded, but have not established the boundedness of $T\Omega - \Omega T$.

<u>Proof:</u> We only need to show that (2.4) holds. Observe that from (i), (ii); and writing

$$T(a_0 + a_1) = T(a_0 + a_1) - T(a_0) + T(a_0),$$

we easily see that for all t

$$K(t,Tf;\bar{B}) \leq cK(t,f;\bar{A}).$$

Now if $f \in A_0 \cap A_1$, then

$$\|TD_K(t)f - D_K(t)Tf\|_{B_0} \leq c_1\|D_K(t)f\|_{A_0} + K(t,Tf;\bar{B}) \leq cK(t,f;\bar{A})$$

and,

$$t\|TD_K(t)f - D_K(t)Tf\|_{B_1} \leq t\|TD_K(t)f - Tf\|_{B_1} + t\|Tf - D_K(t)Tf\|_{B_1}$$

$$\leq tc_1\|D_K(t))f - f\|_{B_1} + cK(t,f;\bar{A})$$

$$\leq cK(t,f;\bar{A})$$

and (2.4) follows.

Operators which satisfy the hypotheses are common. For example, suppose the couples $\bar{A}$ and $\bar{B}$ are lattices of measurable functions and T is pointwise positive, i.e. for any f, $Tf(x) \geq 0$ a.e.. If T is bounded from $\bar{A}$ to $\bar{B}$ and satisfies the pointwise estimate:

for f,g in $A_0 + A_1$, $f - g$ in A_1,

(ii)′ $$|Tf(x) - Tg(x)| \leq c\,|T(f - g)(x)|.$$

then the hypotheses of the theorem are easily checked. (ii)′ holds for T which are obtained as the pointwise supremum of positive sublinear operators. Hence the theorem applies to most classical maximal functions, sharp functions, area functions, g functions, etc.

VIII: Some questions: There are many open problems related to these topics. Here are some we find especially interesting.

1. We have seen that commutator estimates arise in a natural way both for the complex and real method of interpolation, and that these methods often produce the same estimates. This suggests there might be a more general point of view. A natural problem is to find an abstract set-up for these commutator estimates (for instance in terms of the maximal and minimal methods of Aronszajn-Gagliardo, see [J]; or perhaps in the parameterized families of interpolations spaces presented in [CS]). A related problem is to give a conceptual (rather than computational) proof of the equivalence of the various domain spaces described in Theorem 4.2.

2. Is there an analog of the Hille-Yosida Theorem which would tell which operators could arise as Ω's and how to use the Ω's to construct an associated interpolation couple?

3. Although there is clearly a substantial relation, we don't understand very well how Kalton's techniques and results mesh with those described in the previous sections. In particular, Kalton makes important use of ideas from abstract interpolation theory, but he also makes substantial use of the pointwise product of functions. We don't see how that part of his analysis can be put in our abstract context.

4. Is $T\Omega - \Omega T$ bounded when T is nonlinear? What is the analog for the complex method of the estimates for nonlinear operators; Theorem 7.15?

5. Is it possible to give a proof of Zafran's result (Lemma 3.4) and Proposition 3.7 without using function theory? Are similar results true for quasinormed spaces?

6. For the complex method of interpolation the space $A\ {}_{\Omega}\oplus A$ has an alternative, and more intrinsic definition. (Here A denotes one of the spaces $A_\theta = [A_0, A_1]_\theta$ of the complex

interpolation method.) $A \,_\Omega\!\oplus A$ can be identified with the quotient space of $\mathcal{F}$ by the subspace of those functions F in $\mathcal{F}$ which satisfy $F(\theta) = F'(\theta) = 0$. Alternatively, $A \,_\Omega\!\oplus A$ is the space $\{(F(\theta), F'(\theta)) : F \in \mathcal{F}\}$ with the natural norm. This space is called $A^{(2)}$ in [RW] where it is shown that the norm on that space is equivalent to the quasinorm described for the twisted direct sum. It would be interesting to have such a natural characterization of $A \,_\Omega\!\oplus A$ as a normed space in the case of real interpolation. For instance, it may be true that $A \,_\Omega\!\oplus A$ can be identified with the set of pairs

$$(a,b) = \left(\int_0^\infty u(t)\, \frac{dt}{t},\ \int_0^\infty \log t\, u(t)\, \frac{dt}{t}\right)$$

where $u(t)$ is a strongly measurable $A_0 \cap A_1$ valued functions and we measure the size of the pair by

$$\inf \left[\int_0^\infty (t^{-\theta} J(t, u(t), \bar{A}))^q \frac{dt}{t}\right]^{1/q}$$

where, as usual, the infimum is over $u(t)$ which give the same (a,b).

7. Do analogs of Theorems 3.15 and 3.19 hold for the Ω's associated to the E_α functionals?

8. (Probably easy.) Is there an analog of Theorem 4.2 for the range spaces?

9. Is Theorem 5.14 true for couples which are not quasilinearizable?

10. Let A_w be the set of all $a \in \Sigma(\bar{A})$ such that $\int_0^\infty w(t) K(t, a, \bar{A}) \frac{dt}{t} < \infty$. Under what conditions on w_1 and w_2 does the inclusion $A_{w_1} \subset A_{w_2}$ force $A_0 = A_1$?

References

[BL] J. Bergh, J. Löfström, Interpolation Spaces: An Introduction, Springer-Verlag-Berlin-Heidelberg-New York, (1976).

[C] S. Chanillo, A note on commutators, Ind. U. Math. J. 31 (1982), 7-17.

[CRW] R. Coifman, R. Rochberg, G. Weiss, Factorization theorems for Hardy spaces in several variables, Ann. Math. 103 (1976), 611-635.

[CS] R. Coifman, S. Semmes, Interpolation of Banach spaces, Perron Processes, and Yang-Mills, Preprint, 1988.

[CJM] M. Cwikel, B. Jawerth, M. Milman, The domain spaces of quasilogarithmic operators, preprint.

[DRS] R. A. DeVore, S. D. Riemenschneider, and R. Sharpley, Weak interpolation in Banach spaces, J. Functional Anal, 33 (1979), 58-94.

[F] G. Feissner, Hypercontractive semigroups and Sobolev's inequality, Trans. Amer. Math. Soc. 210 (1975), 51-62.

[G] L. Gross, Logarithmic Sobolev inequalities, Amer. J. Math. 97 (1975), 1061-1083.

[J] S. Janson, Minimal and maximal methods of interpolation, J. Funct. Anal. 44 (1981), 50-73.

[JNP] S. Janson, P. Nilsson, and J. Peetre, Notes on Wolff's note on interpolation spaces, Proc. Lond. Math. Soc. 3 48 (1984), 283-299.

[JRW] B. Jawerth, R. Rochberg, G. Weiss, Commutator and other second order estimates in real interpolation theory, Ark. Mat. 24(1986), 191-219.

[K1] N. Kalton, Non-linear commutators in interpolation theory, preprint 1987.

[K2] ________, Trace class operators and commutators, preprint 1987.

[LT] J. Lindenstrauss and L. Tzafiri, Classical Banach Spaces II, Springer-Verlag, Berlin-Heidleberg-New York, 1979.

[RW] R. Rochberg, G. Weiss, Derivatives of analytic families of Banach spaces, Ann. Math. 118(1983), 315-347.

[Sc] M. Schechter, Complex Interpolation, Compositio Math. 18 (1967), 117-147.

[S] B. Simon, Schrödinger semigroups, Bull. Amer. Math. Soc. 7 (1982),447-526.

[SH-K] B. Simon and R. Hoegh-Krohn, Hypercontractive semigroups and two dimensional self-coupled Bose fields, J. Funct. Anal. 9 (1972), 121-180.

[Z] M. Zafran, Spectral theory and interpolaltion of operators, J. Funct. Anal. 36 (1980), 185-204.

A Survey of Nest Algebras

Kenneth R. Davidson†

University of Waterloo

Every linear map on $\mathbb{C}^n$ has an upper triangular form; and for a fixed basis, the set of upper triangular matrices is a tractable object. For operators on Hilbert space, the notion of triangular form is replaced by the search for a maximal chain of invariant subspaces. This has been a rather intensive search, but the Invariant Subspace Problem remains, and is likely to remain for some time. The study of nest algebras takes the other point of view: fix a complete chain of closed subspaces (a nest) and study the algebra of all operators leaving each element of the nest invariant. That is, we study all operators with a given triangular form.

This sub-discipline of operator theory is about twenty five years old. It has reached a stage where there are many nice results, and a fairly satisfactory theory. Yet there are still interesting and compelling problems remaining. In these lectures, I will attempt to describe some of the results and to state some of these open questions.

Closely related to nest algebras are the so called CSL algebras. A CSL is a complete lattice $\mathcal{L}$ of *commuting* projections. The associated algebra $Alg\,\mathcal{L}$ consists of all operators leaving the ranges of $\mathcal{L}$ invariant. That is, all operators A such that $P^{\perp}AP = 0$ for all P in L. These algebras are closely related to nest algebras, and indeed are the intersection of nest algebras with "commuting" nests. Nevertheless, they are much less well understood. The results for nest algebras naturally lead to open questions about CSL algebras. We will try to describe some of the these problems as well.

I would like to thank Earl Berkson and David Berg at the University of Illinois for inviting me to give this series of lectures; and to thank the NSF for supporting it.

† This research partially supported by a grant from NSERC.

Lecture 1 Compact Operators

A complete chain of subspaces will be called a *nest*. For each element N of a nest $\mathcal{N}$, let N_- denote the sup (closed span) of all its predecessors $(N'<N, N' \in \mathcal{N})$. It is not difficult to check that $\mathcal{N}$ is maximal if and only if $\dim N/N_- \leq 1$ for all N in $\mathcal{N}$. A subspace $E = N \ominus N'$ is called an *interval* of $\mathcal{N}$ $(N'<N \in \mathcal{N})$. The minimal intervals of $\mathcal{N}$ are the *atoms* $N \ominus N_-$, when $N_- < N$. Before considering triangular forms, let us look at a few examples of nests.

Example 1. Let $\{e_n, n \in \mathbb{N}\}$ be an orthonormal basis for $\mathcal{H}$. Let $P_k = span\{e_n, n \leq k\}$ for k in $\mathbb{N}$. $\mathcal{P} = \{\{0\}, P_k, k \in \mathbb{N}, \mathcal{H}\}$ is a maximal nest. The atoms $A_k = span\{e_k\} = P_k \ominus P_{k-1}$ span $\mathcal{H}$, so $\mathcal{P}$ is *atomic*.

Example 2. Let $\mathcal{H} = L^2(0,1)$. For $0 \leq t \leq 1$, let N_t be the set of functions in $L^2(0,1)$ supported on $[0,t]$. The *Volterra nest* is the nest $\mathcal{N} = \{N_t, 0 \leq t \leq 1\}$. This is maximal and has no atoms (a *continuous* nest).

Example 3. Let $\mathcal{H} = \ell^2(\mathbb{Q})$. For each t in $\mathbb{R}$, let Q_t^+ and Q_t^- denote all functions in $\ell^2(\mathbb{Q})$ supported on $(-\infty, t]$ and $(-\infty, t)$ respectively. Note that $Q_t^+ = Q_t^-$ when $t \notin \mathbb{Q}$, and $Q_r^+ \ominus Q_r^- = A_r = span\{\delta_r\}$; where δ_r is the characteristic function of r. Let $\mathcal{Q} = \{Q_t^{\mp}, t \in \mathbb{R}\}$. This nest is atomic, maximal, and uncountable.

Let $\mathcal{T}(\mathcal{N})$ denote the algebra of all operators T such that each N in $\mathcal{N}$ is invariant for T. Let $E = N_2 \ominus N_1$ be an interval of $\mathcal{N}$. Then the decomposition of $\mathcal{H} = N_1 \oplus E \oplus N_2^{\perp}$ yields a 3×3 matrix form

$$T = \begin{bmatrix} T_{11} & T_{12} & T_{13} \\ 0 & T_{22} & T_{23} \\ 0 & 0 & T_{33} \end{bmatrix}$$

for each T in $\mathcal{T}(\mathcal{N})$. The map $\Phi_E(T) = P(E)T|E = T_{22}$ compressing T to $\mathcal{B}(E)$ is an *algebra homomorphism*. The converse is also valid [68].

Let K be a fixed compact operator. An old and well known theorem of Aronszajn and Smith [2] and von Neumann (unpublished) is that *every compact operator has a proper invariant subspace*. An easy application of Zorn's lemma yields: *For every compact operator K, there is a maximal nest $\mathcal{N}$ consisting of invariant subspaces of K.* As we have seen above, every atom A of $\mathcal{N}$ is one dimensional. The compression Φ_A of $\mathcal{T}(\mathcal{N})$ to $\mathcal{B}(A) \cong \mathbb{C}$ is a a multiplicative linear functional. Thus the values $\Phi_A(K)$ will be in the spectrum $\sigma_{\mathcal{T}(\mathcal{N})}(K)$ of K as an element of $\mathcal{T}(\mathcal{N})$. The following theorem of Ringrose [65] is a beautiful extension of the finite dimensional theory.

Ringrose's Theorem. *Let K be a compact operator, and let $\mathcal{N}$ be a maximal nest of invariant subspaces of K. Then*

$$\Lambda = \{\Phi_A(K): A \text{ is an atom of } \mathcal{N}\} \cup \{0\}$$

equals $\sigma(K)$, and each non-zero eigenvalue is listed according to its algebraic multiplicity.

Sketch of Proof. The nest $\mathcal{N}$ is complete in the order topology and thus is compact. The map $N \to P(N)$ taking N to the orthogonal projection onto N is continuous if the range is endowed with the strong * operator topology. If A_n is a bounded net converging strong* to A, and K is compact, then $A_nK \to AK$ and $KA_n \to KA$ in norm.

The atoms A_n are pairwise orthogonal and K is compact, so $\lambda_n = \Phi_{A_n}(K)$ converge to zero. Let $\lambda \neq 0$ and let $\epsilon = \frac{1}{2} dist(\lambda, \Lambda \setminus \{\lambda\})$. A compactness argument yields a finite subset

$$\{0\} = N_0 < N_1 < \cdots < N_k = \mathcal{H}$$

such that either $E_j = N_j \ominus N_{j-1}$ is an atom or $\|P(E_j)KP(E_j)\| < \epsilon$ for $1 \leq j \leq k$. Decompose $\mathcal{H} = E_1 \oplus E_2 \oplus \cdots \oplus E_k$. Then K has an upper triangular matrix (K_{ij}) where $K_{ij} = 0$ if $j < i$, and either $\|K_{jj}\| < \epsilon$ or $K_{jj} = \lambda_j$ is a one dimensional scalar. Let $n \geq 0$ be the number of times λ occurs in Λ, and thus $\lambda = \lambda_j$ for exactly n values of j. Let C denote the circle with centre λ and radius ϵ. For each z in C, $zI-K$ is upper triangular with invertible diagonal entries. Thus it is invertible with upper triangular inverse. The Riesz projection

$$E_K\{\lambda\} = \frac{1}{2\pi i} \int_C (zI-K)^{-1} dz$$

is therefore upper triangular with diagonal entries $\frac{1}{2\pi i} \int_C (zI-K_{ii})^{-1} dz$. This entry is zero if $\|K_i\| < \epsilon$ or if λ_j is not inside C (which occurs if $\lambda_j \neq \lambda$). So $E_K\{\lambda\}$ has exactly n ones on the diagonal. It is an idempotent of finite rank equal to n as desired. ■

Now consider the question of how many compact operators belong to $\mathcal{T}(\mathcal{N})$. Let $x \otimes y^*$ denote the rank one operator $x \otimes y^*(z) = (z,y)x$. Fix N in $\mathcal{N}$, and let $A = N \ominus N_-$ (may be $\{0\}$). Decompose $\mathcal{H} = N_- \oplus A \oplus N^\perp$. Every operator T of the form

$$T = \begin{bmatrix} 0 & * & * \\ 0 & * & * \\ 0 & 0 & 0 \end{bmatrix} = P(N)TP(N_-)^{\perp}$$

belongs to $\mathcal{T}(\mathcal{N})$. In particular, $x \otimes y^*$ belongs to $\mathcal{T}(\mathcal{N})$ if $x \in N$ and $y \in (N_-)^{\perp}$. These rank operators are enough to prove:

Lemma. *Every nest is reflexive.*

Proof. Let L be invariant for $\mathcal{T}(\mathcal{N})$. Let N be the least element of $\mathcal{N}$ dominating L. Suppose $M \in \mathcal{N}$ and $M_- < N$, whence $(M_-)^{\perp} \cap L$ contains a nonzero vector y. For every x in M,

$$x \in (x \otimes y^*)(L) \subseteq \mathcal{T}(\mathcal{N})L.$$

So L contains $\vee\{M \in \mathcal{N}: M_- < N\} = N$. ■

In fact, the compact operators in $\mathcal{T}(\mathcal{N})$ are strongly dense in $\mathcal{T}(\mathcal{N})$. Indeed, one has

Erdos Density Theorem [30]: *There is a sequence F_n of finite rank contractions in $\mathcal{T}(\mathcal{N})$ converging strong * to the identity. Hence the unit ball of $\mathcal{T}(\mathcal{N}) \cap K$ is strong * dense in the ball of $\mathcal{T}(\mathcal{N})$.*

Sketch The second statement is immediate from the first. The existence of such a sequence is trivial for atomic nests, and for the atomic part of any nest. So one compresses to the non- atomic part. It is fairly easy to reduce to the case of a nest with a cyclic vector. We illustrate the ideas with the Volterra nest of example 2. Fix n for a moment, and let x_k be the positive multiple of the characteristic function $\chi_{[(k-1)/2^n, k/2^n]}$ with norm one. Let $F_n = \sum_{k=2}^{2^n} x_{k-1} \otimes x_k^*$. This is a finite rank contraction in $\mathcal{T}(\mathcal{N})$. A simple computation shows that $F_n \chi_{[a,b]} \to \chi_{[a,b]}$ for every interval $[a,b] \subseteq [0,1]$. These intervals span $L^2(0,1)$. So $F_n \xrightarrow{s} I$, and likewise $F_n^* \xrightarrow{s} I$. ■

This can be used to give a simple proof [31] (also [61]) of a theorem of Lidskii [51]. Try to give your own proof that doesn't use nests. (It's hard.).

Lidskii's Theorem. *The trace of a trace class operator is equal to the sum of its eigenvalues including multiplicity.*

Proof. This is easy for finite rank operators, as it readily reduces to the familiar matrix case. Let T be a trace class operator. Fix a maximal nest $\mathcal{N}$ of invariant subspaces for T. Define a linear functional on $\mathcal{C}_1 \cap \mathcal{T}(\mathcal{N})$, the trace class operators in $\mathcal{T}(\mathcal{N})$, by $\Phi(X) = \Sigma \Phi_A(X)$ where the sum is taken over all atoms of $\mathcal{N}$. It is straightforward to verify that $\|\Phi\| \leq 1$. By Ringrose's Theorem, $\Phi(X)$ is the sum of the eigenvalues of X including multiplicity. Let F_n be the approximate identity given by Erdös's Density Theorem. It follows easily that $F_n T$ converges to T in the trace class norm. Since each F_n is a finite rank operator $\mathcal{T}(\mathcal{N}), \Phi(F_n T) = tr(T)$. Hence

$$\Phi(T) = \lim_{n \to \infty} \Phi(F_n T) = \lim_{n \to \infty} tr(F_n T) = tr(T). \blacksquare$$

These ideas lead naturally to corresponding questions about the compact operators in CSL algebras. Such algebras need not contain any compact operators at all. For example, the multiplication algebra $\mathcal{M} = \{M_f : f \in L^\infty(0,1)\}$ acting on $L^2(0,1)$ has a commutative lattice, but no compact operators. Nevertheless, all commutative subspaces are reflexive [4,18]. The following question remains open:

Problem 1. *Characterize those CSL lattices such that $(Alg\, \mathcal{L}) \cap \mathcal{K}$ is weak * dense in $Alg\, \mathcal{L}$.*

Some interesting partial results exist. A CSL is called *completely distributive* if the infinite distributive law

$$\bigwedge_{x \in X} \bigvee_{\alpha \in \Lambda_x} L_\alpha = \bigvee_{f \in \Pi \Lambda_x} \bigwedge_{x \in X} L_{f(x)}$$

holds for every collection of subsets $\{\Lambda_x : x \in X\}$ of $\mathcal{L}$. Fortunately, this horrible condition has a number of more tractable formulations [52,64]. Using a difficult result of Arveson [4], Laurie and Longstaff [50] and Hopenwasser, Laurie and Moore [49] prove:

Theorem *Let $\mathcal{L}$ be a CSL on a separable space. Then the following are equivalent:*

1) *$\mathcal{L}$ is completely distributive*
2) $\overline{Alg(\mathcal{L}) \cap \mathcal{C}_2}^{WOT} = Alg\, \mathcal{L}$
3) $\overline{span\{\text{rank one operators in } Alg\, \mathcal{L}\}}^{w^*} = Alg(\mathcal{L})$

*Here, $\mathcal{C}_2$ is the set of Hilbert-Schmidt operator, WOT and w^* are the weak operator and weak * topologies respectively.*

Froelich [33] has shown that CSL algebras may contain compact operators and even C_p operators for $p>2$ and still fail to have C_2 operators. But the compact operators are not dense in these examples. Wagner [70] shows that the density implies that $\mathcal{L}$ is compact in its strong topology. This in turn implies a certain lattice theoretic condition which is strictly weaker than complete distributivity for arbitrary distributive lattices. No example is known, however, which is strongly compact and not completely distributive. So a weaker version of problem 1 is:

Problem 1′: *Does* $\overline{\mathcal{K} \cap Alg\,\mathcal{L}}^{w^*} = Alg\,\mathcal{L}$ *imply that* $\mathcal{L}$ *is completely distributive?*

Returning to nests again, we consider the question of generators. Every nest algebra $\mathcal{T}(\mathcal{N})$ contains the von Neumann algebra $\mathcal{N}' = \{P(N): N \in \mathcal{N}\}'$. This has abelian communtant, and thus contains a maximal abelian von Neumann subalgebra (masa) $\mathcal{M}$. In the separable case, $\mathcal{M}$ is always singly generated as an algebra by some self adjoint operator M. The nest algebra is not abelian, so is not singly generated as an algebra. However, it is singly generated as a weak * $\mathcal{M}$ bi-module [3,50]. The generator T can be taken to be trace class. It follows that $\{A,T\}$ generate $\mathcal{T}(\mathcal{N})$ as a weak * closed algebra.

The key to proving this is a result of Arveson [4] which extends a theorem of Radjavi and Rosenthal [63] from the *WOT* topology to the weak * topology (non-trivial!):

Theorem (Arveson): *Let* $\mathcal{S}$ *be a weak * closed subalgebra of* $\mathcal{B}(\mathcal{H})$ *containing a masa* $\mathcal{M}$ *such that* $Lat\,\mathcal{S} = \mathcal{N}$ *is a nest. Then* $\mathcal{S} = \mathcal{T}(\mathcal{N})$.

Now one constructs an operator T such that $Lat(T) \cap Lat(\mathcal{M}) = \mathcal{N}$. The weak * closed algebra generated by $\{A,T\}$ satisfies the hypothesis of the theorem, and thus equals $\mathcal{T}(\mathcal{N})$.

A much more difficult question is:

Problem 2. *Is every (separably acting) maximal nest equal to the invariant subspace lattice of a single operator?*

Donoghue [28] showed that the weighted backward unilateral shift given by $Se_1=0, Se_{n+1}=2^{-n}e_n$ for $n\geq 1$ has $Lat\,S = \mathcal{P}$ of Example 1. Dixmier [26] showed that the Volterra operator $Vf(x) = \int_x^1 f(t)dt$ has the Volterra nest of Example 2 as its lattice of invariant subspaces. Such operators are called *unicellular,* and are rather difficult to come by.

Recently, some progress has been made. Domar [27] showed that the doubly infinite nest on $l^2(\mathbb{Z})$ equals $Lat(W)$ for certain weighted shifts. Harrison and Longstaff [34] succeed in gluing two copies of Example 1 together. Barria [9] generalized this, and later Barria and Davidson [10] construct unicellular operators for all ordinals and certain related nests. A useful test case is

Problem 2′: *Is the nest on $\ell^2(\mathbb{Q})$ of Example 3 the invariant subspace lattice of a single operator?*

Lecture 2. Spatial Structure

It is a persistent aim in mathematics to classify all objects up to their natural isomorphisms in some class. We consider the classification of nests under unitary equivalence and similarity. There is a natural parallel with the classification of self-adjoint operators. The similarities and differences are both instructive. We restrict out attention to the separable case.

The von Neumann algebra $\mathcal{N}'' = \{P(N): N \in \mathcal{N}\}''$ is abelian and hence has a separating vector x of norm one. The map $\Phi(N) = \|P(N)x\|^2$ is an order preserving homeomorphism of $\mathcal{N}$ onto a compact subset of ω of $[0,1]$. We will call ω (up to an order preserving homeomorphism of $[0,1]$) the *order type* of $\mathcal{N}$. Using ω to parametrize $\mathcal{N}$, define a spectral measure

$$E(a,b] = P(N_b) - P(N_a).$$

This extends to a regular Borel measure on $[0,1]$ supported on ω. The Hahn-Hellinger theory determines $E(\cdot)$ up to unitary equivalence by its measure class - the set of scalar measures $[\mu]$ mutually absolutely continuous with $E(\cdot)$, and its multiplicity function - $m: \omega \to \mathbb{N} \cup \{\infty\}$. The set ω splits into a disjoint union of Borel sets $\omega_\infty, \omega_1, \omega_2, \ldots$ and the measure μ decomposes as $\Sigma \mu_n$ where $\mu_n = \mu|\omega_n$. The spectral measure $E(\cdot)$ is unitarily equivalent to the standard spectral measure on $\sum_{n=1}^{\infty} \oplus L^2(\omega_n)^{(n)} \oplus L^2(\omega_\infty)^{(\infty)}$ where $\mathcal{H}^{(n)}$ indicates the direct sum of n copies of $\mathcal{H}$.

Theorem (Erdos [29]): *Two nests $\mathcal{N}$ and $\mathcal{M}$ are unitarily equivalent if and only if there exists an order isomorphism of $\mathcal{N}$ onto $\mathcal{M}$ which preserves the measure class and multiplicity of the spectral measure.*

The connection with Hermitian operators is very direct. Let E be the spectral measure of $\mathcal{N}$, and define $A = \int_0^1 tE(dt)$. It is fairly easy to see that the spectral subspaces $E_A[0,t]$ are precisely $\mathcal{N}$. If θ is an order isomorphism of $\mathcal{N}$ onto another nest $\mathcal{M}$, define $\Psi(M) = \Phi(\theta^{-1}(M))$ and let F be the corresponding spectral measure. Set $B = \int_0^1 tF(dt)$. Then θ is implemented by a unitary operator if and only if B is unitarily equivalent to A.

The problem remains open however: how do you determine if two nests are unitarily equivalent? For example, suppose $\mathcal{N}$ and $\mathcal{M}$ are both continuous nests of multiplicity one. Their spectral measure can be equivalent to any scalar measure with support [0,1]. But Kadison and Singer [41] used the Halmos-von Neumann theorem (any two nonatomic finite Borel measures are equivalent) and to show that all continuous nests of multiplicity one are unitarily equivalent. Essentially, the nest is reparametrized so that the spectral measure is Lebesgue measure. The general problem reduces to a technical problem in measure theory.

What about similarity? The two self-adjoint operators are similar if and only if they are unitarily equivalent. Ringrose asked if multiplicity was a similarity invariant for nests. Recently, Larson [47,48] showed that it is not, nor is the measure class! The key is another equivalence relation valid for both operators and nests. Two operators A and B are *approximately unitarily equivalent* ($A \underset{a}{\sim} B$) if there is a sequence of unitaries U_n such that $B - U_n^* A U_n$ is compact, and $\lim_{n\to\infty} \|B - U_n^* A U_n\| = 0$. Two nests $\mathcal{N}$ and $\mathcal{M}$ are *approximately unitarily equivalent* ($\mathcal{N} \underset{a}{\sim} \mathcal{M}$) if there is an order isomorphism θ of $\mathcal{N}$ onto $\mathcal{M}$ and a sequence of unitaries U_n such that $f_n(N) = P(\theta(N)) - U_n^* P(N) U_n$ are continuous compact valued functions converging uniformly to zero.

For both operators and nests, these notions turn out to be equivalent to weaker versions of themselves, dropping the conditions of compactness and norm continuity. For Hermitian operators A and B, $A \underset{a}{\sim} B$ if and only if $\sigma(A) = \sigma(B)$ and their isolated eigenvalues have the same multiplicity. For nests, the key step is the case of a continuous nest due to Andersen [1]. Other proofs are given in [6,21].

Andersen's Theorem. *Let $\mathcal{N}$ and $\mathcal{M}$ be (separably acting) continuous nests, and let θ be an order isomorphism of $\mathcal{N}$ onto $\mathcal{M}$. For any $\epsilon > 0$, there is a unitary U such that $f(N) = P(\theta(N)) - UP(N)U^*$ is continuous, compact valued and of norm at most ϵ.*

Larson [48] used Andersen's theorem to prove that multiplicity need not be preserved by similarity. Then he used this to show that any two continuous nests are similar. Davidson [19] extended Larson's theorem to arbitrary nests. If S is an invertible operator and $\mathcal{N}$ is a nest, $S\mathcal{N} = \{SN: N \in \mathcal{N}\}$ is a nest and there is a natural order isomorphism $\theta_S(N) = SN$ of $\mathcal{N}$ onto $S\mathcal{N}$. The spatial nature of θ_S forces

$dim\theta_S(N)/\theta_S(N') = dimN/N'$ for every $N' < N$ in $\mathcal{N}$ (i.e. θ_S *preserves dimension*). An order isomorphism θ of $\mathcal{N}$ onto $\mathcal{M}$ compares with the equal spectra $\sigma(A) = \sigma(B)$ of Hermitian operators. Preserving dimension should be compared with multiplicity of isolated eigenvalues. This is particularly appropriate because θ preserves dimension exactly when it preserves the dimension of atoms.

Similarity Theorem [19]. *Let $\mathcal{N}$ and $\mathcal{M}$ be separably acting nests. Then the following are equivalent:*

1) *There is an order isomorphism of $\mathcal{N}$ onto $\mathcal{M}$ which preserves dimension.*
2) *$\mathcal{N}$ and $\mathcal{M}$ are approximately unitarily equivalent.*
3) *$\mathcal{N}$ and $\mathcal{M}$ are similar.*

Moreover, given such an order isomorphism θ and an $\epsilon > 0$, there is a unitary U and a compact operator K with $\|K\| < \epsilon$ such that invertible operator $S = U+K$ implements θ (i.e. $\theta = \theta_S$).

It follows from this theorem that neither multiplicity nor measure class is preserved by similarity. The uncountable atomic nest on $l^2(\mathbb{Q})$ of Example 3 is similar to a nest which is not atomic. To see this, consider $\mathcal{H} = l^2(\mathbb{Q}) \oplus L^2(\mathbb{R})$. Define M_t to be the functions in $L^2(\mathbb{R})$ supported on $(-\infty,t]$. Let $R_t^{\pm} = Q_t^{\pm} \oplus M_t$. The nest $\mathcal{R} = \{R_t^{\pm}, t \in \mathbb{R}\}$ is order isomorphic to $\mathcal{Q}$, and this order isomorphism preserves dimension because they have the same (one dimensional) atoms. Hence they are similar.

No explicit operator is known to act on any nest and change either the multiplicity or the spectral measure. An example might be illuminating if it were simple enough. One reason this is difficult is because a result of Gohberg and Krein [36] shows that $(I+K)\mathcal{N}$ is unitarily to $\mathcal{N}$ if K belongs to the the Mac'caev ideal. The Ma'caev ideal consists of all compact operators whose singular values satisfy $\sum_{n=1}^{\infty} s_n(K)/n < +\infty$. This contains all the C_p classes, and hence most tractable compact operators.

The notion of similarity is a poor one for CSL's because similarity does not preserve orthogonality. One can however extend the class of lattices to those projection lattices contained in a bounded, σ-complete Boolean algebra. This class is invariant under similarity [20]. The notions of unitary equivalence and approximate unitary equivalence do preserve CSL's, so are more pertinent to the present discussion. Arveson [4] found a nice representation for CSL's which is very useful in developing the structure of the associated algebras. He did not find a complete set of unitary invariants, but the present situation is good enough for most purposes. It is reasonably clear that the Hellinger-Hahn theory should yield a version of Erdös's Theorem for CSL's. The question of approximate unitary equivalence is likely to be of greater interest. Arveson [6] has a version of Andersen's Theorem valid for many "homogeneous" nonatomic CSL's. The general problem is wide open.

Problem 3. *Classify CSL lattices up to approximate unitary equivalence, or at least identify non-trivial invariants that unify the case of nests and Boolean algebras.*

A tool of major importance in the nest case is a distance formula due to Arveson. Note that if A is any operator, T belongs to $\mathcal{T}(\mathcal{N})$, and N is in $\mathcal{N}$, one has

$$\|P(N)^{\perp}AP(N)\| = \|P(N)^{\perp}(A-T)P(N)\| \leq \|A-T\|.$$

Hence $\sup_{N\in\mathcal{N}}\|P(N)^{\perp}AP(N)\| \leq dist(A,\mathcal{T}(\mathcal{N}))$. Arveson [5] proved the converse. Other proofs are given in [45,60].

Arveson's Distance Formula. For A in $\mathcal{B}(\mathcal{H})$ and $\mathcal{N}$ a nest

$$dist(A,\mathcal{T}(\mathcal{N}))) = \sup_{N\in\mathcal{N}}\|P(N)^{\perp}AP(N)\|.$$

Sketch [60]. Given operators A,C and D, Davis, Kahane and Weinberger [25] (also [56]) explicitly write down an operator X so that

$$\left\|\begin{bmatrix} A & X \\ C & D \end{bmatrix}\right\| = \max\left\{\left\|\begin{bmatrix} A \\ C \end{bmatrix}\right\|, \left\|\begin{bmatrix} C, D \end{bmatrix}\right\|\right\}.$$

We will refer to this as the "filling in" lemma. Suppose $\mathcal{F}$ is a finite nest $\{0\} = F_0 < F_1 < \cdots < F_n = \mathcal{H}$, and decompose $\mathcal{H} = \mathcal{H}_1 \oplus \cdots \oplus \mathcal{H}_n$ where $\mathcal{H}_k = F_k \ominus F_{k-1}$. Any operator A in $\mathcal{B}(\mathcal{H})$ has an operator matrix (A_{ij}). The operators in $\mathcal{T}(\mathcal{F})$ are precisely those for which $A_{ij} = 0$ if $j < i$. Thus $dist(A, \mathcal{T}(\mathcal{F}))$ becomes the problem of changing the upper triangle entries of (A_{ij}) to minimize the norm. The term $P(F_j)^{\perp}AP(F_j)$ is a rectangular matrix in the "lower hand corner" of A. Consider

$$\begin{bmatrix} 0 & 0 & 0 & 0 & 0 \\ A_{21} & X_{22} & X_{23} & X_{2n+1} & 0 \\ A_{31} & A_{32} & X_{33} & & \\ A_{41} & A_{42} & A_{43} & & \\ & & & X_{n-1n-1} & 0 \\ A_{n1} & A_{n2} & A_{n3} & A_{nn-1} & 0 \end{bmatrix}$$

Let $L = \max\|P(F_j)^{\perp}AP(F_j)\|$. Consider the lower left $(n-1)\times 2$ rectangle as a 2×2 matrix with X_{22} undefined. Use the "filling in" lemma to define X_{22} so that the rectangle has norm at most L. Then consider the lower left $(n-1)\times 3$ rectangle and use the lemma to define $\begin{pmatrix} X_{23} \\ X_{33} \end{pmatrix}$. Eventually the matrix is completed and the norm is at most L.

The proof is now completed by taking a limit over the net of all finite sub-nests of $\mathcal{N}$. The compactness of the unit ball of $\mathcal{T}(\mathcal{N})$ in the weak operator topology is crucial. ■

This distance formula is essential at several points in the proof of the Similarity Theorem. The same kind of distance formula is valid up to a constant for AF von Neumann algebras (those algebras which are the weak * closure of an increasing net of finite dimensional subalgebra) including type I algebras.

Theorem (Christensen) [15]. *Let $\mathcal{A}$ be an AF von Neumann algebra. For every T in $\mathcal{B}(\mathcal{H})$,*

$$dist(T,\mathcal{A}) \leq 4 \sup \{\|P^{\perp}TP\| : P \in Lat\, \mathcal{A}\}.$$

For abelian and commutants of abelian algebras, the constant can be reduced to 2 [68]. An algebra $\mathcal{A}$ for which there is a constant K such that $dist(T,\mathcal{A}) \leq K \sup \{\|P^{\perp}TP\| : P \in Lat\, \mathcal{A}\}$ for all T in $\mathcal{T}(\mathcal{N})$ is called and the optimal K is called ***the distance constant.*** Other than the examples above, very few such algebras are known [22,42,43]. One interesting example is $W(S)$, the weakly closed algebra generated by the unilateral shift [22]. For CSL algebras, there is some negative information. Davidson and Power [23] show that many CSL algebras including all infinite tensor products of proper nests fail to have a distance formula. Larson [49] generalized this construction to produce more pathology. One hope remains. A CSL lattice is called *width* n if it is generated by n nests, and *finite width* if it is width n for $n < \infty$. For example, consider the lattice $\mathcal{L} = \mathcal{N} \otimes \mathcal{N}$

on $L^2([0,1]^2)$ generated by the nests $\mathcal{N} \otimes I = \{N_t \otimes I: N_t \in \text{Volterra nest}\}$ and $\{I \otimes N_t: N_t \in \text{Volterra nest}\}$.

Problem 4. *Is there a distance constant for $Alg(\mathcal{N}\otimes\mathcal{N})$? More generally, is there a distance constant for width n algebras?*

Though not directly related to CSL algebras, we mention another interesting class: *Is there a distance constant for the weakly closed algebra $W(S)$ generated by a subnormal operator?* A theorem of Olin and Thomson [55] generalizing the work of Scott Brown [11] shows that $W(S)$ is reflexive, and yields quantitative control over the predual of $W(S)$. For a distance constant, one needs the same kind of control over the preannihilator [46].

The distance between two subspaces X,Y of a Banach space is defined by

$$d(X,Y) = \max\{\sup_{x\in X_1} dist(x,Y), \sup_{y\in Y_1} dist(y,X)\}$$

where X_1 denote the unit ball. This is equivalent to (but slightly different from) the Hausdorff metric on the unit balls. Similarly, one can define a distance between lattices as the Hausdorff distance between the corresponding sets of projections. In the following, interpret "small" and "close" as being $O(\epsilon)$.

Theorem (Lance)[45]. *Let $\mathcal{N}$ and $\mathcal{M}$ be two nests. The following are equivalent:*

1) *$\mathcal{T}(\mathcal{N})$ and $\mathcal{T}(\mathcal{M})$ are close.*
2) *$\mathcal{N}$ and $\mathcal{M}$ are close.*
3) *There is an invertible operator S such that $S\mathcal{N} = \mathcal{M}$ and $\|S-I\|$ is small.*

Theorem (Christensen) [14]. *Let $\mathcal{A}$ and $\mathcal{B}$ be type I von Neumann algebras. The following are equivalent:*

1) *$\mathcal{A}$ and $\mathcal{B}$ are close.*
2) *$\mathcal{A}'$ and $\mathcal{B}'$ are close (whence Lat $\mathcal{A}$ and Lat $\mathcal{B}$ are close).*
3) *There is a unitary operator U such that $U\mathcal{A}U^* = \mathcal{B}$ and $\|U-I\|$ is small.*

In both these results, proximity of the algebras or even of the lattices forces the algebras to be spatially isomorphic. This is not the case of CSL algebras. In [23], it is shown that the lattices may be close yet any similarity is far from the identity. Christensen and Choi [12] have shown that close C^* need not be isomorphic. Even in finite dimensions, very peculiar things may occur for arbitrary algebras [13]. On the positive side, one has

Theorem (Davidson) [20]. *Let $\mathcal{A}$ be a CSL algebra. Suppose that $\mathcal{B}$ is an operator algebra with $d(\mathcal{A},\mathcal{B}) = \epsilon < .01$. Then there is a lattice isomorphism θ of Lat $\mathcal{B}$ onto Lat $\mathcal{A}$ such that $\|\theta - id\| \leq 4\epsilon$.*

Problem 5. *Is there an example of two CSL algebras which are close but not similar? or even close but not isomorphic?*

Lecture 3. Algebraic Structure

There is an intriguing problem about nest algebras which remains open even for the easiest nest $\mathcal{P} = \{P_n, n \geq 1\}$ where P_n is an increasing sequence of subspaces such that $dim P_n = n$ and $\bigcup_{n \geq 1} P_n$ is dense in $\mathcal{H}$.

Problem 6. *Is $\mathcal{T}(\mathcal{P})^{-1}$ connected?*

The answer is probably no. One reason for expecting this to be the case is the analogy with Toeplitz operators. The Toeplitz operators in $\mathcal{T}(\mathcal{P})$ are precisely $\{T_h^*: h \in H^\infty\}$. The invertibles in H^∞ are not connected. For in any abelian Banach algebra $\mathcal{A}$, $\mathcal{A}_0^{-1} = e^{\mathcal{A}}$ consists of all elements with logarithms in $\mathcal{A}$. Take h to be the conformal map of the disc onto a region spiralling out to the unit circle, but circling around infinitely often. This h will not have a bounded logarithm. Is T_h^* connected to I via path of upper triangular invertible operators?

It is tempting to try to use K-theory to attack this problem. Since $\mathcal{T}(\mathcal{N}) \otimes \mathcal{M}_n \cong \mathcal{T}(\mathcal{N}^{(n)})$ where $\mathcal{N}^{(n)}$ is the n-fold ampliation of $\mathcal{N}$, questions about $n \times n$ matrices over a nest algebra are just questions about another nest algebra. Pitts [59] has computed $K_0(\mathcal{T}(\mathcal{N}))$ by showing that it is equal to $K_0(\mathcal{D}_a(\mathcal{N}))$ where $\mathcal{D}_a(\mathcal{N})$ is the atomic part of the diagonal. Moreover, in the special case of $\mathcal{T}(\mathcal{P})$, every idempotent in $\mathcal{T}(\mathcal{P})$ is similar (in $\mathcal{T}(\mathcal{P})$) to a diagonal projection. The analogous result for K_1 might solve Problem 7 in the affirmative because the unitary group of a von Neumann algebra is connected.

One of the first questions asked about any Banach algebra is in regard to irreducible representations, and hence about the Jacobson radical. For the upper triangular algebra $\mathcal{T}_n$ of $n \times n$ matrices, one readily identifies the radical as the strictly upper triangular algebra. The quotient

$\mathcal{T}_n/rad\,\mathcal{T}_n$ is isomorphic to the diagonal algebra $\mathcal{D}_n$. If $\mathcal{F}$ is a finite nest, one obtains in precisely the same way that $rad\ \mathcal{T}(\mathcal{F})$ consists of the block strictly upper triangular opertors and $\mathcal{T}(\mathcal{F})/rad\ \mathcal{T}(\mathcal{F})$ is isomorphic to the diagonal algebra $\mathcal{D}(\mathcal{F}) = \{P(F): F \in \mathcal{F}\}'$. There is a (unique) expectation $\Delta_{\mathcal{F}}$ of $\mathcal{B}(\mathcal{H})$ onto $\mathcal{D}(\mathcal{F})$, and $rad\ \mathcal{T}(\mathcal{F}) = ker\,\Delta_{\mathcal{F}} \cap \mathcal{T}(\mathcal{F})$.

Now consider an arbitrary nest $\mathcal{N}$. The C^*-algebra $C^*(\mathcal{N}) = \overline{span}\{P(N): N \in \mathcal{N}\}$ is commutative. Each function ϕ in its maximal ideal space $M_{\mathcal{N}}$ is determined by the restriction of ϕ to $\mathcal{N}$. This is an increasing function from $\mathcal{N}$ onto $\{0,1\}$. Thus $\phi|\mathcal{N}$ is an element of $Hom(\mathcal{N},2)$, the lattice homomorphisms into the two element lattice $2 = \{0,1\}$. Moreover, this identification is a homeomorphism from $M_{\mathcal{N}}$ onto $Hom(\mathcal{N},2)$ with its order topology. Each ϕ gives rise to a seminorm on $\mathcal{T}(\mathcal{N})$:

$$\|T\|_\phi = \inf\{\|ETE\|: E \text{ is an interval of } \mathcal{N}, \phi(E) = 1\}$$

Since $\Phi_E(T) = ETE$ is a contractive homomorphism, the set $I_\phi = \{T \in \mathcal{T}(\mathcal{N}): \|T\|_\phi = 0\}$ is a closed two sided ideal.

For example, if A is an atom of $\mathcal{N}$, define $\delta_A(P(N)) = 1$ if $P(N) \geq A$ and $\delta_A(P(N)) = 0$ otherwise. This extends by linearity to a multiplicative linear functional on $C^*(\mathcal{N})$. The seminorm $\|T\|_{\delta_A} = \|ATA\|$. In this case, $\mathcal{T}(\mathcal{N})/I_\phi$ is isomorphic to $\mathcal{B}(A\mathcal{H})$.

As a second example, suppose that $N = N_-$. So $N = \sup N_k$ where N_k is an increasing sequence in $\mathcal{N}$. Define $\phi(M) = 1$ if $M \geq N$ and 0 otherwise, and extend to $C^*(\mathcal{N})$ by linearity. The seminorm becomes

$$\|T\|_\phi = \inf\|P(N \ominus N_k)TP(N \ominus N_k)\| = \lim_{k\to\infty}\|P(N \ominus N_k)TP(N \ominus N_k)\|.$$

The quotient algebra $\mathcal{D}_\phi = \mathcal{T}(\mathcal{N})/I_\phi$ does not seem to have an easy description in this case.

Ringrose [66] showed that the quotient algebras $\mathcal{D}_\phi$ are semisimple, and thereby obtained the following characterization of the radical of a nest algebra.

Theorem (Ringrose). *Let $\mathcal{N}$ be a nest. For T in $\mathcal{T}(\mathcal{N})$, the following are equivalent:*

1) $T \in rad\,\mathcal{T}(\mathcal{N})$
2) $T \in \cap\{I_\phi: \phi \in M_{\mathcal{N}}\}$
3) $T \in \overline{\cup\{rad\ \mathcal{T}(\mathcal{F}): \mathcal{F} \text{ finite subnest of } \mathcal{N}\}}$
4) *For every $\epsilon > 0$, there is a finite subnest $\mathcal{F}$ of $\mathcal{N}$ such that $\|\Delta_{\mathcal{F}}(T)\| < \epsilon$.*

Sketch. 1) $\Rightarrow$ 2) is the difficult step. It follows from the semisimplicity of $\mathcal{D}_\phi$. 2) $\Rightarrow$ 4) is a compactness argument. Since $\|T\|_\phi = 0$, there is an interval E_ϕ such that $\phi(E_\phi) = 1$ and $\|E_\phi T E_\phi\| < \epsilon$. The set $O_\phi = \{\psi \in M_\mathcal{N}: \psi(E_\phi) = 1\}$ is open. A finite subcover yields intervals $E_1, \ldots, E_n$ with $\sum_{i=1}^{n} E_i \geq I$ and $\|E_i T E_i\| < \epsilon$. The finite nest $\mathcal{F}$ given by the end points of the E_i's is the desired subnest. 4) $\Rightarrow$ 3) is trivial. 3) $\Rightarrow$ 1) is easy. For $rad\, \mathcal{T}(\mathcal{F}) \subseteq \mathcal{T}(\mathcal{N}) \subseteq \mathcal{T}(\mathcal{F})$, from whence one obtains $rad\, \mathcal{T}(\mathcal{F}) \subseteq rad\, \mathcal{T}(\mathcal{N})$. The radical is always closed. ∎

For general Banach algebras, every algebraically irreducible representation is equivalent to a continuous representation. However, it is not known whether every topologically irreducible representation has the same kernel as an algebraically irreducible one, or even if it contains the radical. For nest algebras, the situation is better.

Proposition. *The kernel of each topolgically irreducible representation of $\mathcal{T}(\mathcal{N})$ contains exactly one ideal I_ϕ.*

Sketch. Let π be the representation. A computation shows that $\pi(P(N))$ is invariant and hence equals 0 or I. It is clearly increasing, so determines an element ϕ in $Hom(\mathcal{N},2)$. If E is any interval with $\phi(E) = 1$, one has $\|\pi(T)\| = \|\pi(ETE)\| \leq \|\pi\|\|ETE\|$. Hence $\|\pi(T)\| \leq \|\pi\|\|T\|_\phi$ and $ker\,\pi$ contains I_ϕ. Uniqueness follows by showing that $I_\phi + I_\psi = \mathcal{T}(\mathcal{N})$ for $\phi \neq \psi$. ∎

Form the algebra $\mathcal{D} = \Sigma \oplus \mathcal{D}_\phi$ with the sup norm. Let $\Phi(T) = (T + I_\phi)$ be the natural homomorphism of $\mathcal{T}(\mathcal{N})$ into $\mathcal{D}$. Denote the range of Φ by $\mathcal{D}_\mathcal{N}$. This map factors through $\mathcal{T}(\mathcal{N})/rad\,\mathcal{T}(\mathcal{N})$ and turns out to be isometrically isomorphic to this quotient algebra. Lance [44] showed that each $\mathcal{D}_\phi$ has trivial centre, and used this to obtain a pretty description of the centre of $\mathcal{D}_\mathcal{N}$.

Theorem (Lance). *For T in $\mathcal{T}(\mathcal{N})$, the following are equivalent:*

1) $T \in C^*(\mathcal{N}) + rad\,\mathcal{T}(\mathcal{N})$.
2) $T + rad\,\mathcal{T}(\mathcal{N})$ *belongs to the centre of* $\mathcal{T}(\mathcal{N})/rad\,\mathcal{T}(\mathcal{N})$.
3) $T + I_\phi$ *is a scalar for all* ϕ *in* $M_\mathcal{N}$.
4) $\Phi(T) = (T + I_\phi)$ *is a continuous, scalar valued function on* $M_\mathcal{N}$.

Furthermore, Lance showed that each $\mathcal{D}_\phi$ has an isometric representation as an algebra of operators on Hilbert space. It is not known, however, if $\mathcal{D}_\phi$ is primitive (i.e. does $\mathcal{D}_\phi$ have a faithful, irreducible representation?).

The problem of determining the radical of a CSL algebra is unresolved. Hopenwasser and Larson [38,39] observed that $Hom(\mathcal{L},2)$ or equivalently the maximal ideal space of $C^*(\mathcal{L})$ gives rise to a family of ideals I_ϕ as in the nest case. They show that direct analogues of properties 2,3,4 are all equivalent in this more general context. Moreover, the kernel of every irreducible representation of $Alg\,\mathcal{L}$ contains exactly one I_ϕ. Hence $rad(Alg\,\mathcal{L})$ is contained in $\cap\{I_\phi\colon \phi \in Hom(\mathcal{L},2)\}$.

Problem 7. *If $\mathcal{L}$ is a CSL, is $rad(Alg\,\mathcal{L}) = \cap\{I_\phi\colon \phi \in Hom(\mathcal{L},2)\}$?*

From here, one branches out to more general representations. Recently, Paulsen, Power and Ward [57,58] have successfully analyzed contractive, weak* continuous representations of nest algebras. First they show that nest algebras are *semi discrete* [58] by showing that the identity map is the strong limit of contractive maps which factor through finite dimensional nest algebras. McAsey and Muhly [54] have shown that contractive representations of finite dimensional nest algebras are completely contractive (i.e. $\pi \otimes id_n$ on $\mathcal{T}(\mathcal{N}) \otimes \mathcal{M}_n$ is contractive for all $n \geq 1$). Thus it follows that contractive representations of all nest algebras are completely contractive.

Arveson [3] showed that completely contractive representations of an operator algebra $\mathcal{A}$ are precisely the ones which can be dilated to a * representation of the enveloping C^*-algebra. The weak * continuity allows an extension to the W^*-algebra generated by $\mathcal{T}(\mathcal{N})$, namely $\mathcal{B}(\mathcal{H})$. The weak * continuity of the dilation forces this to be an ampliation $A \mapsto A^{(\infty)}$. This dilation is obtained in the finite dimensional case by Ball and Gohberg [8]. Paulsen and Power [57] exploit the analogy with the Sz-Nagy dilation theory. They generalize Ando's theorem which gives an unitary dilation for two commuting contractions to a dilation of two contractive representations of nest algebras.

Does this have any bearing on Problem 4?

Next, let us look at algebra isomorphisms. Ringrose proved another very nice theorem. He showed:

Theorem (Ringrose [67]). *Every algebra isomorphism of one nest algebra onto another has the form $\alpha(T) = STS^{-1}$ for some invertible operator S. In particular, α is continuous.*

Sketch. Every nest algebra contains a maximal abelian von Neumann algebra (masa), say $\mathcal{A}$. Thus $\alpha(\mathcal{A})$ is maximal abelian in the image and so is closed. Moreover, $\|\alpha(A)\| \geq spr(\alpha(A)) = spr(A) = \|A\|$ for all A in $\mathcal{A}$. So $\alpha^{-1}|\alpha(\mathcal{A})$ is continuous, whence $\alpha|\mathcal{A}$ is continuous. The unitary group of $\mathcal{A}$ is abelian and thus amenable. So one can use an averaging argument to produce a similarity which makes $\alpha(\mathcal{A})$ self adjoint, hence a masa. The standard theory of abelian von Neumann algebras shows that the two isomorphic masas are unitarily equivalent. This reduces the problem to an automorphism of a nest algebra containing a fixed masa $\mathcal{A}$ such that $\alpha|\mathcal{A} = id$.

The second step of the argument is to characterize the rank one operators in $\mathcal{T}(\mathcal{N})$ algebraically. They are determined by the property: *For all A,B in* $\mathcal{T}(\mathcal{N})$, $ATB = 0$ *implies* $AT = 0$ *or* $TB = 0$. So algebra isomorphisms preserve rank one operators. Now some vector chasing produces the similarity that agrees with α on the finite rank operators. There are enough of them (Erdös Density Theorem) and they form an ideal, so one deduces that α agrees with the similarity on all of $\mathcal{T}(\mathcal{N})$. ∎

An automorphism α determines an automorphism of the nest. For if $\alpha = AdS$, then $\mathcal{T}(\mathcal{N}) = S\mathcal{T}(\mathcal{N})S^{-1} = \mathcal{T}(S\mathcal{N})$. So θ_S is a dimension preserving automorphism of $\mathcal{N}$. If $\theta_S = id$, then $SN = N$ for all N in $S\mathcal{N}$. Hence S belongs to $\mathcal{T}(\mathcal{N})^{-1}$ and α is inner. By the Similarity Theorem, every θ in $Aut(\mathcal{N})$ is implemented by an invertible operator. Define the outer automorphism group to be $Out\,\mathcal{T}(\mathcal{N}) = Aut\,\mathcal{T}(\mathcal{N})/Inn\,\mathcal{T}(\mathcal{N})$, the quotient of the automorphism group by the inner automorphism.

Corollary [24]. $Out\,\mathcal{T}(\mathcal{N}) \cong Aut(\mathcal{N})$.

Gilfeather and Moore [35] have generalized these results to show that every algebra isomorphism of CSL algebras is a continuous map between similar algebras. But not all such isomorphisms are spatial. This is related to problems about derivations and cohomology. Every deviation of $\mathcal{T}(\mathcal{N})$ into itself, or into a normal bimodule is inner [16], as is every derivation into the compact operators [17]. Indeed, all cohomology groups with coefficients in a normal bimodule are zero [45]. However, even finite dimensional CSL algebras have non-trivial derivations [34].

Finally, we explore a connection between CSL-algebras and harmonic analysis. The structure of weak* closed ideals in or even bimodules of a nest algebra were described by Erdos and Power [32] in a very simple way. Let $\mathcal{S}$ be a $\mathcal{T}(\mathcal{N})$ bimodule. For each N in $\mathcal{N}$, $\overline{SN} = \Phi_S(N)$ is invariant for $\mathcal{T}(\mathcal{N})$ and thus belongs to $\mathcal{N}$. Φ is increasing and left continuous. Conversely, given such a Φ,

$$S_\Phi = \{T \in B(\mathcal{H}) : TN \subseteq \Phi(N)\}$$

is a $\mathcal{T}(\mathcal{N})$ bimodule. In [32], it is shown that these are the only ones (i.e., $S = S_{\Phi_S}$).

This is not the case for arbitrary CSL algebras:

Theorem (Arveson): *Let $\mathcal{L}$ be a CSL lattice. There is a weak* closed algebra $\mathcal{A}_{\min}(\mathcal{L})$ with Lat $\mathcal{A}_{\min}(\mathcal{L}) = \mathcal{L}$ such that if $\mathcal{A}$ is a weak* closed algebra containing a masa with Lat $\mathcal{A} = \mathcal{L}$, then $\mathcal{A}_{\min}(\mathcal{L}) \subseteq \mathcal{A} \subseteq Alg(\mathcal{L})$.*

If $\mathcal{L}$ is finite width or completely distributive, then $\mathcal{A}_{\min}(\mathcal{L}) = Alg(\mathcal{L})$. A lattice with $\mathcal{A}_{\min}(\mathcal{L}) = Alg(\mathcal{L})$ is called *synthetic*. The reason for this was an example constructed by Arveson showing that the failure of spectral synthesis for S^2 in $\mathbb{R}^3$ lead to an example of a non-synthetic lattice. The connection has been made even more directly by Froelich [33]. Let (X,m) be a compact metric space with a regular Borel measure m. An operator T on $L^2(X,m)$ is said to have support $\Sigma \subseteq X \times X$ if the complement Σ^c is the union of all open rectangles $U \times V$ such that $P_U T P_V = 0$ where P_U is multiplication by χ_U. Let G be a locally compact abelian group and let E be a closed subset of G. Define an algebra $\mathcal{A}_E$ on $L^2(G,m) \oplus L^2(G,m)$, (m is Haar measure), as the set of all operators $\begin{pmatrix} M_1 & T \\ 0 & M_2 \end{pmatrix}$ such that M_i are multiplication operators in $L^\infty(G)$, and T is an operator supported in $\{(x,y): x-y \in E\}$.

Theorem (Froelich). *Lat $\mathcal{A}_E$ is synthetic if and only if E is a set of spectral synthesis in G.*

Froelich [33] has exploited this connection to construct a variety of CSL algebras with various pathologies carried over from pathology in harmonic analysis. At the present time, the operator theoretic side is bogged down by measure theoretic difficulties.

Problem 8. *Find an operator theoretic approach to the study of synthetic lattices that avoids measure theoretic technicalities.*

This question is admittedly vague. The test of a good new approach should be a clearer idea of which lattices are synthetic. For example, if $\mathcal{L}_1$ and $\mathcal{L}_2$ are commuting synthetic lattices, is $\mathcal{L}_1 \vee \mathcal{L}_2$ synthetic? I would hope that the flow of information from harmonic analysis could be reversed. If a result in operator theory leads to a new result in harmonic analysis, you will know that you are onto an interesting idea.

References

[1] Andersen, N.T., Compact perturbations of reflexive algebras, J. Func. Anal. 38(1980), 366-400

[2] Aronszajn, N, and Smith, K.T., Invariant subspaces of completely continuous operators, Ann. Math. 60(1954), 345-350.

[3] Arveson, W.G., Subalgebras of C*-algebras, Acta Math. 123 (1969), 141-224.

[4] Arveson, W.G., Operator algebras and invariant subspaces, Ann. Math. (2) 100(1974), 433-532.

[5] Arveson, W.G., Interpolation problems in nest algebras, J. Func. Anal. 3(1975), 208-233.

[6] Arveson, W.G., Perturbation theory for groups and lattices, J. Func. Anal. 53(1983), 22-73.

[7] Arveson, W.G., Ten lectues on operator algebras, CBMS series 55, Amer. Math. Soc., Providence 1983.

[8] Ball, J.A. and Gohberg, I., A commutant lifting theorem for triangular matrices with diverse applications, Int. Operators Operator Thy. 8(1985), 205-267.

[9] Barria, J., The invariant subspaces of a Volterra operator, J. Operator Thy. 6(1981), 341-349.

[10] Barria, J. and Davidson, K.R., Unicelluar operators, Trans. Amer. Math. Soc. 284(1984), 229-246.

[11] Brown, S., Some invariant subspaces for subnormal operators, J. Int. Equations Op. Theory 1(1978), 310-333.

[12] Choi, M.D. and Christensen, E., Completely order isomorphic and close C*-algebras need not be *-isomorphic, Bull. London Math. Soc. 15(1983), 604-610.

[13] Choi, M.D. and Davidson, K.R., Perturbations of matrix algebras, Mich. Math. J. 33(1986), 273-287.

[14] Christensen, E., Perturbations of type I von Neumann algebras, J. London Math. Soc. (2), 9(1975), 395-405.

[15] Christensen, E., Perturbations of operator algebras II, Indiana Math. J. 26(1977), 891-904.

[16] Christensen, E., Derivations of nest algebras, Math. Ann. 229 (1977), 155-161.

[17] Christensen, E. and Peligrad, C., Commutants of nest algebras modulo the compact operators, Inv. Math. 56(1980), 113-116.

[18] Davidson, K.R., Commutative Subspaces Lattices, Ind. Math. J. 27(1978), 479-490.

[19] Davidson, K.R., Similarity and Compact Perturbations of Nest Algebras, J. reine angew. Math. 348(1984), 72-87.

[20] Davidson, K.R., Perturbations of reflexive operator algebras, J. Operator Theory, 15(1986), 289-305.

[21] Davidson, K.R., Approximate unitary equivalences of continuous nests, Proc. Amer. Math. Soc. 97(1986), 655-660.

[22] Davidson, K.R., The distance to the analytic Toeplitz operators, Illinois J. Math., 31(1987), 265-273.

[23] Davidson, K.R. and Power, S.C., Failure of the distance formula, J. London Math. Soc. (2) 32(1984), 157-165.

[24] Davidson, K.R. and Wagner, B., Automorphisms of quasitriangular algebras, J. Func. Anal. 59(1984), 612-627. (M486c #47061)

[25] Davis, C., Kahan, W.M., and Weinberger, W.F., Norm preserving dilations and their applications to optimal error bounds, SIAM J. Numer. Anal. 19(1982), 445-469.

[26] Dixmier, J., Les opèrateurs permutables á l'opérateur integral, Fas. 2 Portugal, Math. 8(1949), 73-84.

[27] Domar, Y., Translation invariant subspaces and weighted ℓ^p and L^p spaces, Math. Scand. 49(1981), 133-144.

[28] Donoghue, W.F., The lattice of invariant subspaces of a completely continuous quasinilpotent transformation, Pacific J. Math. 7(1957), 1031-1035.

[29] Erdos, J.A., Unitary invariants for nests, Pacific J. Math. 23(1967), 229-256.

[30] Erdos, J.A., Operators of finite rank in nest algebras, J. London Math. Soc. 43(1968), 391-397.

[31] Erdos, J.A., On the trace of a trace class operator, Bull. London Math. Soc. 6(1974), 47-50.

[32] Erdos, J.A. and Power,S.C., Weakly closed ideals of nest algebras, J. Operator Thy. 7(1982), 219-235.

[33] Froelich, J., Compact operators, invariant subspaces and spectral synthesis, Ph.D. thesis, Univ. Iowa, 1984.

[34] Gilfeather, F., Derivations of certain CSL algebras, J. Operator Theory 11(1984), 145-156.

[35] Gilfeather, F. and Moore, R.L., Isomorphisms of certain CSL algebras, J. Func. Anal. 67(1986), 264-291.

[36] Gohberg, I.C. and Krein, M.G., Theory and applications of Volterra operators in Hilbert space, "Nauka", Moscow, 1967; English transl., Transl. Math. Monographs, 24, Amer. Math. Soc., Providence, R.I., 1970.

[37] Harrison, K.J. and Longstaff, W.E., A invariant subspace lattice of order type $\omega+\omega+1$, Proc. Amer. Math. Soc. 79(1980), 45-49.

[38] Hopenwasser, A., The radical of a reflexive operator algebra, Pacific J. Math. 65(1976), 375-392.

[39] Hopenwasser, A. and Larson, D.R., The carrier space of a reflexive operator algebra, Pacific J. Math. 81(1979), 417-434.

[40] Hopenwasser, A., Laurie, C., and Moore, R., Reflexive algebras with completely distributive subspace lattices, J. Operator Theory, 11(1984), 91-108.

[41] Kadison, R.V. and Singer, I.M., Triangular operator algebras, Amer. J. Math. 82(1960), 227-259.

[42] Kraus, J. and Larson, D.R., Some applications of a technique for constructing reflexive operator algebras, J. Operator Theory, 13(1985), 227-236.

[43] Kraus, J. and Larson, D.R., Reflexivity and distance formulae, J. London Math. Soc., to appear.

[44] Lance, E.C., Some properties of nest algebras, Proc. London Math. Soc. (3) 19(1969), 45-68.

[45] Lance, E.C., Cohomology and perturbations of nest algebras, Proc. London Math. Soc. (3) 43(1981), 334-356.

[46] Larson, D.R., Annihilators of operator algebras, Topics in Modern Operator Theory 6, p.119-130, Birhauser Verlag Basel 1982.

[47] Larson, D.R., A solution to a problem of J.R. Ringrose, Bull. Amer. Math. Soc. 7(1982), 243-246.

[48] Larson, D.R., Nest algebras and similarity transformations, Ann. Math. 121(1985), 409-427.

[49] Larson, D.R., Hyperreflexivity and a dual product construction, Trans. Amer. Math. Soc.

[50] Laurie, C. and Longstaff, W., A note on rank one operators in reflexive algebras, Proc. Amer. Math. Soc. 89(1983), 293-297.

[51] Lidskii, V.B., Nonselfadjoint operators with trace, (Russian) Dokl. Akad. Nauk SSR 125(1959), 485-487; English transl., Transl. Amer. Math. Soc. (2) 47(1965), 43-46.

[52] Longstaff, W.E., Strongly reflexive lattices, J. London Math. Soc. (2) 11(1975), 491-498.

[53] Longstaff, W.E., Generators of reflexive algebras, J. Australian Math. Soc. ser.A 20(1975), 159-164.

[54] McAsey, M.J. and Muhly, P.S., Representations of nonselfadjoint crossed products, Proc. London Math. Soc. (3) 47(1983), 128-144.

[55] Olin, R.F. and Thomson, J.E., Algebras of subnormal operators, J. Func. Anal. 37(1980), 271-301.

[56] Parrott, S.K., On a quotient norm and the Sz-Nagy-Foias lifting theorem, J. Func. Anal. 30(1978), 311-328.

[57] Paulsen, V.I. and Power, S.C., Lifting theorems for nest algebra, preprint.

[58] Paulsen, V.I., Power, S.C. and Ward, J., Semidiscreteness and dilation theory for nest algebras, J. Func. Anal., to appear.

[59] Pitts, D.R., On the K_o groups of nest algebras, preprint.

[60] Power, S.C., The distance to upper triangular operators, Math. Proc. Camb. Phil. Soc. 88(1980), 327-329.

[61] Power, S.C., Another proof of Lidskii's theorem on the trace, Bull. London Math. Soc. 15(1983), 146-148.

[62] Power, S.C., Analysis in nest algebras, Surveys of Recent Results in Operator Theory, Editor J. Conway, Pitman Research Notes in Mathematics, Longman, to appear.

[63] Radjavi, H. and Rosenthal, P., On invariant subspaces and reflexive algebras, Amer. J. Math. 91(1969), 683-692.

[64] Raney, G.N., Completely distributive lattices, Proc. Amer. Math. Soc. 3(1952), 677-680.

[65] Ringrose, J.R., Superdiagonal forms for compact linear operators, Proc. London Math. Soc. (3) 12(1962), 367-384.

[66] Ringrose, J.R., On some algebras of operators, Proc. London Math. Soc. (3) 95(1965), 61-83.

[67] Ringrose, J.R., On some algebras of operators II, Proc. London Math. Soc. (3) 16(1966), 385-402.

[68] Rosenoer, S., Distance estimates for von Neumann algebras, Proc. Amer. Math. soc. 86(1982), 248-252.

[69] Sarason, D., On spectral sets having connected complement, Acta Sci. Math. (Szeged) 26(1965), 289-299.

[70] Wagner, B., Automorphisms and derivations of certain operator algebras, Ph.D. dissertation, U. Calif., Berkeley, 1982.

Some Notes on Non-commutative Analysis

Richard V. Kadison

Dedicated to the memory of Michel Sirugue

Abstract.

An unbounded non-commutative monotone convergence result (Theorem 9) is proved. A new derivation of the Friedrichs extension is given. The basics of the Takesaki cones are studied.

1. Introduction.

We study certain aspects of the theory of von Neumann algebras that emphasize its interpretation as non-commutative measure theory. In this interpretation, which is direct and unmistakable, the projections in the algebra are the characteristic functions of the (non-commuting) measurable sets (which do not appear!), the elements of $\mathcal{R}$ are the bounded measurable functions, and the (normal) states of $\mathcal{R}$ are the probability measures on the underlying (non-commutative) measure space (which, again, does not appear). An important result, in the early stages of the theory, states that if $\{A_n\}$ is a monotone increasing sequence of self-adjoint operators, bounded above (by some multiple of the identity operator I), then there is a bounded self-adjoint operator A such that $A_n x \to Ax$ for each x in the underlying Hilbert space. If each $A_n \in \mathcal{R}$, then, of course, $A \in \mathcal{R}$ (cf. [6; Lemma 5.1.4]). This is a primitive non-commutative monotone convergence theorem. In Section 3, the restriction that the sequence $\{A_n\}$ be bounded above is removed; the limit A is now an appropriate unbounded self-adjoint operator affiliated with $\mathcal{R}$. (We write $A \; \eta \; \mathcal{R}$ to indicate this affiliation.) The extension of the classical bounded non-commutative monotone convergence result to this unbounded version (Theorem 9) seems not to be routine. A simpler unbounded monotone convergence assertion (Proposition 7),that assumes the presence of a "well-placed" separating vector for $\mathcal{R}$, is also proved in Section 3.

The Friedrichs extension is a self-adjoint extension of a positive closed (densely defined) symmetric operator [4]. Subject to a certain domain condition, this extension is unique. The Friedrichs extension of a symmetric operator affiliated with a von Neumann algebra is also affiliated with that algebra, as follows from the uniqueness. This extension plays a somewhat hidden, but important, role in the theory of von Neumann algebras. In the first of the monumental series of papers by Murray and von Neumann, it supplies the basic ingredient of a crucial comparison result (cf. [7; Lemma 9.3.3]). the Friedrichs extension is vital in establishing the basic properties of the Takesaki cones [8; pp. 101–106].

This research was carried out with partial support of NSF (USA).

A new proof for the existence of the Friedrichs extension is given in Section 2. In Section 4, the Friedrichs extension is used to give a readily accessible account of the fundamental properties of the Takesaki cones. The view of these cone properties as a broad non-commutative Radon-Nikodým theorem is described.

2. The Friedrichs Extension.

Throughout this section, A_0 is a closed linear operator with domain $\mathcal{D}(A_0)$ dense in a Hilbert space $\mathcal{H}$ and $0 \le \langle A_0 x, x\rangle$ for each x in $\mathcal{D}(A_0)$. The lemma that follows is needed primarily for establishing the uniqueness of the Freidrichs extension. The conditions and relations that the extension must fulfill are explored in this lemma, and that points the way to defining the extension.

Lemma 1. *If A is a positive self-adjoint extension of A_0, then $A+I$ is one-to-one with range $\mathcal{H}$. The inverse B to $A+I$ is a positive operator, everywhere defined, $\|B\| \le 1$, and*

$$(*) \qquad \langle x, y\rangle = \langle (A_0+I)x, By\rangle \qquad (x \in \mathcal{D}(A_0), y \in \mathcal{H}).$$

The range of B is contained in $\mathcal{D}(A_0^*)$.

Proof. Since $\langle A_0 x, x\rangle$ is real for each x in $\mathcal{D}(A_0)$, polarization (cf. [6; Prop. 2.1.7 and 2.4 (3)]) yields that $\langle A_0 x, y\rangle = \langle x, A_0 y\rangle$, when x and y are in $\mathcal{D}(A_0)$. Hence $A_0 \subseteq A_0^*$, that is, A_0 is symmetric. As $\langle (A_0+I)x, x\rangle \ge 0$ for each x in $\mathcal{D}(A_0)$ $(= \mathcal{D}(A_0+I))$, A_0+I is symmetric. In addition, A_0+I is closed, since A_0 is closed. By the same token, $A+I$ is closed and thus, self-adjoint. Moreover,

$$\|(A+I)x\|\,\|x\| \ge \langle (A+I)x, x\rangle = \langle Ax, x\rangle + \|x\|^2 \ge \|x\|^2 \ge 0$$

for each x in $\mathcal{D}(A)$ $(= \mathcal{D}(A+I))$. Thus $A+I$ is a positive self-adjoint operator with null space (0), and $\|x\| \le \|(A+I)x\|$ for each x in $\mathcal{D}(A)$. As the closure of the range of $A+I$ $(= (A+I)^*)$ is the orthogonal complement of the null space of $A+I$ (cf. [6; Exercise 2.8.45]), $A+I$ has range dense in $\mathcal{H}$.

If $\{x_n\}$ is a sequence in $\mathcal{D}(A)$ such that $\{(A+I)x_n\}$ tends to y, then $\|x_n - x_m\| \le \|(A+I)(x_n - x_m)\|$, and $\{x_n\}$ is a Cauchy sequence in $\mathcal{H}$. It follows that $\{x_n\}$ converges to x. Since $A+I$ is closed, $x \in \mathcal{D}(A+I)$ and $(A+I)x = y$. Hence A+I has a closed range. From our earlier conclusion, this range is dense. Thus $A+I$ has range $\mathcal{H}$.

If B is the mapping inverse to $A+I$ and $y = (A+I)x$, then

$$0 \le \|By\|^2 = \|x\|^2 \le \langle x, (A+I)x\rangle = \langle By, y\rangle \le \|By\|\,\|y\|.$$

Thus $\|B\| \le 1$ and $B \ge 0$. For each x in $\mathcal{D}(A_0+I)$, $B(A_0+I)x = B(A+I)x = x$. Hence, with y in $\mathcal{H}$,

$$\langle x, y\rangle = \langle B(A_0+I)x, y\rangle = \langle (A_0+I)x, By\rangle.$$

It follows that $By \in \mathcal{D}((A_0+I)^*)$ and $(A_0+I)^* By = y$. Since $(A_0+I)^* = A_0^* + I$ (more generally, $(T+S)^* = T^* + S^*$ when S is bounded), $By \in \mathcal{D}(A_0^*)$. ∎

For the lemmas that follow, we define a positive definite inner product on $\mathcal{D}(A_0)$ by

$$\langle u, v\rangle' = \langle (A_0 + I)u, v\rangle$$

and denote by $\mathcal{D}'$ the completion of $\mathcal{D}(A_0)$ relative to this inner product.

Lemma 2. *The identity mapping of $\mathcal{D}(A_0)$ onto itself has a (unique) bounded extension ι mapping $\mathcal{D}'$ into $\mathcal{H}$, ι is one-to-one, and $\|\iota\| \le 1$. For each y in $\mathcal{H}$, the functional $x \to \langle x, y\rangle$ on $\mathcal{D}(A_0)$ extends to a bounded linear functional on $\mathcal{D}'$ whose norm does not exceed $\|y\|$. There is a (unique) vector By in $\mathcal{D}(A_0^*)$ and in $\iota(\mathcal{D}')$ satisfying*

$$\langle x, y\rangle = \langle (A_0 + I)x, By\rangle \qquad (x \in \mathcal{D}(A_0))$$

$$\langle x, y\rangle = \langle x, \iota^{-1}(By)\rangle' \qquad (x \in \mathcal{D}(A_0)).$$

Proof. With x in $\mathcal{D}(A_0)$,

$$\|x\|^2 = \langle x, x\rangle \le \langle x, x\rangle + \langle A_0 x, x\rangle = \langle x, x\rangle' = \|x\|'^2.$$

Thus the identity mapping of $\mathcal{D}(A_0)$ onto itself has a (unique) bounded extension ι mapping $\mathcal{D}'$ into $\mathcal{H}$ and $\|\iota\| \le 1$. To see that ι is one-to-one, choose x_n in $\mathcal{D}(A_0)$ tending to z' in $\mathcal{D}'$ and note that

$$\|x_n - \iota(z')\| = \|\iota(x_n) - \iota(z')\| \le \|x_n - z'\|' \to 0,$$

whence $\|x_n\| \to 0$ when $\iota(z') = 0$. Thus, for each m,

$$\begin{aligned}\langle z', x_m\rangle' &= \lim_n \langle x_n, x_m\rangle' \\ &= \lim_n \langle (A_0 + I)x_n, x_m\rangle = \lim_n \langle x_n, (A_0 + I)^* x_m\rangle = 0,\end{aligned}$$

since $x_m \in \mathcal{D}(A_0 + I) \subseteq \mathcal{D}((A_0 + I)^*)$. But,

$$\langle z', z'\rangle' = \lim_m \langle z', x_m\rangle' = 0,$$

whence $z' = 0$ and ι is one-to-one.

Since $|\langle x, y\rangle| \le \|x\| \, \|y\| \le \|x\|' \, \|y\| \quad (x \in \mathcal{D}(A_0), y \in \mathcal{H})$ (as we have just shown), the functional $x \to \langle x, y\rangle$ on $\mathcal{D}(A_0)$ has bound not exceeding $\|y\|$ relative to the norm $x \to \|x\|'$. This functional extends (uniquely) to a linear functional of norm not exceeding $\|y\|$ on $\mathcal{D}'$. From this and Riesz's representation of functionals on Hilbert space, there is a (unique) vector z' in $\mathcal{D}'$ such that $\langle x, y\rangle = \langle x, z'\rangle'$ for each x in $\mathcal{D}(A_0)$. Let By be $\iota(z')$. We can choose x_n in $\mathcal{D}(A_0)$ tending to z' in $\mathcal{D}'$. Then, as before,

$$\|x_n - \iota(z')\| = \|\iota(x_n) - \iota(z')\| \le \|x_n - z'\|' \to 0.$$

It follows that, for each x in $\mathcal{D}(A_0)$,

$$\begin{aligned}\langle x, y\rangle = \langle x, z'\rangle' &= \lim_n \langle x, x_n\rangle' = \lim_n \langle (A_0 + I)x, x_n\rangle \\ &= \langle (A_0 + I)x, \iota(z')\rangle = \langle (A_0 + I)x, By\rangle,\end{aligned}$$

whence $By \in \mathcal{D}((A_0 + I)^*) = \mathcal{D}(A_0^*)$ (cf. the proof of Lemma 1 where it is noted that $(A_0 + I)^* = A_0^* + I$). At the same time, $\iota^{-1}(By) = z'$, so that $\langle x, y\rangle = \langle x, \iota^{-1}(By)\rangle'$. ∎

Lemma 3. *With B as in Lemma 2, $B \in \mathcal{B}(\mathcal{H})$, $B \geq 0$, $\|B\| \leq 1$, and B is one-to-one.*

Proof. Choose y in $\mathcal{H}$ and let z' be the (unique) vector in $\mathcal{D}'$ found in the proof of Lemma 2, such that $\langle x, y\rangle = \langle x, z'\rangle'$ for each x in $\mathcal{D}(A_0)$ and in terms of which By was defined to be $\iota(z')$. Then

$$\|By\| = \|\iota(z')\| \leq \|z'\| \leq \|y\|.$$

Hence $\|B\| \leq 1$. Choose x_n in $\mathcal{D}(A_0)$ tending to z' in $\mathcal{D}'$. From Lemma 2,

$$\|x_n - By\| = \|\iota(x_n - z')\| \leq \|x_n - z'\|' \to 0$$

and

$$\begin{aligned}\langle By, y\rangle &= \lim_n \langle x_n, y\rangle = \lim_n \langle x_n, z'\rangle' \\ &= \lim_n \langle x_n, \iota^{-1}(By)\rangle' = \langle z', \iota^{-1}(By)\rangle' = \|\iota^{-1}(By)\|'^2 \geq 0.\end{aligned}$$

Thus $B \geq 0$.

If y is a non-zero element of $\mathcal{H}$, then for some x in (the dense linear manifold) $\mathcal{D}(A_0)$,

$$0 \neq \langle x, y\rangle = \langle x, \iota^{-1}(By)\rangle'.$$

Hence $\iota^{-1}(By) \neq 0$. From Lemma 2, ι is one-to-one, whence $By \neq 0$, and B is one-to-one. ∎

We use the notation ι and B, in the theorem that follows, to describe the Friedrichs extension.

Theorem 4. *If A_1 is the mapping inverse to B, then $A_1 - I$ $(= A)$ is a positive self-adjoint extension (the Friedrichs extension) of A_0, and $\mathcal{D}(A) \subseteq \iota(\mathcal{D}')$. Moreover, A is the unique positive self-adjoint extension of A_0 satisfying $\mathcal{D}(A) \subseteq \iota(\mathcal{D}')$.*

Proof. Suppose $x = Bu$ and $y = Bv$. Then $x \in \mathcal{D}(A_1) = \mathcal{D}(A)$, and

$$\begin{aligned}\langle Ax, y\rangle &= \langle (A_1 - I)x, y\rangle = \langle u - x, Bv\rangle \\ &= \langle B(u - x), v\rangle = \langle x - Bx, v\rangle \\ &= \langle x, (I - B)v\rangle = \langle x, v - y\rangle \\ &= \langle x, (A_1 - I)y\rangle = \langle x, Ay\rangle.\end{aligned}$$

Thus $y \in \mathcal{D}(A^*)$ and $A^*y = Ay$. It follows that $A \subseteq A^*$.

With v in $\mathcal{H}$, let $x = Bv$. If $z \in \mathcal{D}(A^*)$, then

$$\begin{aligned}\langle v, BA^*z\rangle &= \langle Bv, A^*z\rangle = \langle x, A^*z\rangle \\ &= \langle Ax, z\rangle = \langle (A_1 - I)Bv, z\rangle \\ &= \langle v - Bv, z\rangle = \langle v, (I - B)z\rangle.\end{aligned}$$

Since this equality holds for all v in $\mathcal{H}$,

$$(I - B)z = BA^*z.$$

Thus $z = B(I + A^*)z$, z is in the range of B, and $z \in \mathcal{D}(A)$. Thus $A = A^*$.

If x and u are in $\mathcal{D}(A_0)$, then from Lemmas 2 and 3,

$$\langle u, x\rangle' = \langle (A_0 + I)u, x\rangle = \langle u, (A_0 + I)x\rangle = \langle u, \iota^{-1}(B(A_0 + I)x)\rangle'.$$

Since $\mathcal{D}(A_0)$ is dense in $\mathcal{D}'$, $x = \iota^{-1}(B(A_0 + I)x)$, whence

$$x = \iota(x) = B(A_0 + I)x.$$

Hence $x \in \mathcal{D}(A_1)$ and $A_1 x = (A_0 + I)x$. Thus $Ax = (A_1 - I)x = A_0 x$, and A is a self-adjoint extension of A_0.

The range of B is $\mathcal{D}(A_1)$ $(= \mathcal{D}(A))$. From Lemma 2, the range of B is contained in $\iota(\mathcal{D}')$. Thus $\mathcal{D}(A) \subseteq \iota(\mathcal{D}')$. For each y in $\mathcal{H}$,

$$\langle ABy, By\rangle = \langle (A_1 - I)By, By\rangle = \langle (I - B)y, By\rangle \geq 0,$$

since $I - B$ and B are positive, commuting, bounded operators on $\mathcal{H}$. Thus A is positive.

The restrictions on (that is "properties of") a positive self-adjoint extension of A_0 are noted in Lemma 1. Suppose now that A' is such an extension and that $\mathcal{D}(A') \subseteq \iota(\mathcal{D}')$. Let B' be the operator arising from A' with the properties corresponding to those of B in Lemma 1. Then

$$\langle (A_0 + I)x, (B - B')y\rangle = 0 \qquad (x \in \mathcal{D}(A_0), y \in \mathcal{H}).$$

Since $\mathcal{D}(A') \subseteq \iota(\mathcal{D}')$, there is a vector u' in $\mathcal{D}'$ such that $\iota(u') = (B - B')y$. Let $\{x_n\}$ be a sequence of vectors in $\mathcal{D}(A_0)$ tending to u' (in $\mathcal{D}'$). Then

$$\|x_m - (B - B')y\| = \|x_m - \iota(u')\| \leq \|x_m - u'\|' \to 0$$

and

$$\begin{aligned}\langle x_n, u'\rangle' &= \lim_m \langle x_n, x_m\rangle' \\ &= \lim_m \langle (A_0 + I)x_n, x_m\rangle \\ &= \langle (A_0 + I)x_n, (B - B')y\rangle \\ &= 0.\end{aligned}$$

Hence $\langle u', u'\rangle' = \lim_n \langle x_n, u'\rangle' = 0$, and $u' = 0$. It follows that $\iota(u') = (B - B')y = 0$ and $B = B'$. Since B and B' are the mappings inverse to A and A', respectively, $A = A'$. ∎

Corollary 5. *Let $\mathcal{R}$ be a von Neumann algebra acting on a Hilbert space $\mathcal{H}$ and A_0 be a closed, densely defined, symmetric operator affiliated with $\mathcal{R}$. Suppose $\langle A_0 x, x\rangle \geq 0$ for each x in $\mathcal{D}(A_0)$ and A is the Friedrichs extension of A_0. Then $A \,\eta\, \mathcal{R}$.*

Proof. Let V' be a unitary operator in $\mathcal{R}'$. Then $V'AV'^*$ is a positive self-adjoint extension of $V'A_0V'^*$ and $\mathcal{D}(V'AV'^*) \subseteq V'(\iota(\mathcal{D}'))$. Since $A_0 \,\eta\, \mathcal{R}$, $V'A_0V'^* = A_0$. From uniqueness of the Friedrichs extension, it remains to show that $V'(\iota(\mathcal{D}')) \subseteq \iota(\mathcal{D}')$.

Suppose $z \in \iota(\mathcal{D}')$ and $\iota(z') = z$ (with z' in $\mathcal{D}'$). Then $\{x_n\}$ tends to z' for some sequence $\{x_n\}$ in $\mathcal{D}(A_0)$. Since $A_0 \,\eta\, \mathcal{R}$, $V'(\mathcal{D}(A_0)) = \mathcal{D}(A_0)$ and $V'x_n \in \mathcal{D}(A_0)$. Now

$$\begin{aligned}\|V'x_n - V'x_m\|'^2 &= \langle (A_0 + I)V'(x_n - x_m), V'(x_n - x_m)\rangle \\ &= \langle (A_0 + I)(x_n - x_m), (x_n - x_m)\rangle \\ &= \|x_n - x_m\|'^2 \to 0\end{aligned}$$

as $m, n \to \infty$ since $\{x_n\}$ converges in $\mathcal{D}'$. Thus $\{V'x_n\}$ converges in $\mathcal{D}'$ to some u' and $\{V'x_n\}$ converges in $\mathcal{H}$ to $\iota(u')$. Since $\{x_n\}$ tends to z in $\mathcal{H}$, $\{V'x_n\}$ tends to $V'z$ in $\mathcal{H}$. Thus $V'z = \iota(u') \in \iota(\mathcal{D}')$ and $V'(\iota(\mathcal{D}')) \subseteq \iota(\mathcal{D}')$. ∎

3. Monotone Convergence.

We prove an unbounded non-commutative monotone convergence result (Theorem 9) and use it to give a proof of the Murray-von Neumann "BT-Lemma" [7; Lemma 9.2.1] (cf. [6; Theorem 7.2.1']). A more easily proven unbounded monotone convergence result, with the assumption of a separating vector, is found in Proposition 7 and Corollary 8. The lemma that follows, blending weak and norm convergence of nets of vectors in a Hilbert space, will be useful throughout this section.

Lemma 6. *Suppose $\{y_a\}_{a\in\mathsf{A}}$ is a bounded net of vectors in a Hilbert space $\mathcal{H}$.*

(i) Suppose $\{x_b\}_{b\in\mathsf{B}}$ converges in norm to a vector x and y is a vector in $\mathcal{H}$ such that $\lim_a\langle y_a, x_b\rangle = \langle y, x_b\rangle$ for each b in B. Then $\lim_a\langle y_a, x\rangle = \langle y, x\rangle$ and $\lim_{a,b}\langle y_a, x_b\rangle = \langle y, x\rangle$.

(ii) Suppose $\mathcal{D}$ is a dense linear submanifold of $\mathcal{H}$ such that $\lim_a\langle y_a, x\rangle$ converges for each x in $\mathcal{D}$. Then $\{y_a\}$ converges weakly to some y in $\mathcal{H}$.

Proof. (i) Choose k such that $\|y\| \le k$ and $\|y_a\| < k$ for each a in A. Given a positive ϵ, choose b' in B such that $\|x - x_b\| < \epsilon/6k$ when $b \ge b'$. Now choose a' in A such that $|\langle y_a - y, x_{b'}\rangle| < \epsilon/6$ when $a \ge a'$. Then

$$\begin{aligned}|\langle y_a, x\rangle - \langle y, x\rangle| &\le |\langle y_a, x - x_{b'}\rangle| + |\langle y_a - y, x_{b'}\rangle| + |\langle y, x_{b'} - x\rangle| \\ &\le \|y_a\|\epsilon/6k + \epsilon/6 + \|y\|\epsilon/6k < \epsilon/2\end{aligned}$$

when $a \ge a'$. Thus $\lim_a\langle y_a, x\rangle = \langle y, x\rangle$. At the same time, when $a \ge a'$ and $b \ge b'$,

$$\begin{aligned}|\langle y_a, x_b\rangle - \langle y, x\rangle| &\le |\langle y_a, x_b\rangle - \langle y_a, x\rangle| + |\langle y_a, x\rangle - \langle y, x\rangle| \\ &< \|y_a\|\epsilon/6k + \epsilon/2 < \epsilon.\end{aligned}$$

Thus $\lim_{a,b}\langle y_a, x_b\rangle = \langle y, x\rangle$.

(ii) The mapping $x \to \lim_a\langle y_a, x\rangle$ is a conjugate-linear functional on $\mathcal{D}$. Since $|\langle y_a, x\rangle| \le \|y_a\|\,\|x\|$ and $\{\|y_a\|\}_{a\in\mathsf{A}}$ is bounded, this functional is bounded and extends, without change of norm, to a bounded conjugate-linear functional on $\mathcal{H}$. From Riesz's representation of such functionals, there is a vector y in $\mathcal{H}$ such that $\lim_a\langle y_a, x\rangle = \langle y, x\rangle$ for each x in $\mathcal{D}$. With z in $\mathcal{H}$, there is a sequence $\{x_n\}$ in $\mathcal{D}$ that converges in norm to z. Since $\lim_a\langle y_a, x_n\rangle = \langle y, x_n\rangle$, the condition (i) is satisfied and $\lim_a\langle y_a, z\rangle = \langle y, z\rangle$. Thus $\{y_a\}_{a\in\mathcal{A}}$ converges weakly to y. ∎

For the purposes of the following proposition, we define $H \le K$ for positive symmetric operators H and K to mean that $\mathcal{D}(K) \subseteq \mathcal{D}(H)$ and $\langle Hx, x\rangle \le \langle Kx, x\rangle$ for each x in $\mathcal{D}(K)$.

Proposition 7. *Let $\{H_a\}_{a\in\mathbb{A}}$ be a monotone increasing net of positive symmetric operators affiliated with a von Neumann algebra $\mathcal{R}$ acting on a Hilbert space $\mathcal{H}$ and x_0 be a separating unit vector for $\mathcal{R}$ in the domain of each H_a. If either of the following two conditions is satisfied,*

(i) $\{H_a x_0\}_{a\in\mathbb{A}}$ converges weakly in $\mathcal{H}$ to some vector ($\{H_a\}_{a\in\mathbb{A}}$ need not be required to be increasing in this case),

(ii) the net $\{\|H_a x_0\|\}_{a\in\mathbb{A}}$ is bounded,

then there is a positive self-adjoint operator H affiliated with $\mathcal{R}$ such that $\{H_a T' x_0\}_{a\in\mathbb{A}}$ converges weakly to $HT'x_0$ for each T' in $\mathcal{R}'$.

Proof. (i) By assumption $\{H_a x_0\}_{a\in\mathbb{A}}$ converges weakly to some vector $H_0 x_0$ in $\mathcal{H}$. Since bounded operators on $\mathcal{H}$ are continuous on $\mathcal{H}$ in its weak topology, $\{T' H_a x_0\}_{a\in\mathbb{A}}$ $(= \{H_a T' x_0\}_{a\in\mathbb{A}})$ converges to the vector $T' H_0 x_0$ in $\mathcal{H}$. We define $H_0 T' x_0$ to be $T' H_0 x_0$. Then H_0 is linear with (dense) domain $\mathcal{R}' x_0$. Moreover,

$$\langle H_0 T' x_0, T' x_0\rangle = \lim_a \langle H_a T' x_0, T' x_0\rangle \ge 0$$

since each H_a is positive. It follows that H_0 is symmetric and, therefore, has a closure H_1 that is positive. With V' a unitary operator in $\mathcal{R}'$, $H_1 V' T' x_0 = V' T' H_0 x_0 = V' H_1 T' x_0$. From [6; Remark 5.6.3], $H_1 \,\eta\, \mathcal{R}$ since $\mathcal{R}' x_0$ is a core for H_1. The Friedrichs extension H of H_1 is affiliated with $\mathcal{R}$ (Corollary 5) and has the properties required.

(ii) Suppose k is a bound for the net $\{\|H_a x_0\|\}_{a\in\mathbb{A}}$. Since the ball of radius k with centre 0 is weakly compact in $\mathcal{H}$, some cofinal subnet of $\{H_a x_0\}_{a\in\mathbb{A}}$ converges weakly to a vector in that ball. From (i), there is a positive self-adjoint H affiliated with $\mathcal{R}$ such that that cofinal subnet of $\{H_a T' x_0\}$ converges weakly to $HT' x_0$ for each T' in $\mathcal{R}'$. Since $\{H_a\}_{a\in\mathbb{A}}$ is monotone and that subnet is cofinal, $\{\langle H_a T' x_0, T' x_0\rangle\}_{a\in\mathbb{A}}$ converges to $\langle HT' x_0, T' x_0\rangle$ (over $\mathbb{A}$) for each T' in $\mathcal{R}'$. Polarizing, we have that $\{\langle H_a T' x_0, S' x_0\rangle\}_{a\in\mathbb{A}}$ converges to $\langle HT' x_0, S' x_0\rangle$ for all T' and S' in $\mathcal{R}'$. From Lemma 6 (ii), $\{H_a T' x_0\}_{a\in\mathbb{A}}$ converges weakly to some vector y in $\mathcal{H}$. Thus $\langle y - HT' x_0, S' x_0\rangle = 0$ for all S' in $\mathcal{R}'$. Since $\mathcal{R}' x_0$ is dense in $\mathcal{H}$, $y = HT' x_0$. Thus $\{H_a T' x_0\}_{a\in\mathbb{A}}$ converges weakly to $HT' x_0$ for each T' in $\mathcal{R}'$. ∎

Corollary 8. *Let $\{H_n\}$ be a sequence of positive symmetric operators affiliated with a von Neumann algebra $\mathcal{R}$ acting on a Hilbert space $\mathcal{H}$ and x_0 be a separating unit vector for $\mathcal{R}$ in the domain of each H_n^2. Suppose $\{H_n^2\}$ is monotone increasing. If some subsequence of $\{H_n x_0\}$ converges weakly, then there is a positive self-adjoint operator H affiliated with $\mathcal{R}$ such that $\{H_n T' x_0\}$ converges weakly to $HT' x_0$ for each T' in $\mathcal{R}$.*

Proof. The weakly convergent subsequence of $\{H_n x_0\}$ is bounded from [6; Theorem 1.8.10]. If $n \le m$ then $\|H_n x_0\|^2 = \langle H_n^2 x_0, x_0\rangle \le \langle H_m^2 x_0, x_0\rangle = \|H_m x_0\|^2$. Thus $\{H_n x_0\}$ is bounded, (ii) of Proposition 7 applies and completes the argument. ∎

Theorem 9. *Suppose $\{H_a\}_{a\in\mathbb{A}}$ is a monotone increasing net of operators in a von Neumann algebra $\mathcal{R}$ acting on a Hilbert space $\mathcal{H}$. Let $\mathcal{F}$ be the family of vectors x in $\mathcal{H}$ such that $\{\langle H_a x, x\rangle : a \ge a_0\}$ is bounded for some a_0 in $\mathbb{A}$ and E be the projection with range $[\mathcal{F}]$. Then $E \in \mathcal{R}$, and there is a self-adjoint operator H affiliated with $\mathcal{R}$ such that $EH \subseteq HE$, $H(I-E) = 0$, $\mathcal{D}(H) \ominus (I-E)(\mathcal{H}) \subseteq \mathcal{F}$, and $\{\langle H_a x, x\rangle\}_{a\in\mathbb{A}}$ converges to $\langle Hx, x\rangle$ for each x in $\mathcal{D}(H) \cap \mathcal{F}$.*

Proof. We may replace $\{H_a\}_{a\in A}$ by $\{H_a\}_{a\ge a'}$ for some a' in $\mathbb{A}$. Since $-\|H_{a'}\|I \le H_{a'}$, we may assume that $rI \le H_a$ for each a in $\mathbb{A}$ and some (fixed) real number r. Let K_a be $H_a + (1-r)I$. Then $\{K_a\}$ is monotone increasing, $I \le K_a$, and $x \in \mathcal{F}$ if and only if $\{\langle K_a x, x\rangle\}_{a\in\mathbb{A}}$ is bounded. If we find a self-adjoint operator K affiliated with $\mathcal{R}$ such that $EK \subseteq KE$, $K(I-E) = 0$, $\mathcal{D}(K) \ominus (I-E)(\mathcal{H}) \subseteq \mathcal{F}$, and $\{\langle K_a x, x\rangle\}$ converges to $\langle Kx, x\rangle$ for each x in $\mathcal{D}(K) \cap \mathcal{F}$, then $K - (1-r)E$ will serve as the required H.

Since $\langle H_a U'x, U'x\rangle = \langle U'^* U' H_a x, x\rangle$ for each unitary operator U' in $\mathcal{R}'$, $U'x \in \mathcal{F}$ when $x \in \mathcal{F}$. Thus $E \in \mathcal{R}'' = \mathcal{R}$. From [6; Proposition 4.2.8], $\{K_a^{-1}\}_{a\in\mathbb{A}}$ is monotone decreasing. Now $0 \le K_a^{-1}$, whence $\{K_a^{-1}\}_{a\in\mathbb{A}}$ converges to its greatest lower bound S in the strong-operator topology. We have that $S \in \mathcal{R}$ and $0 \le S \le I$. Since $I \le K_b \le K_a$ when $b \le a$, we have that $0 \le K_a^{-1/2} K_b K_a^{-1/2} \le I$. As $\{K_a^{-1/2} K_b K_a^{-1/2}\}_{a\in\mathbb{A}}$ tends to $S^{1/2} K_b S^{1/2}$ in the strong-operator topology, we have that $0 \le S^{1/2} K_b S^{1/2} \le I$ and $\|K_b^{1/2} S^{1/2}\| \le 1$ for each b in $\mathbb{A}$. (See [6; Remark 2.5.10, Proposition 5.3.2].) It follows that $\|K_a^{1/2} S^{1/2} z\| \le \|z\|$ for each z in $\mathcal{H}$. Thus $S^{1/2}z$ and Sz $(= S^{1/2}S^{1/2}z)$ are in $\mathcal{F}$, so that $F \le E$, where F is the range projection of S.

Suppose $Fy = 0$. Then $Sy = 0$, since $S = S^*$, and $S^{1/2}y = 0$. From [6; Proposition 4.2.8], $\{K_a^{-1/2}\}_{a\in\mathbb{A}}$ is monotone decreasing with strong-operator limit $S^{1/2}$. Hence $\{\|K_a^{-1/2}y\|\}_{a\in\mathbb{A}}$ tends to 0. Let z_a be $K_a^{-1/2}y$ so that $y = K_a^{1/2} z_a$. With x in $\mathcal{F}$, we have that

$$|\langle y, x\rangle| = |\langle K_a^{\frac{1}{2}} z_a, x\rangle| = |\langle z_a, K_a^{\frac{1}{2}} x\rangle| \le \|z_a\|\,\|K_a^{\frac{1}{2}} x\| \underset{a}{\rightarrow} 0$$

since $\{\|K_a^{1/2}x\|\}_{a\in\mathbb{A}}$ $(= \{\langle K_a x, x\rangle^{1/2}\}_{a\in\mathbb{A}})$ is bounded and $\|z_a\| \to 0$. Thus y is orthogonal to $\mathcal{F}$, and $I - F \le I - E$. Combining this with the inequality $F \le E$, established in the preceding paragraph, we conclude that $F = E$.

Define $K(Sx+y)$ to be x when $x \in E(\mathcal{H})$ and $y \in (I-E)(\mathcal{H})$. Since S is one-to-one on $E(\mathcal{H})$, K is well defined. The range of S is dense in $E(\mathcal{H})$, whence K is densely defined. Moreover,

$$\langle K(Sx+y), Sx+y\rangle = \langle x, Sx+y\rangle = \langle x, Sx\rangle \ge 0.$$

As in the proof of Lemma 1, K is symmetric and positive. Moreover, $EK(Sx+y) = Ex = x = KE(Sx+y)$ so that $EK \subseteq KE$, $K(I-E) = 0$, and $\mathcal{D}(K) \ominus (I-E)(\mathcal{H}) \subseteq \mathcal{F}$.

At this point, several possibilities present themselves for proceeding with the argument. Theorem 4 and Corollary 5 apply to yield a positive self-adjoint extension of K affiliated with $\mathcal{R}$. By using the Borel function f, defined at 0 as 0 and at a positive real t as $1/t$, and the Borel function calculus of [6; Section 5.6] (see, especially [6; Theorem 5.6.26]) to form $f(S)$, we arrive at a self-adjoint operator extending K (inverse to S on $E(\mathcal{H})$). If we restrict S to $E(\mathcal{H})$, then S is invertible in the sense of the discussion of p. 595 of [6], and the argument of the last paragraph of p. 596 applies to show that $f(S)$ restricted to $E(\mathcal{H})$ is the mapping inverse to the restriction of S. Thus $f(S) = K$, and K is a positive self-adjoint operator affiliated with $\mathcal{R}$. This last conclusion can also be argued directly, without appeal to the Borel function calculus, as presented at the top of p. 467 of [6] (with K in place of T_0).

By construction of K, $\mathcal{D}(K) \ominus (I - E)(\mathcal{H})$ is the range of S, which is contained in $\mathcal{F}$, as we have shown. Thus $\mathcal{D}(K) \cap \mathcal{F}$ is the range of S. With x in $E(\mathcal{H})$, $\{K_a^{1/2} S^{1/2} x\}_{a \in \mathbb{A}}$ lies in the closed ball in $\mathcal{H}$ of radius $\|x\|$ with center 0. Since this ball is weakly compact, some cofinal subnet $\{K_{a'}^{1/2} S^{1/2} x\}_{a' \in \mathbb{A}'}$ of $\{K_a^{1/2} S^{1/2} x\}_{a \in \mathbb{A}}$ converges weakly to a vector u. We shall show that $Eu = x$. Given z in $\mathcal{H}$, from Lemma 6 (i),

$$\begin{aligned}\langle u, S^{\frac{1}{2}} z\rangle &= \lim_{a',a}\langle K_{a'}^{\frac{1}{2}} S^{\frac{1}{2}} x, K_a^{-\frac{1}{2}} z\rangle \\ &= \lim_{a'}\langle K_{a'}^{\frac{1}{2}} S^{\frac{1}{2}} x, K_{a'}^{-\frac{1}{2}} z\rangle \\ &= \lim_{a'}\langle S^{\frac{1}{2}} x, z\rangle \\ &= \langle S^{\frac{1}{2}} x, z\rangle.\end{aligned}$$

Thus $S^{1/2}u = S^{1/2}x$, and

$$Eu = KSEu = KS^{\frac{1}{2}}S^{\frac{1}{2}}u = KS^{\frac{1}{2}}S^{\frac{1}{2}}x = x.$$

Suppose, now, that $\|x\| = 1$. Given a positive ϵ, we can choose a' in $\mathbb{A}'$ such that

$$|\langle K_{a'}^{\frac{1}{2}} S^{\frac{1}{2}} x, x\rangle - \langle u, x\rangle| < \epsilon.$$

Then

$$\begin{aligned}1 - \epsilon = \langle x, x\rangle - \epsilon = \langle Eu, x\rangle - \epsilon &= \langle u, x\rangle - \epsilon \\ &< |\langle K_{a'}^{\frac{1}{2}} S^{\frac{1}{2}} x, x\rangle| \le \|K_{a'}^{\frac{1}{2}} S^{\frac{1}{2}} x\|\,\|x\| \\ &= \|K_{a'}^{\frac{1}{2}} S^{\frac{1}{2}} x\| \le \|x\| = 1.\end{aligned}$$

Thus, if $a \ge a'$,

$$\begin{aligned}(1-\epsilon)^2 \le \|K_{a'}^{\frac{1}{2}} S^{\frac{1}{2}} x\|^2 &= \langle S^{\frac{1}{2}} K_{a'} S^{\frac{1}{2}} x, x\rangle \\ &\le \langle S^{\frac{1}{2}} K_a S^{\frac{1}{2}} x, x\rangle \le \|x\|^2 = 1.\end{aligned}$$

It follows that $\{\langle S^{1/2} K_a S^{1/2} x, x\rangle\}_{a \in \mathbb{A}}$ tends to 1. As $\{S^{1/2} K_a S^{1/2}\}_{a \in \mathbb{A}}$ is monotone increasing and bounded above by E, this net converges to some positive A ($\subseteq E$) in the strong-operator topology. From what we have proved $\langle Ax, x\rangle = 1$ for each x of norm 1 in $E(\mathcal{H})$. Thus $A = E$ and

$$\begin{aligned}\langle K_a Sx, Sx\rangle &= \langle S^{\frac{1}{2}} K_a S^{\frac{1}{2}} S^{\frac{1}{2}} x, S^{\frac{1}{2}} x\rangle \\ &\underset{a}{\to} \langle ES^{\frac{1}{2}} x, S^{\frac{1}{2}} x\rangle = \langle x, Sx\rangle = \langle KSx, Sx\rangle.\end{aligned}$$

∎

Theorem 10. *Let $\mathcal{R}$ be a von Neumann algebra acting on a Hilbert space $\mathcal{H}$, x_0 be a vector in $\mathcal{H}$ and z_0 be a vector in $[\mathcal{R}x_0]$. Then there is a B in $\mathcal{R}$ and a positive self-adjoint T affiliated with $\mathcal{R}$ such that $BTx_0 = z_0$.*

Proof. We may assume the $\|z_0\| = 1$. We first choose T_0 in $\mathcal{R}$ such that $\|z_0 - T_0x_0\| \le 4^{-1}$ and $\|T_0x_0\| \le \|z_0\| = 1$. We then choose T_1 in $\mathcal{R}$ such that $\|z_0 - T_0x_0 - T_1x_0\| \le 4^{-2}$ and $\|T_1x_0\| \le \|z_0 - T_0x_0\|$ $(\le 4^{-1})$. Continuing in this way, we choose T_n in $\mathcal{R}$ such that

$$\|z_0 - T_0x_0 - T_1x_0 - \cdots - T_{n-1}x_0 - T_nx_0\| \le 4^{-(n+1)}, \quad \|T_nx_0\| \le 4^{-n}.$$

Then $\sum_{n=0}^{\infty} T_nx_0$ converges to z_0.

Let V_nH_n be the polar decomposition of T_n. Then $\|H_nx_0\| = \|T_nx_0\| \le 4^{-n}$. Let K_m be $(I + \sum_{k=0}^{m} 4^kH_k^2)^{1/2}$. Then

$$\|K_mx_0\|^2 = \langle K_m^2x_0, x_0\rangle = \|x_0\|^2 + \sum_{k=0}^{m} 4^k\|H_kx_0\|^2 \le \|x_0\|^2 + \sum_{k=0}^{m} 4^{-k}.$$

Thus $\{K_mx_0\}$ is bounded. Let u be a weak limiting point of $\{K_mx_0\}$ in $\mathcal{H}$. We apply Theorem 9 with $\{K_m\}$ in place of $\{K_a\}_{a\in\mathbb{A}}$. Let T be the positive self-adjoint operator affiliated with $\mathcal{R}$ such that $\{\langle K_mx, x\rangle\}$ converges to $\langle Tx, x\rangle$ for each x in $\mathcal{D}(T) \cap \mathcal{F}$, S be the strong-operator limit of the monotone decreasing sequence $\{K_m^{-1}\}$, and z be a vector in $\mathcal{H}$. From Lemma 6, we have

$$\begin{aligned}\langle u, Sz\rangle &= \lim_{m'}\langle K_{m'}x_0, Sz\rangle = \lim_{m'}\lim_{n}\langle K_{m'}x_0, K_n^{-1}z\rangle \\ &= \lim_{m',n}\langle K_{m'}x_0, K_n^{-1}z\rangle = \lim_{m'}\langle K_{m'}x_0, K_{m'}^{-1}z\rangle \\ &= \langle x_0, z\rangle. \quad (\{K_{m'}x_0\} \text{ a subnet of } \{K_mx_0\})\end{aligned}$$

Thus $Su = x_0$. It follows that $x_0 \in \mathcal{D}(T) \cap \mathcal{F}$ and that $Eu = TSu = Tx_0$. Thus $x_0 = Su = SEu = STx_0$.

Since $4^nH_n^2 \le I + \sum_{k=0}^{m} 4^kH_k^2 = K_m^2$ when $n \le m$, we have that $K_m^{-1}H_n^2K_m^{-1} \le 4^{-n}I$. As the product of operators is jointly strong-operator continuous on bounded subsets of $\mathcal{B}(\mathcal{H})$, $\{K_m^{-1}H_n^2K_m^{-1}\}$ tends to SH_n^2S in the strong-operator topology as m tends to ∞. Thus $0 \le SH_n^2S \le 4^{-n}I$ and

$$\|T_nS\|^2 = \|ST_n^*T_nS\| = \|SH_n^2S\| \le 4^{-n}.$$

It follows that $\sum_{n=0}^{\infty} T_nS$ converges in norm to an operator B in $\mathcal{R}$. Moreover,

$$BTx_0 = \left(\sum_{n=0}^{\infty} T_nS\right)Tx_0 = \sum_{n=0}^{\infty} T_nSTx_0 = \sum_{n=0}^{\infty} T_nx_0 = z_0.$$

∎

4. Cones and States.

In this section, we study the Takesaki cones $\mathcal{V}_u^0$ and $\mathcal{V}_u^{1/2}$ associated with a von Neumann algebra $\mathcal{R}$ acting on a Hilbert space $\mathcal{H}$ and a separating and generating vector u. (See [8], where these cones were first introduced and investigated.) The notation and results of Tomita's modular theory [8,9], as described in [6; section 9.2], will be used. We recall that conjugate-linear operators S_0 and F_0, with dense domains $\mathcal{R}u$ and $\mathcal{R}'u$, respectively, are defined by $S_0Au = A^*u$ and $F_0A'u = A'^*u$ for A in $\mathcal{R}$ and A' in $\mathcal{R}'$. It is easy to show that $F_0 \subseteq S_0^*$ and $S_0 \subseteq F_0^*$, so that S_0 and F_0 have closures S and F, respectively. The polar decomposition $J\Delta^{1/2}$ of S, where $\Delta = S^*S$, supplies the main elements, J and Δ, of Tomita's theory.

With x a vector in $\mathcal{H}$, define functionals ϕ_x on $\mathcal{R}$ and ϕ'_x on $\mathcal{R}'$ by $\phi_x(A) = \langle Au, x\rangle$ and $\phi'_x(A') = \langle A'u, x\rangle$. The vectors x that give rise to positive functionals ϕ'_x form a cone $\mathcal{V}_u^0$ in $\mathcal{H}$. Symmetrically, those vectors x that give rise to positive functionals ϕ_x form a cone $\mathcal{V}_u^{1/2}$ in $\mathcal{H}$. The Friedrichs extension plays a key role in establishing that $x \in \mathcal{V}_u^0$ if and only if $x = Hu$ for some positive self-adjoint operator H affiliated with $\mathcal{R}$. This result is a crucial step in the exposition of Takesaki's results that follows.

Theorem 11. *The functional ϕ'_x on $\mathcal{R}'$ is positive if and only if $x = Hu$ for some positive self-adjoint H affiliated with $\mathcal{R}$.*

Proof. Suppose $x \in \mathcal{D}(F_0^*)$ and $F_0^*x = x$. With A' self-adjoint in $\mathcal{R}'$, we have that

$$\langle A'u, x\rangle = \langle F_0A'u, x\rangle = \langle F_0^*x, A'u\rangle = \langle x, A'u\rangle;$$

whence $\phi'_x(A')$ is real and ϕ'_x is hermitian.

Assume, now, that ϕ'_x is hermitian and $T' \in \mathcal{R}'$. Then

$$\langle F_0T'u, x\rangle = \langle T'^*u, x\rangle = \phi'_x(T'^*) = \overline{\phi'_x(T')} = \overline{\langle T'u, x\rangle} = \langle x, T'u\rangle,$$

whence $x \in \mathcal{D}(F_0^*)$ and $x = F_0^*x$. (That $F_0^* = S$ is proved in [6; Corollary 9.2.30].)

Suppose $x = Hu$, where H is a positive self-adjoint operator affiliated with $\mathcal{R}$. Let $\{E_\lambda\}$ be the resolution of the identity for H (cf. [6; pp. 310, 311]), and let H_n be HE_n for each positive integer n. Then $H_n \in \mathcal{R}$ and $E_nH \subseteq H_n$. Hence $H_nu = E_nHu \to Hu = x$ as n tends to ∞ (since E_n is strong-operator convergent to I). Now $\langle A'u, H_nu\rangle = \langle H_nA'u, u\rangle \geq 0$, when A' is a positive operator in $\mathcal{R}'$. Thus

$$0 \leq \lim_n \langle A'u, H_nu\rangle = \langle A'u, Hu\rangle = \langle A'u, x\rangle,$$

and $\phi'_x \geq 0$.

Suppose that $\phi'_x \geq 0$ for an x in $\mathcal{H}$. Then, in particular, ϕ'_x is hermitian. From what we have proved, $x \in \mathcal{D}(F_0^*)$ and $F_0^*x = x$. Let $L_x^0A'u$ be $A'x$ for each A' in $\mathcal{R}'$. Then L_x^0 is a linear operator with (dense) domain $\mathcal{R}'u$, and

$$\begin{aligned}\langle L_x^0A'u, B'u\rangle &= \langle x, A'^*B'u\rangle = \langle x, F_0B'^*A'u\rangle \\ &= \langle A'u, B'F_0^*x\rangle = \langle A'u, B'x\rangle,\end{aligned}$$

for all A' and B' in $\mathcal{R}'$. Thus $B'u \in \mathcal{D}(L_x^{0*})$ and $L_x^{0*}B'u = B'x$. It follows that the domain of L_x^{0*} is dense, whence L_x^0 has a closure L_x. (See [6; Theorem 2.7.8 (ii)].) If $T' \in \mathcal{R}'$,

$$\langle L_xT'u, T'u\rangle = \langle T'x, T'u\rangle = \overline{\phi'_x(T'^*T')} \geq 0,$$

since T'^*T' is a positive operator in $\mathcal{R}'$. But $\mathcal{R}'u$ is a core for L_x, whence L_x is a positive symmetric operator. Moreover, with U' a unitary operator in $\mathcal{R}'$,

$$U'L_xT'u = U'T'x = L_xU'T'u.$$

Since $\mathcal{R}'u$ is a core for L_x, $L_x \,\eta\, \mathcal{R}$ from [6; Remark 5.6.3]. Theorem 4 and Corollary 5 apply; L_x has a positive self-adjoint extension H, its Friedrichs extension, affiliated with $\mathcal{R}$. Finally, $x = L_xu = Hu$. ∎

With $\mathbf{X}$ a linear space and $\mathbf{X}^\#$ its dual space, cones $\mathcal{V}$ in $\mathbf{X}$ and $\mathcal{V}^\#$ in $\mathbf{X}^\#$ are said to be dual cones when $x \in \mathcal{V}$ if and only if $\phi(x) \geq 0$ for each ϕ in $\mathcal{V}^\#$, and $\eta \in \mathcal{V}^\#$ if and only if $\eta(y) \geq 0$ for each y in $\mathcal{V}$. In a sense, $\mathcal{V}$ and $\mathcal{V}^\#$ are dual when each "determines" the ordering induced by the other. This concept applies in various contexts of linear space and dual space. If X is a compact Hausdorff space and $C(X)$ is the algebra of continuous complex-valued functions on X, the cone of positive functions and the cone of positive linear functionals on X are dual. As interpreted in a Hilbert space $\mathcal{H}$, and using the fact that $\mathcal{H}$ can be identified with its dual, cones $\mathcal{V}$ and $\mathcal{V}'$ in $\mathcal{H}$ are said to be dual cones when $x \in \mathcal{V}$ if and only if $\langle x, x'\rangle \geq 0$ for all x' in $\mathcal{V}'$, and $y' \in \mathcal{V}'$ if and only if $\langle y, y'\rangle \geq 0$ for all y in $\mathcal{V}$.

Proposition 12. *The Takesaki cones $\mathcal{V}_u^0$ and $\mathcal{V}_u^{1/2}$ are norm-closed dual cones in $\mathcal{H}$.*

Proof. If A' is a positive operator in $\mathcal{R}'$, then $0 \leq \langle A'u, ax + y\rangle$ when $a \geq 0$ and $x, y \in \mathcal{V}_u^0$. Thus $ax + y \in \mathcal{V}_u^0$. If v and $-v$ are in $\mathcal{V}_u^0$, then $\langle A'u, v\rangle = 0$ for each positive A' in $\mathcal{R}'$. Since each operator T' in $\mathcal{R}'$ is a linear combination of (four) positive operators in $\mathcal{R}'$, $\langle T'u, v\rangle = 0$. Since $[\mathcal{R}'u] = \mathcal{H}$, $v = 0$. Thus $\mathcal{V}_u^0$ and, symmetrically, $\mathcal{V}_u^{1/2}$ are cones in $\mathcal{H}$.

If $\{x_n\}$ is a sequence of vectors in $\mathcal{V}_u^0$ tending to x in norm and A' is a positive operator in $\mathcal{R}'$, then $0 \leq \langle A'u, x_n\rangle \to \langle A'u, x\rangle$. Hence $x \in \mathcal{V}_u^0$, and $\mathcal{V}_u^0$ is norm closed. Symmetrically, $\mathcal{V}_u^{1/2}$ is norm closed.

If $v \in \mathcal{V}_u^{1/2}$, then $\langle Au, v\rangle \geq 0$ for each positive A in $\mathcal{R}$. If $w \in \mathcal{V}_u^0$, then from Theorem 11, $w = Hu$ for some positive self-adjoint H affiliated $\mathcal{R}$. With $\{E_\lambda\}$ the resolution of the identity for H,

$$0 \leq \langle H_nu, v\rangle = \langle E_nHu, v\rangle \to \langle Hu, v\rangle = \langle w, v\rangle,$$

where $H_n = HE_n \in \mathcal{R}$.

If $\langle w, v\rangle \geq 0$ for each v in $\mathcal{V}_u^{1/2}$, then

$$0 \leq \langle w, A'u\rangle = \langle A'u, w\rangle$$

for each positive A' in $\mathcal{R}'$, since $A'u \in \mathcal{V}_u^{1/2}$ (from Theorem 11 applied with $\mathcal{R}'$ in place of $\mathcal{R}$). Hence $\phi'_w \geq 0$ and $w \in \mathcal{V}_u^0$. Thus $w \in \mathcal{V}_u^0$ if and only if $\langle w, v\rangle \geq 0$ for each v in $\mathcal{V}_u^{1/2}$. Symmetrically, $v \in \mathcal{V}_u^{1/2}$ if and only if $\langle w, v\rangle \geq 0$ for each w in $\mathcal{V}_u^0$. ∎

We recall the notation $\mathcal{R}^+$ and $\mathcal{R}'^+$ for the sets of positive operators in $\mathcal{R}$ and $\mathcal{R}'$, respectively.

Proposition 13. *The Takesaki cones $\mathcal{V}_u^0$ and $\mathcal{V}_u^{1/2}$ are the respective norm closures of $\mathcal{R}^+u$ and $\mathcal{R}'^+u$. Moreover, $\Delta^{1/2}\mathcal{R}^+u = \mathcal{R}'^+u$ and $\Delta^{-1/2}\mathcal{R}'^+u = \mathcal{R}^+u$, whence $\mathcal{V}_u^0$ and $\mathcal{V}_u^{1/2}$ are the respective norm closures of $\Delta^{-1/2}\mathcal{R}'^+u$ and $\Delta^{1/2}\mathcal{R}^+u$.*

Proof. From Theorem 11, $\mathcal{R}^+u \subseteq \mathcal{V}_u^0$. If $x \in \mathcal{V}_u^0$, then $x = Hu$ for some positive self-adjoint H affiliated with $\mathcal{R}$, again from Theorem 11. With $\{E_\lambda\}$ the resolution of the identity for H, $H_n = HE_n \in \mathcal{R}^+$ and $H_nu = E_nHu \to Hu = x$. Thus x is in the norm closure of $\mathcal{R}^+u$. It follows that $\mathcal{V}_u^0$ is the norm closure of $\mathcal{R}^+u$ and, symmetrically, $\mathcal{V}_u^{1/2}$ is the norm closure of $\mathcal{R}'^+u$.

Let $\Phi(A)$ be JA^*J for A in $\mathcal{R}$. The mapping Φ is a *anti-isomorphism of $\mathcal{R}$ onto $\mathcal{R}'$ [6; p. 591]. Thus $\Phi(\mathcal{R}^+) = \mathcal{R}'^+$. With A in $\mathcal{R}^+$, $Au \in \mathcal{D}(S) = \mathcal{D}(\Delta^{1/2})$, and

$$\Delta^{1/2}Au = JSAu = JA^*u = JA^*Ju = \Phi(A)u.$$

Hence $\Delta^{1/2}\mathcal{R}^+u = \mathcal{R}'^+u$. Symmetrically, with A' in $\mathcal{R}'^+$, $A'u \in \mathcal{D}(F) = \mathcal{D}(\Delta^{-1/2})$, and

$$\Delta^{-1/2}A'u = JFA'u = JA'^*u = JA'^*Ju = \Phi^{-1}(A')u.$$

Thus $\Delta^{-1/2}\mathcal{R}'^+u = \mathcal{R}^+u$. ∎

The notation we are using for the Takesaki cones is motivated by Proposition 13, which identifies $\mathcal{V}_u^0$ and $\mathcal{V}_u^{1/2}$ with the norm closures of $\mathcal{R}^+u$ $(= \Delta^0\mathcal{R}^+u)$ and $\Delta^{1/2}\mathcal{R}^+u$, respectively. This notation is Araki's who introduces [1,2] and subjects to a deep and penetrating analysis, the one-parameter family of cones $\mathcal{V}_u^a$ defined as the norm closures of $\Delta^a\mathcal{R}^+u$, where $a \in [0, 1/2]$. Araki shows that $\mathcal{V}_u^a$ and $\mathcal{V}_u^{a'}$ are dual cones, where $a' = 1/2 - a$. The cone $\mathcal{V}_u^{1/4}$, which is "self-dual," exhibits surprising and useful properties. Independently, and at the same time, Connes [3] introduces and studies the self-dual cone, proving an important order characterization result for von Neumann algebras. Haagerup, in an unpublished note, studies the self-dual cone at about this same time. (His clever techniques are incorporated in the solution to [6; Exercises 9.6.62-4].) In [5], Haagerup extends the scope of his self-dual-cone techniques and results to the non-countably decomposable case.

Theorem 14. *With ω a normal state of $\mathcal{R}$, there is a unique vector v in $\mathcal{V}_u^0$ such that $\omega_v \mid \mathcal{R} = \omega$. Moreover,*

$$\|v - u\| = \inf\{\|z - u\| : \omega_z \mid \mathcal{R} = \omega\}.$$

Proof. From [6; Theorem 7.2.3], there is a unit vector z in $\mathcal{H}$ such that $\omega = \omega_z \mid \mathcal{R}$ (as $\mathcal{R}$ admits the separating vector u). From [6; Theorem 7.3.2], there is a partial isometry V' in $\mathcal{R}'$ such that ω' is a positive normal linear functional on $\mathcal{R}'$, where $\omega'(A') = \phi'_z(V'A')$ for each A' in $\mathcal{R}'$, and such that $\phi'_z(A') = \omega'(V'^*A')$. Now

$$\omega'(A') = \phi'_z(V'A') = \langle V'A'u, z\rangle = \langle A'u, V'^*z\rangle,$$

whence $(v =)\, V'^*z \in \mathcal{V}_u^0$. In addition,

$$\langle A'u, z\rangle = \phi'_z(A') = \omega'(V'^*A') = \phi'_z(V'V'^*A') = \langle A'u, V'V'^*z\rangle.$$

Since u is generating for $\mathcal{R}'$, $z = V'V'^*z$. Thus

$$\omega_v \mid \mathcal{R} = \omega_z \mid \mathcal{R} = \omega.$$

If H is a positive self-adjoint operator affiliated with $\mathcal{R}$ and $u \in \mathcal{D}(H)$, then $u \in \mathcal{D}(H^{1/2})$, $H^{1/2}u \in \mathcal{D}(H^{1/2})$, $H^{1/2}H^{1/2}u = Hu$, and $A'H^{1/2}u = H^{1/2}A'u$ for each A' in $\mathcal{R}'$. Thus, if V' is a partial isometry in $\mathcal{R}'$.

$$\begin{aligned} |\langle V'Hu, u\rangle| &= |\langle V'H^{\frac{1}{2}}u, H^{\frac{1}{2}}u\rangle \\ &\le \|V'H^{\frac{1}{2}}u\| \, \|H^{\frac{1}{2}}u\| \\ &\le \|H^{\frac{1}{2}}u\|^2 \\ &= \langle Hu, u\rangle \end{aligned} \tag{1}$$

and

$$\operatorname{Re}\langle V'Hu, u\rangle \le \langle Hu, u\rangle. \tag{2}$$

Suppose z is a unit vector in $\mathcal{H}$ such that $\omega_z \mid \mathcal{R} = \omega_{Hu} \mid \mathcal{R}$. The mapping $AHu \to Az$ $(A \in \mathcal{R})$ extends to a partial isometry W' in $\mathcal{R}'$, with initial space $[\mathcal{R}Hu]$, such that $W'Hu = z$. From (2),

$$\operatorname{Re}\langle z, u\rangle = \operatorname{Re}\langle W'Hu, u\rangle \le \langle Hu, u\rangle,$$

so that

$$\|Hu - u\|^2 = 2 - 2\operatorname{Re}\langle Hu, u\rangle \le 2 - 2\operatorname{Re}\langle z, u\rangle = \|z - u\|^2. \tag{3}$$

From Theorem 11, there is a positive self-adjoint operator H, affiliated with $\mathcal{R}$, such that $v = Hu$. From (3),

$$\|v - u\| = \inf\{\|z - u\| : \omega_z \mid \mathcal{R} = \omega\}.$$

If v' is another vector in $\mathcal{V}_u^0$ such that $\omega_{v'} \mid \mathcal{R} = \omega$, then

$$\|v - u\| = \inf\{\|z - u\| : \omega_z \mid \mathcal{R} = \omega\} = \|v' - u\|,$$

and $v' = V'Hu$ for some partial isometry V' in $\mathcal{R}'$ with Hu $(= v)$ in its initial space. Hence

$$\operatorname{Re}\langle V'Hu, u\rangle = Re\langle v', u\rangle = \operatorname{Re}\langle v, u\rangle = \langle Hu, u\rangle,$$

and the inequality of (2) is equality in the present case. It follows that

$$\langle V'Hu, u\rangle = |\langle V'Hu, u\rangle| = \operatorname{Re}\langle V'Hu, u\rangle = \langle Hu, u\rangle,$$

so that

$$\langle V'H^{\frac{1}{2}}u, H^{\frac{1}{2}}u\rangle = \|V'H^{\frac{1}{2}}u\| \, \|H^{\frac{1}{2}}u\| = \|H^{\frac{1}{2}}u\|^2.$$

Thus $V'H^{1/2}u = H^{1/2}u$ and

$$v' = V'Hu = V'H^{\frac{1}{2}}(H^{\frac{1}{2}}u) = H^{\frac{1}{2}}V'H^{\frac{1}{2}}u = Hu = v.$$

Combining Theorems 11 and 14 yields a far reaching non-commutative Radon-Nikodým result. In effect, the normal state ω is "absolutely continuous" with respect to $\omega_u \mid \mathcal{R}$, and $\omega = \omega_{Hu} \mid \mathcal{R}$, where $H^* = H \geq 0$ and $H \ \eta \ \mathcal{R}$. Loosely, $\omega(A) = \omega_u(HAH)$ for each A in $\mathcal{R}$ (although HAH is not bounded, in general, so that, in fact $\omega(HAH)$ is not defined). Thus H^2 is the Radon-Nikodým derivative of ω with respect to ω_u. In the non-commutative context, "HAH" rather than "AH^2" is the appropriate formulation.

References.

1. H. Araki, *Some properties of modular conjugation operator of von Neumann algebras and a non-commutative Radon-Nikodým theorem with a chain rule*, Pac. J. Math. **50**(1974), 309-354.
2. H. Araki, *Positive cones for von Neumann algebras*, in "Operator Algebras and Applications" (Proc. of Symposia in Pure Math. Vol. 38 Part 2, R. Kadison, ed., 1980), pp. 5-15. Amer. Math. Soc., Providence, 1982.
3. A. Connes, *Caractérisation des espaces vectoriels ordonnés sous-jacents aux algèbres de von Neumann*, Ann. Inst. Fourier., Grenoble **24**(1974), 121-155.
4. K. Friedrichs, *Spektraltheorie halbbeschränkter Operatoren*, Math. Ann. **109**(1934), 465-487.
5. U. Haagerup, *The standard form of von Neumann algebras*, Math. Scand. **37**(1975), 271-283.
6. R. Kadison and J. Ringrose, "Fundamentals of the Theory of Operator Algebras," Vols. I, II. Academic Press, New York, Orlando, 1983, 1986.
7. F. Murray and J. von Neumann, *On rings of operators*, Ann. of Math. **37**(1936), 116-229.
8. M. Takesaki, "Tomita's Theory of Modular Hilbert Algebras and Its Applications," Lecture Notes in Mathematics, Vol. 128. Springer-Verlag, Heidelberg, 1970.
9. M. Tomita, *Standard forms of von Neumann algebras*, Fifth Functional Analysis Symposium of the Math. Soc. of Japan, Sendai, 1967.

Mathematics Department
University of Pennsylvania

Some remarks on interpolation of families of quasi-Banach spaces

N. J. KALTON[1]

UNIVERSITY OF MISSOURI-COLUMBIA

Abstract. We study some questions raised in theory of complex interpolation of quasi-Banach spaces. In particular we give a criterion for the interpolated space to be locally convex.

1. Introduction. In [1] and [2], Coifman, Cwikel, Rochberg, Saghar and Weiss introduced and studied complex interpolation of families of Banach spaces. Recently, Tabacco [11],[12] and Rochberg [10] have studied the extension of these ideas to the non-locally convex quasi-Banach case.

We let $\mathbf{T}$ denote the unit circle in the complex plane and λ denote normalized Haar measure on $\mathbf{T}$, i.e. $d\lambda = (2\pi)^{-1}d\theta$. Δ denotes the unit disk, $\{z : |z| < 1\}$. We then suppose that we are given a family of quasinormed spaces X_w for $w \in \mathbf{T}$ and define interpolation spaces X_z for $z \in \Delta$. The precise details of the construction are given in Section 2.

In this paper we prove two main results on interpolation of analytic families of quasi-Banach spaces. In Theorem 4, we answer a question of Rochberg [10] by giving a condition for the interpolated space to be locally convex. We use here the notion of (Rademacher) type. A quasi-Banach space X is of type p where $0 < p \leq 2$ if there is a constant C so that if $x_1, \ldots, x_n \in X$ then

$$\mathcal{E}\left(\|\epsilon_1 x_1 + \cdots + \epsilon_n x_n\|^p\right) \leq C^p\left(\|x_1\|^p + \cdots + \|x_n\|^p\right)$$

[1]Supported by NSF-grant DMS-8601401

where the signs $\epsilon_k = \pm 1$ are chosen at random. In fact if $p < 1$ then type p is equivalent to p-normability [5], but there are type one spaces which are not locally convex (e.g. the Lorentz spaces $L(1,p)$ where $1 < p < \infty$, or the Ribe space [5]). Now if X_w is type $p(w)$ for $w \in \mathbf{T}$, and a mild separability assumption holds, then the interpolated space at the origin X_0 is locally convex provided

$$\int p(w)^{-1} d\lambda(w) < 1.$$

Simple examples show that equality does not guarantee the conclusion, although under mild hypotheses it will force X_0 to be type one. By conformal transformation, similar results can be given for any interpolated space X_z where $|z| < 1$.

In Theorem 9 we show, again under some mild additional assumptions, that the interpolated space X_0 is unchanged if we replace the quasinorm on each X_w by the largest plurisubharmonic (semi-)quasinorm it dominates. It follows that the interpolated space X_0 collapses to the trivial zero space if X_w is A-trivial in the sense of [7] on a set of positive measure; an example of an A-trivial space is L_p/H_p where $p < 1$. The proof of Theorem 9 is similar to arguments used in [7].

Theorem 9 suggests that the interpolated space always has a plurisubharmonic quasinorm. This, however, is false as we show by example. We also show that the Iteration Theorem of [1] has no non-locally convex extension, at least in an isometric sense. We leave open, however, the question of whether it is true isomorphically.

Let us note that we have attempted to avoid unnecessary continuity assumptions on the given quasinorms, since in many naturally arising spaces these quasinorms are not continuous. This does result in some tedious measurability problems.

2. Some basic interpolation results. Suppose X is a complex linear space and that we are given a function $H : \mathbf{T} \times X \to [0, \infty]$ with the property that the restriction of H to

$\mathbf{T} \times V$ is a Borel function for each finite-dimensional subspace V of X. Suppose further that for each $x \in X$ we have that $\log^+ H(w,x) \in L_1(\mathbf{T})$.

We define $\mathcal{A}$ to be the set of all analytic maps $g : \Delta \to X$ of the form

$$g(z) = \sum_{j=1}^{N} \phi_j(z) b_j$$

where $\phi_j \in N^+$ for $1 \leq j \leq N$. Here N^+ denotes the Smirnov class (see, for example, [3] p. 25).

Then for $|z| < 1$ we define $H_z : X \to [0,\infty)$ by

$$H_z(x) = \inf\{\int_{\mathbf{T}} H(w, g(w))\, d\lambda(w) : \; g \in \mathcal{A},\; g(z) = x\}$$

We shall also write $H(z,x)$ for $H_z(x)$.

Now suppose in addition that H satisfies the condition:

$$H(w, \alpha x) = |\alpha| H(w,x) \qquad \alpha \in \mathbf{C}$$

Under these conditions we say that H is homogeneous. It then follows as in the work of Tabacco [12] (see Theorem 1.8) that H_z can also be obtained from the formulas:

$$H(z,x) = \inf\{\Big(\int_{\mathbf{T}} P(z,w) H(w, g(w))^p\, d\lambda(w)\Big)^{1/p} : \; g \in \mathcal{A},\; g(z) = x\}$$
$$H(z,x) = \inf\{\operatorname{ess\,sup} H(w, g(w)) : \; g \in \mathcal{A},\; g(z) = x\}$$
$$H(z,x) = \inf\{\exp \int_{\mathbf{T}} P(z,w) \log(H(w,g(w)) d\lambda(w) : \; g \in \mathcal{A},\; g(z) = x\}$$

where $0 < p < \infty$ and $|z| < 1$. Here P denotes the Poisson kernel

$$P(z,w) = \frac{1 - |z|^2}{|w - z|^2}.$$

Note now that our hypotheses force $H(z,x) < \infty$ for $|z| < 1$ and $x \in X$.

Next we further suppose the existence of a Borel function $c = c_H$ so that

$$H(w, x_1 + x_2) \leq c(w)(H(w,x_1) + H(w,x_2)) \qquad x_1, x_2 \in X.$$

Then for each w the space $X_w = \{x \in X : H(w,x) < \infty\}$ is a linear space and $\|x\|_w = H(w,x)$ is a semi-quasinorm on X. In future we will use the term quasinorm with the understanding that positive-definiteness is not assumed. If, in addition, we assume that $\log c \in L_1(\mathbf{T})$ then for $|z| < 1$, $\|x\|_z = H_z(x)$ defines a quasinorm on X and indeed

$$\|x_1 + x_2\|_z \leq c(z)(\|x_1 + x_2\|_z)$$

where

$$c(z) = \exp\left(\int_{\mathbf{T}} P(z,w)\log c(w)\, d\lambda(w)\right).$$

For details, see Tabacco [12], Propositions 1.9 and 2.7. Under these hypotheses, we refer to X_w as an analytic family of quasinormed spaces, and say that H defines the analytic family X_w. The interpolated space X_z is the completion of the Hausdorff quotient of $(X, \| \|_z)$.

We shall concentrate on the case $z = 0$ since a conformal transformation then gives the corresponding results for all $|z| < 1$.

Suppose then as above H defines an analytic family of quasinormed spaces. Let $p : \mathbf{T} \to (0,\infty)$ be a measurable function such that $1/p \in L_1(\mathbf{T})$. Then we may consider H^p in place of H and Rochberg [10] shows that

$$H_0^p = e^{\bar{p}\Lambda}(H_0)^{\bar{p}}$$

where

$$\bar{p}^{-1} = \int_{\mathbf{T}} p(w)^{-1}\, d\lambda(w)$$

and

$$\Lambda = \bar{p}^{-1}\log(\bar{p}^{-1}) - \int_{\mathbf{T}} p^{-1}\log(p^{-1})\, d\lambda.$$

Rochberg calls this the Power Theorem for Complex Interpolation.

Let us derive two simple conclusions from these results. In order to state our results economically we shall sometimes impose a further natural conditon on the analytic family

X_w. We shall say this family is *uniformly separable* if there is a countable set D in X such that for each $w \in \mathbf{T}$ the set $D \cap X_w$ is dense in X_w. This will avoid certain cumbersome measurability assumptions, and is satisfied in most examples.

PROPOSITION 1. *Let $H : \mathbf{T} \times X \to [0,\infty]$ and $K : \mathbf{T} \times Y \to [0,\infty]$ induce analytic families of quasinormed spaces X_w and Y_w. Suppose $T : X \to Y$ is a linear map and define for $w \in \mathbf{T}$,*

$$\|T\|_w = \sup\{\|Tx\|_w : \|x\|_w \leq 1\}.$$

Then:

(i) If η is a Borel measurable function such that $\eta(w) \geq \|T\|_w$ for all w and $\log_+ \eta \in L_1$ then

$$\|T\|_0 \leq \exp\left(\int_{\mathbf{T}} \log \eta(w) d\lambda(w)\right)$$

(ii) If X is uniformly separable and both X_w and Y_w are continuously quasinormed then $w \to \|T\|_w$ is Borel measurable.

(iii) If X is uniformly separable, but we do not assume that the families are continuously quasinormed, then there exists a Borel function η such that

$$\|T\|_w \leq \eta(w) \leq c_H(w)^6 c_K(w)^6 \|T\|_w$$

for $w \in \mathbf{T}$.

PROOF: (i) is a special case of a theorem of Tabacco [12], Theorem 1.13. (ii) is an immediate consequence of uniform separability. For (iii) let (x_n) be any sequence whose intersection with each X_w is dense. Define $\eta(w) = c_H(w)^2 c_K(w)^4 \sup\{\|Tx_n\|_w : \|x_n\|_w \leq 1\}$. Note that on X_w there is an equivalent continuous quasinorm $|\ |$ with $\|x\|_w \leq |x| \leq c_H(w)^2 \|x\|_w$ (see [8], Lemma 1.1 and Theorem 1.2). Similarly there is an continuous equivalent quasinorm $|\ |$ on Y_w with $\|y\|_w \leq |y| \leq c_K(w)^2 \|y\|_w$. Now if $|x_n| \leq 1$ then $|x_n|_w \leq 1$ and so $|T| \leq \sup\{|Tx_n| : \|x_n\|_w \leq 1\}$. Thus $\|T\|_w \leq c_H(w)^2 c_K(w)^2 |T| \leq \eta(w)$. On the other hand, $\eta(w) \leq c_H(w)^2 c_K(w)^4 \sup\{|Tx_n| : |x_n| \leq c_H(w)^2\} \leq c_H(w)^4 c_K(w)^4 |T| \leq c_H(w)^6 c_K(w)^6 \|T\|_w$.

Now let Ω be a finite set and let μ be a finite strictly positive measure on Ω. Denote by X^Ω the space of all maps $f : \Omega \to X$.

PROPOSITION 2. *Let $H : \mathbf{T} \times X \to [0, \infty]$ induce an analytic family of quasinormed spaces on X. Let $p : \mathbf{T} \to (0, \infty)$ be a measurable map such that $1/p \in L_1$. Define $K : \mathbf{T} \times X^\Omega \to [0, \infty]$ by*

$$K(w, f) = \left(\int_\Omega H(w, f)^{p(w)} \, d\mu \right)^{1/p(w)}.$$

Then K also induces an analytic family with

$$c_K(w) \leq \gamma(w) c_H(w)$$

where $\gamma(w) = \max(2^{\frac{1}{p}-1}, 1)$ and further

$$\|f\|_0 = K_0(f) = \left(\int_\Omega \|f(\omega)\|_0^{\bar{p}} \, d\mu(\omega) \right)^{1/\bar{p}}$$

where $\bar{p} = \int p^{-1} \, d\lambda$.

PROOF: We omit the simple proof that K satisfies the hypotheses to induce an analytic family, with c_K as given. Observe then that K^p is given by

$$K^p(w, f) = \int_\Omega H^p(w, f) \, d\mu$$

and since Ω is finite it then easily follows that

$$K_0^{\bar{p}}(f) = \int_\Omega H_0^{\bar{p}}(f(\omega)) \, d\mu.$$

The result now follows from the Power Theorem of Rochberg quoted above.

Either as a limiting case or by a simple direct argument we can also easily show that:

PROPOSITION 3. *Under the hypotheses of Proposition 2 define $J : \mathbf{T} \times X^{\Omega} \to [0,\infty]$ by*

$$J(w,f) = \max_{\omega} H(w, f(\omega)).$$

Then J also defines an analytic family with $c_J = c_H$ and

$$\|f\|_0 = J_0(f) = \max_{\omega} \|f(\omega)\|_0.$$

3. Convexity of the interpolated space. We now prove a theorem which provides an answer to a question of Rochberg [10].

THEOREM 4. *Suppose $H : \mathbf{T} \times X \to [0,\infty]$ induces a uniformly separable analytic family of quasinormed spaces. Suppose that $p : \mathbf{T} \to (0,\infty)$ is a measurable function such that each X_w is of type $p(w)$ and*

$$\int p(w)^{-1}\, d\lambda < 1.$$

Then X_0 is locally convex, i.e. $\| \ \|_0$ is equivalent to a seminorm on X.

PROOF: We set $p_0 = \int 1/p\, d\lambda$. For any $n \in \mathbf{N}$ let Ω be a finite set (depending on n), let μ be a probability measure on Ω and let $\epsilon_1, \ldots, \epsilon_n$ be a sequence of independent random variables on (Ω, μ) such that $\mu(\epsilon_k = 1) = \mu(\epsilon_k = -1) = \frac{1}{2}$ for $k = 1, 2, \ldots, n$. Let $d_n(w)$ be the least constant such that for every $x_1, \ldots, x_n$ we have

$$\left(\int_{\Omega} \|\epsilon_1 x_1 + \cdots + \epsilon_n x_n\|_w^{p(w)}\, d\mu \right)^{1/p(w)} \leq d_n(w) \max(\|x_1\|_w, \ldots, \|x_n\|_w).$$

Similarly $d_n(0)$ is the least constant such that

$$\left(\int_{\Omega} \|\epsilon_1 x_1 + \ldots + \epsilon_n x_n\|_0^{p_0}\, d\mu \right)^{1/p_0} \leq d_n(0) \max(\|x_1\|_0, \ldots, \|x_n\|_0).$$

A crude estimate on d_n when $2^{m-1} < n \leq 2^m$ is given by

$$d_n(w) \leq (2c(w))^m$$

so that we have

$$\log d_n(w) \le \frac{\log n + \log 2}{\log 2}(\log 2 + \log c(w)).$$

Define $T_n : X^n \to X^\Omega$ by

$$T_n(x_1, \ldots, x_n) = \epsilon_1 x_1 + \cdots + \epsilon_n x_n.$$

If we equip X^n with the quasinorm $\|(x_1, \ldots, x_n)\|_w = \max \|x_k\|_w$ and X^Ω with the quasinorm $\|f\|_w = (\int \|f\|_w^{p(w)} \, d\mu)^{1/p(w)}$ then $d_n(w) = \|T_n\|_w$ and so it follows from Propositions 1, 2 and 3 that

$$d_n(0) = \|T_n\|_0 \le \exp\left(\int_{\mathbf{T}} \log \eta_n(w) \, d\lambda(w)\right),$$

where η is any Borel measurable function satisfying $\eta_n(w) \ge d_n(w)$. We may further suppose that $\eta_n(w) \le \gamma(w)^6 c(w)^{12} d_n(w)$ where $\gamma(w) = \max(2^{\frac{1}{p(w)}-1}, 1)$.

Now from the calculation above we have

$$\frac{\log d_n(w)}{\log n} \le A(\log 2 + \log c(w))$$

for suitable constant A. Thus

$$\frac{\log \eta_n(w)}{\log n} \le A_1(\log 2 + \log c(w) + \log \gamma(w))$$

for suitable A_1. Furthermore for each w then exists a constant B_w such that

$$\left(\int_\Omega \|\epsilon_1 x_1 + \cdots + \epsilon_n x_n\|_w^{p(w)} \, d\mu\right)^{1/p(w)} \le B_w (\|x_1\|_w^{p(w)} + \cdots + \|x_n\|_w^{p(w)})^{1/p(w)}$$

whenever $x_1, \cdots, x_n \in X_w$. Thus

$$d_n(w) \le B_w n^{1/p}.$$

It follows that

$$\limsup \frac{\log d_n(w)}{\log n} \le \frac{1}{p(w)}$$

and so

$$\limsup \frac{\log \eta_n(w)}{\log n} \leq \frac{1}{p(w)}.$$

Now by the Dominated Convergence Theorem,

$$\limsup_n \int_{\mathbf{T}} \frac{\log \eta_n(w)}{\log n} d\lambda \leq \int_{\mathbf{T}} \frac{1}{p(w)} d\lambda$$

Hence

$$\limsup \frac{\log d_n(0)}{\log n} < 1.$$

It now follows from Theorem 2.5 of [4] that X_0 is locally convex.

In the above theorem we did not assume any bound on the type $p(w)$ constants B_w. If we assume some control on these constants then we can make more precise statements.

THEOREM 5. *Suppose $H : \mathbf{T} \times X \to [0, \infty]$ induces a uniformly separable family of continuously quasinormed spaces. Suppose that $p : \mathbf{T} \to (0, \infty)$ and $B : \mathbf{T} \to (0, \infty)$ are measurable functions such that each X_w is of type $p(w)$ and further:*

(a)
$$\int p(w)^{-1} d\lambda = p_0^{-1}$$

(b)
$$\int \log B(w) d\lambda = \log B_0 < \infty$$

(c) $$\left(\int_\Omega \|\epsilon_1 x_1 + \cdots + \epsilon_n x_n\|_w^{p(w)} d\mu\right)^{1/p(w)} \leq B(w)(\|x_1\|_w^{p(w)} + \cdots + \|x_n\|_w^{p(w)})^{1/p(w)}$$

whenever $x_1, \ldots x_n \in X_w$ and $\epsilon_1, \ldots \epsilon_n$ is a sequence of independent random variables defined on a probability space (Ω, μ) with $\mu(\epsilon_k = 1) = \mu(\epsilon_k = -1) = \frac{1}{2}$ for $1 \leq k \leq n$.

Then X_0 is type p_0 with type constant at most B_0 i.e. for $x_1, \ldots, x_n \in X$ we have

$$\left(\int_\Omega \|\epsilon_1 x_1 + \cdots \epsilon_n x_n\|_0^{p_0} d\mu\right)^{1/p_0} \leq B_0(\|x_1\|_0^{p_0} + \cdots + \|x_n\|_0^{p_0})^{1/p_0}.$$

REMARK: If we relax the conditions that the quasinorms are continuous then it is still possible to conclude that X is type p_0, but without the precise bound on the type p_0 constant.

We omit the proof of Theorem 5 in view of its similarity to Theorem 4. Let us note that these results can be applied to complex interpolation of two spaces (E,F) by regarding this as a special case of a family. Let $H(w,x) = \|x\|_E$ if $0 < \arg w < 2\pi\theta$ and $H(w,x) = \|x\|_F$ otherwise; then $X_0 = (E,F)_\theta$. The separability assumptions now become unnecessary. Suppose E is p-normable where $p < 1$ and F is a Banach space of type q where $q > 1$. Then $(E,F)_\theta$ is isomorphic to a Banach space provided $\frac{\theta}{p} + \frac{1-\theta}{q} < 1$. If θ is chosen at the critical value i.e. $\frac{\theta}{p} + \frac{1-\theta}{q} = 1$ then $(E,F)_\theta$ is type one, but need not be locally convex. An easy example is given by interpolating between weak $L_{1/2}$ and L_2.

Let us note here that if we interpolate in the last example between $L_{1/2}$ and L_2 then the critical space is L_1 and is locally convex. This result is related to the lattice structure. The notion of lattice p-convexity studied in [6] interpolates nicely and a lattice 1-convex space is locally convex. The distinction here is that weak $L_{1/2}$ is $\frac{1}{2}$-normable but not lattice $\frac{1}{2}$-convex.

Finally let us note that if E is p-normable and F is a Hilbert space then if we choose θ to satisfy $\frac{\theta}{p} + \frac{1-\theta}{2} = 1$ then $(E,F)_\theta$ can be renormed to be type one with constant one. This point is discussed, for Banach spaces, in the context of the Clarkson inequalities by Pisier [9].

4. A-convexity. Let X be a quasinormed space with the property that the quasinorm is a Borel function on each finite-dimensional subspace. Then we can define $H : \mathbf{T} \times X \to [0,\infty]$ by $H(w,x) = \|x\|$ for all w. In this case the interpolated quasinorm $\| \ \|_0$ on X is given by

$$\|x\|_0 = \inf \int_{\mathbf{T}} \|g(w)\| d\lambda(w)$$

where $g \in \mathcal{A}$ runs through all maps with with $g(0) = x$. We shall denote the interpolated quasinorm $\| \ \|_0$ in this case by $\| \ \|_A$. This corresponds, when the original quasinorm is continuous, with the terminology introduced in [7]. It is shown in [7] that $\| \ \|_A$ is plurisubharmonic at least when the original quasinorm is p-subadditive for some $p > 0$. This result will be reproved, for separable spaces, and extended below. We say that X is A-convex if $\| \|_A$ is equivalent to the original quasinorm; A-convexity may be reformulated in terms of the Maximum Modulus Principle or the existence of an equivalent plurisubharmonic quasinorm, as in [7]. Conversely X is A-trivial if $\|x\|_A = 0$ for every $x \in X$. Tabacco [12] uses the A-trivial space L_p/H_p for $0 < p < 1$ to show that in general complex interpolation of quasi-Banach spaces can yield trivial spaces.

Let us now describe our general setup. Suppose $H : \mathbf{T} \times X \to [0, \infty]$ defines a uniformly separable family of quasinormed spaces, with the additional property that the associated quasinorms $\| \|_w$ are upper-semi-continuous . Then we define

$$H'(w, x) = \inf_{y \in X} \int_{\mathbf{T}} H(w, x + zy) d\lambda(z).$$

Then it is readily verified that H' also defines a uniformly separable family of quasinormed spaces (with the constants $c(w)$ unchanged) and has the property that the associated quasinorms $\| \|'_w$ are upper-semi-continuous on $(X_w, \| \|_w)$. Furthermore the spaces X'_w coincide with the spaces X_w as linear subspaces of X.

LEMMA 6. *Under the hypotheses above, we have* $H'_0 = H_0$.

PROOF: Suppose $x \in X$ and that $g \in \mathcal{A}$ is such that $g(0) = x$. Next suppose $y_1, \ldots, y_k \in X$ and that $\phi_1, \ldots, \phi_k$ are polynomials. Define $h(z) = \phi_1(z)y_1 + \cdots + \phi_k(z)y_k$. For any $m \in \mathbf{N}$ we have

$$H_0(x) \leq \int_{\mathbf{T}} H(w, g(w) + w^m h(w)) d\lambda(w).$$

Now for almost every fixed w we have for every continuous function ψ on $\mathbf{T}$

$$\lim_{N \to \infty} \frac{1}{N} \sum_{j=1}^{N} \psi(w^j) = \int_{\mathbf{T}} \psi d\lambda.$$

For such w let ψ be any continuous function on $V_w = [g(w), y_1, \ldots, y_k]$ with $\psi(u) \geq H(w,u)$ for $u \in X$. Then

$$\lim_{N\to\infty} \frac{1}{N}\sum_{j=1}^{N} \psi(g(w) + w^j h(w)) = \int_{\mathbf{T}} \psi(g(w) + zh(w))d\lambda(z).$$

By taking infima over all such ψ we conclude that

$$\limsup_{N\to\infty} \frac{1}{N}\sum_{j=1}^{N} H(w, g(w) + w^j h(w)) \leq \int_{\mathbf{T}} H(w, g(w) + zh(w))d\lambda(z).$$

Now by integration we obtain that

$$H_0(x) \leq \int\int H(w, g(w) + zh(w))d\lambda(z)d\lambda(w).$$

Now if $B_1, \ldots, B_k$ are any disjoint Borel sets we can find uniformly bounded sequences of polynomials $\phi_j^{(n)}$ so that the functions $|\phi_j^{(n)}|$ converge almost everywhere to the characteristic functions χ_{B_j}. Define $h^{(n)}(w) = \sum \phi_j^{(n)}(w)y_j$. Suppose $w \in B_j$; except for w in a set of measure zero, we may pick α_n so that $|\alpha_n| = 1$ and $\alpha_n\phi_j^{(n)}(w) \to 1$, while $\alpha_n\phi_r^{(n)}(w) \to 0$ for $r \neq j$. Then

$$\int H(w, g(w) + zh^{(n)}(w))d\lambda(z) = \int H(w, g(w) + \alpha_n zh^{(n)}(w))d\lambda(z).$$

Letting $n \to \infty$ we have (for almost every $w \in B_j$)

$$\limsup \int H(w, g(w) + zh^{(n)}(w))d\lambda(z) \leq \int H(w, g(w) + zy_j)d\lambda(z).$$

Thus

$$H_0(x) \leq \sum_{j=1}^{N} \int_{B_j}\int_{\mathbf{T}} H(w, g(w) + zy_j)d\lambda(z)d\lambda(w).$$

Note that equation holds for any choice of $y_1, \ldots, y_k$ and $B_1, \ldots, B_k$. In particular let $(y_n,\ n \in \mathbf{N})$ be any sequence in X whose intersection with each X_w is dense. Clearly we have for each $k \in \mathbf{N}$,

$$H_0(x) \leq \int_{\mathbf{T}} \inf_{j\leq k} \left(\int_{\mathbf{T}} H(w, g(w) + zy_j)d\lambda(z)\right) d\lambda(w).$$

Letting $k \to \infty$ since each $H(w,.)$ is upper-semi-continuous we obtain

$$H_0(x) \leq \int_{\mathbf{T}} H'(w, g(w))d\lambda(w).$$

Now by allowing g to vary we have $H_0(x) \leq H_0'(x)$.

We then define a sequence by $H^{(n)} = H^{(n-1)\prime}$ for $n > 0$ with $H^{(0)} = H$. This sequence is monotone decreasing, and so we can define a limit by $H^a(w,x) = \lim H^{(n)}(w,x)$. It is again the case that H^a defines a uniformly separable family of quasinorms. Also each $H^{(n)}(w,.)$ and $H^a(w,.)$ is upper-semi-continuous on $(X_w, \| \ \|_w)$. It is easy to verify that $H_0^a(x) = \lim H_0^{(n)}(x)$ and hence by the preceding lemma we also have:

LEMMA 7. $H_0^a = H_0$.

THEOREM 8. *Let X be a separable quasinormed space such that the quasinorm is upper-semi-continuous . Then $\| \ \|_A$ is plurisubharmonic on X. Furthermore if the sequence of quasinorms $\| \ \|^{(n)}$ is defined by $\| \ \|^{(0)} = \| \ \|$ and*

$$\|x\|^{(n)} = \inf_{y \in X} \int_{\mathbf{T}} \|x + zy\|^{(n-1)} d\lambda(z)$$

then for each x we have $\lim_{n\to\infty} \|x\|^{(n)} = \|x\|_A$.

PROOF: Define $H(w,x) = \|x\|$ for all $w \in \mathbf{T}$. Then $H_0(x) = \|x\|_A$ for all $x \in X$. Furthermore $H^{(n)}(w,x) = \|x\|^{(n)}$ for all w, x. Thus $\|x\|_A = H_0(x) = H_0^a(x) \leq H_0^{(n)}(x) \leq \|x\|^{(n)}$ for all n. Now for any x, y we have

$$\begin{aligned}\lim_{n\to\infty} \|x\|^{(n)} &\leq \lim_{n\to\infty} \int_{\mathbf{T}} \|x + zy\|^{(n-1)} d\lambda(z) \\ &= \int_{\mathbf{T}} \lim_{n\to\infty} \|x + zy\|^{(n)} d\lambda(z).\end{aligned}$$

Since the function $\lim \| \ \|^{(n)}$ is upper-semi-continuous we conclude that it is plurisubharmonic. Hence if $g \in \mathcal{A}$ with $g(0) = x$ then

$$\lim_{n\to\infty} \|x\|^{(n)} \leq \int_{\mathbf{T}} \|g(w)\| d\lambda(w)$$

so that $\lim \|x\|^{(n)} \leq \|x\|_A$ and the theorem is proved.

THEOREM 9. *Let $H : \mathbf{T} \times X \to [0,\infty]$ induce a uniformly separable analytic family o upper-semi-continuously quasi-normed spaces. Define $H^*(w,x) = \|x\|_{w,A}$ for $w \in \mathbf{T}$, $x \in X$. Then H^* also induces a uniformly separable analytic family of quasinormed spaces anc $H_0^* = H_0$.*

PROOF: In fact by the above theorem, $H^* = H^a$ and so this reduces to Lemma 7.

COROLLARY 10. *Let H induce a uniformly separable analytic family of upper-semi-continuously quasinormed spaces X_w such that X_w is A-trivial on a set of positive measure. Then $H_0 = 0$.*

PROOF: This is immediate from the preceding theorem.

Theorems 8 and 9 might suggest that we can expect the interpolated quasinorm H_0 itself to be plurisubharmonic. However, this is false and we now present a simple three-dimensional example to show that this is not the case. In fact, our example also shows that there is no iteration theorem in the quasinormed setting, in contrast to the normed setting (see [1], Corollary 4.2).

We will consider $\mathbf{C}^3$ with the ℓ_∞-norm which we denote by $\| \ \|$. Fix $\kappa > 1$ and $0 < p < 1$. Denote by $(e_j, j = 1,2,3)$ the standard basic vectors in $\mathbf{C}^3$ and then for $u, w \in \mathbf{T}$ define $\xi_{u,w} \in \mathbf{C}^3$ by $\xi_{u,w} = e_1 + ue_2 + w\bar{u}e_3$. We then define $\| \ \|_w$ to be the greatest p-subadditive quasinorm such that $\|x\|_w \le \kappa\|x\|$ for $x \in \mathbf{C}^3$ and $\xi_{u,w} \le 1$ for all $u \in \mathbf{T}$. It is immediate from the definition that we have $\|x\| \le \|x\|_w \le \kappa\|x\|$ for all x.

Now define $H(w,x) = \|x\|_w$. It may be checked that H is continuous on $\mathbf{T} \times X$. Thus we may interpolate and we will have for all z with $|z| < 1$, $\|x\| \le H_z(x) = \|x\|_z \le \kappa\|x\|$.

If we set $g(z) = e_1 + ue_2 + z\bar{u}e_3$ where $u \in \mathbf{T}$ is fixed we can see that $\|e_1 + ue_2 + z\bar{u}e_3\|_z \le 1$ and hence since each interpolated quasinorm is p-subadditive,

$$\|e_1 + ue_2\|_z \le (1 + \kappa^p|z|^p)^{1/p}.$$

Now suppose that H_0 is plurisubharmonic. Then using the above equation for $z = 0$ we have $H_0(e_1) = \|e_1\|_0 = 1$. Equally, if the iteration theorem is valid then we would have for any r, $0 < r < 1$,

$$\begin{aligned}\|e_1\|_0 &\le \int_{\mathbf{T}} \|e_1 + we_2\|_{rw}\, d\lambda(w)\\ &\le (1 + \kappa^p r^p)^{1/p}\end{aligned}$$

so that $\|e_1\|_0 \le 1$.

We will show however that $\|e_1\|_0 > 1$. Let us suppose $0 < \epsilon < 1$. In the ensuing argument we will adopt the convention that $\delta = \delta(\epsilon)$ represents a function such that $\delta(\epsilon) \to 0$ as $\epsilon \to 0$; δ may depend on p, κ, and is allowed to vary from line to line.

Suppose $\|e_1\|_0 \le 1$. Then there exists $g \in \mathcal{A}$ such that $g(0) = (1 - \epsilon^3)e_1$ and $\|g(w)\|_w < 1$ for all $w \in \mathbf{T}$. We may write $g(z) = \phi_1(z)e_1 + \phi_2(z)e_2 + \phi_3(z)e_3$ where each ϕ_j is in H_∞ with $\|\phi_j\|_\infty \le 1$. First we observe that

$$\begin{aligned}\int_{\mathbf{T}} |1 - \phi_1(w)|^2 d\lambda(w) &\le 2 - 2\Re \int_{\mathbf{T}} \phi_1(w) d\lambda(w)\\ &\le 2\epsilon^3.\end{aligned}$$

Let A denote the set where $|1 - \phi_1(w)| \le \epsilon$. then $\lambda(A) \ge 1 - 2\epsilon$. If $w \in A$ then we can write

$$g(w) = y + \sum_{n=1}^{\infty} \alpha_n \xi_{u_n, w}$$

where

$$\sum_{n=1}^{\infty} |\alpha_n|^p + \kappa^p \|y\|^p < 1.$$

Thus

$$\phi_1(w) = y_1 + \sum_{n=1}^{\infty} \alpha_n.$$

Hence

$$1 - \epsilon \le |y_1| + \sum_{n=1}^{\infty} |\alpha_n|.$$

Now, since $p < 1 < \kappa$ we conclude that

$$\max_n |\alpha_n| \geq 1 - \delta(\epsilon)$$

and so there exists m such that

$$\|g(w) - \alpha_m \xi_{u_m, w}\| \leq \delta(\epsilon).$$

Considering the first co-ordinate we have

$$|1 - \alpha_m| \leq \delta(\epsilon)$$

and so

$$\|g(w) - \xi_{u_m, w}\| \leq \delta(\epsilon).$$

This in turn implies, for $w \in A$,

$$|\phi_2(w)\phi_3(w) - w| \leq \delta(\epsilon).$$

Hence

$$\int_{\mathbf{T}} |1 - \bar{w}\phi_2(w)\phi_3(w)| d\lambda(w) \leq \delta(\epsilon).$$

However ϕ_2, ϕ_3 both have zeros at the origin and so

$$\int_{\mathbf{T}} \bar{w}\phi_2(w)\phi_3(w) d\lambda(w) = 0.$$

This contradiction completes the example.

References.

1. R. Coifman, M. Cwikel, R. Rochberg, Y. Saghar and G. Weiss, *The complex method for interpolation of operators acting on families of Banach spaces*, 123-153 of Springer Lecture Notes 779, Berlin-Heidelberg-New York, 1980.

2. R. Coifman, M. Cwikel, R. Rochberg, Y. Saghar and G. Weiss, *A theory of complex interpolation for families of Banach spaces,* Adv. Math. 33(1982) 203-229.

3. P. L. Duren, Theory of H^p spaces, Academic press, New York-London 1970.

4. N. J. Kalton, *The three space problem for locally bounded F-spaces,* Comp. Math. 37(1978) 243-276.

5. N. J. Kalton, *Convexity, type and the three space problem,* Studia Math. 69(1981) 247-287.

6. N. J. Kalton, *Convexity conditions for non-locally convex lattices,* Glasgow Math. J. 25(1984) 141-152.

7. N. J. Kalton, *Plurisubharmonic functions on quasi-Banach spaces,* Studia Math., 84(1986) 297-324.

8. N. J. Kalton, N. T. Peck and J. W. Roberts, An F-space sampler, London Math. Soc. Lecture Notes No. 86, Cambridge University Press, 1985.

9. G. Pisier, *Some applications of the complex interpolation method to Banach lattices,* J. d'Analyse Math. 35(1979) 264-281.

10. R. Rochberg, *A generalization of Szego's theorem and the power theorem for complex interpolation,* to appear.

11. A. Tabacco Vignati, Ph. D. Dissertation, Washington University, St. Louis, 1986.

12. A. Tabacco Vignati, *Complex interpolation for families of quasi-Banach spaces,* to appear.

An Application of Edgar's Ordering of Banach Spaces

Lawrence H. Riddle*

Department of Mathematics and Computer Science
Emory University
Atlanta, GA 30322

Abstract

We examine and extend several results about the universal Pettis integral property using a uniform approach via Edgar's ordering structure on Banach spaces. This reduces the problem from one concerning integrable functions to a purely Banach space setting.

*This work was partially supported by Emory University Research Committee Grant 2-50113. The author would also like to thank the Department of Mathematics of The Ohio State University for their hospitality during his visit.

1980 Mathematics Subject Classification. Primary 46B20, 46G10.

1. Introduction

Originally defined in 1938 by B. J. Pettis, the Pettis integral lay dormant for forty years, elusive and seemingly banished to the realm of mathematical curiosities. Since 1978, however, substantial progress has been made, particularly for functions taking values in the dual of a Banach space. There are essentially two basic ways in which to study the integrability properties of functions. One is to concentrate on a particular function $f : \Omega \rightarrow X$ and find conditions on f for which it will be Pettis integrable. This can be done, for example, in terms of an appropriate convergence of simple functions, looking at the core of the function [9], or examining the set $\{ x^*f : ||x^*|| \leq 1 \}$. This approach has successfully lead to various characterizations of the Pettis integral, for which the monograph [16] by M. Talagrand provides an excellent reference. Recent results have also been attained by Riddle-Saab [12] and Andrews [1] on functions that are universally Pettis integrable, that is, functions defined on a compact Hausdorff space and Pettis integrable with respect to all Radon measures on that space.

The other approach is to study the Banach space X and find conditions on X for which *all* functions into X will be Pettis integrable under certain suitable measurability conditions. Edgar's seminal papers [3,4] of 1977 and 1979 introduced the Pettis Integral Property (PIP) for Banach spaces. The space X has the PIP if it has the μ-PIP with respect to all finite measure spaces (Ω, Γ, μ). Later, in 1983, Riddle-Saab-Uhl [13] defined the notion of the universal Pettis integrability property (UPIP) for Banach spaces. A Banach space X is said to have the UPIP if for every compact Hausdorff space K, every bounded function $f : K \rightarrow X$ that is scalarly measurable with respect to every Radon measure on K is Pettis integrable with respect to every Radon measure. It was shown in that paper that the duals of separable Banach spaces have the UPIP and it was asked if the duals of weakly compactly generated spaces have this property. Andrews [1] recently gave several conditions on a Banach space X under which X^* has the UPIP.

In 1983 Edgar [6] published a paper in which he defined and studied an ordering for the Banach spaces. His original interest in this ordering arose from studying the Pettis integral property. At the end of the paper he asked if there is a largest PIP space and if there is a space X_0 such that a Banach space X has the PIP if and only if X precedes X_0. This paper examines an extension of the universal Pettis integral property in terms of Edgar's order, showing how many of the results known about the UPIP can be derived by an application of this ordering.

Notation and terminology concerning the Pettis integral generally match Riddle-Saab-Uhl [13]. If Ω is a measure space and μ a probability measure on a σ-algebra Σ of subsets of Ω, then a bounded function $f : \Omega \to X$ is μ–scalarly measurable if the real-valued function $x^*f(\cdot)$ is μ-measurable. In addition, the function f is called μ–Pettis integrable if for each measurable set E there is an element x_E of X that satisfies

$$x^*(x_E) = \int_E x^*f \, d\mu$$

for every x^* in X^*.

The space X has the μ–Pettis Integral Property (μ–PIP) if every bounded μ–scalarly measurable function with values in X is μ–Pettis integrable. We say that a Banach space X has the Radon Pettis Integral Property if it has the μ–PIP for all Radon probability measures on compact Hausdorff spaces, and that X has the perfect Pettis Integral Property (perfect-PIP) if it has the μ–PIP for all perfect probability measures. A finite measure space (Ω , Σ , μ) is *perfect* if for each measurable function $f : \Omega \to \mathrm{R}$ and each subset F of $\mathbf{R}$ for which $f^{-1}(F) \in \Sigma$, there is a Borel set $G \subset F$ with $\mu f^{-1}(G) = \mu f^{-1}(F)$. Alternatively, for each $\varepsilon > 0$ there is a compact set $K \subset f[X]$ such that $\mu(X \setminus f^{-1}[K]) \leq \varepsilon$. See [14],[16] or [8] for additional information about perfect measures.

The Radon-PIP and perfect-PIP are intermediate properties between the PIP of Edgar and the UPIP of Riddle-Saab-Uhl in the sense that

$$\text{PIP} \Rightarrow \text{perfect-PIP} \Rightarrow \text{Radon-PIP} \Rightarrow \text{UPIP}.$$

The middle implication follows since a finite Radon measure is perfect [16]. We shall restrict our attention to the "well-behaved" perfect measures spaces to avoid the types of measure spaces encountered in the Fremlin and Talagrand proof [8] that l_∞ does not have the PIP. On the other hand, the particular attention to just one perfect measure means we do not need to have our function be universally scalarly measurable as with the UPIP. However, we do not know if perfect-PIP is actually different from either the Radon-PIP or the UPIP. Fremlin and Talagrand asked in [8] which Banach spaces have perfect-PIP or Radon-PIP.

2. Edgar's ordering and the perfect-PIP

Let X and Y be Banach spaces. Then $X < Y$ in Edgar's ordering [6] if and only if

$$X = \bigcap \mathrm{T}^{**-1}(Y),$$

where the intersection is taken over all bounded linear operators $\mathrm{T} : X \to Y$. An alternative way to express this definition is that $X < Y$ if and only if whenever x^{**} is an element of X^{**} such that $\mathrm{T}^{**}(x^{**})$ is in Y for every operator $\mathrm{T} : X \to Y$, then x^{**} belongs to X. The operation $<$ is transitive and reflexive and thus defines a partial order on the equivalence classes obtained by the usual equivalence relation. We shall continue to consider individual spaces, however, rather than their equivalence classes.

An easy adaptation of a proof in Edgar's paper shows that this order preserves the perfect-PIP. For completeness we include the proof here.

THEOREM 1: *Let X,Y be Banach spaces and let $X < Y$. If the space Y has the perfect-PIP, then so does X.*

Proof. Let μ be a perfect measure on Ω and let $f : \Omega \to X$ be a bounded scalarly measurable function with respect to μ. Let A be a measurable subset of Ω and define x^{**} in X^{**} by

$$x^{**}(x^*) = \int_A x^* f \, d\mu.$$

The functional x^{**} is called the Dunford integral of f over A and its existence as an element of X^{**} is an easy application of the closed graph theorem. To show that f is Pettis integrable we must show that x^{**} is actually an element of X.

Suppose $\mathrm{T} : X \to Y$ is a bounded linear operator. Then $\mathrm{T} \circ f$ is a bounded scalarly measurable function into Y. Since Y has the perfect-PIP, there is an element y_A in Y such that

$$y^*(y_A) = \int_A y^* \mathrm{T} \circ f \, d\mu$$

for every y^* in Y^*. Thus

$$\mathrm{T}^{**} x^{**}(y^*) = x^{**}(\mathrm{T}^* y^*) = \int_A \mathrm{T}^* y^* f \, d\mu = \int_A y^*(\mathrm{T} f) \, d\mu = y^*(y_A)$$

for every y^* in Y^* and so we see that $\mathrm{T}^{**}(x^{**}) = y_A$. This says that $\mathrm{T}^{**}(x^{**})$ is an element of Y for every operator T. Therefore x^{**} is an element of X since $X < Y$, and this completes the proof. ■

What is an example of a space with the perfect-PIP? Fremlin and Talagrand [8] have provided an answer contingent on special set axioms ([8], [16]).

Axiom K: There is a cardinal κ such that (i) there is a set of cardinality κ in $[0,1]$ that has Lebesgue outer measure 1, and (ii) the interval $[0,1]$ is not expressible as the union of κ sets of zero Lebesgue measure 0.

Axiom L: The interval [0,1] cannot be covered by fewer than the power of the continuum closed sets of zero Lebesgue measure.

Both of these are known to be consistent. Axiom L is a weakening of the Continuum Hypothesis, which we shall assume for the rest of this paper. Under this assumption, then, the Fremlin-Talagrand results says the following:

THEOREM 2 *(Fremlin-Talagrand): (*K or L*) The space l_∞ has the μ–PIP for every perfect probability measure μ.*

It is not known whether special set axioms are really necessary to prove this result. The work involved in proving the theorem consists in showing that if (f_n) is a sequence of pointwise bounded measurable functions defined on a perfect measure space such that every pointwise cluster point is measurable, then every pointwise cluster point is the almost everywhere limit of some subsequence. If $f : \Omega \rightarrow l_\infty$ is then a bounded μ-scalarly measurable function, one considers the two topologies on $C = \{ x^*f : ||x^*|| \leq 1 \}$ given by convergence in measure, τ_m, and pointwise convergence, τ_p. Then τ_m is regular, τ_p is separable, and the result mentioned above shows that for each countable set $D \subset C$, $\overline{D}^{\tau_p} \subset \overline{D}^{\tau_m}$. These three facts are enough to insure that $\tau_m \subset \tau_p$, and thus that the map sending a function in C to its integral with respect to μ is τ_p-continuous (since it is τ_m-continuous). Then f is Pettis integrable with respect to μ.

The Fremlin-Talagrand result says the l_∞ has the perfect-PIP (under the CH, for example). We immediately obtain the following corollary.

COROLLARY 3: (CH) *If $X < l_\infty$, then X has the perfect-PIP.*

Thus, to show that a space X has the perfect-PIP, it suffices to show that $X < l_\infty$ in Edgar's ordering. A very convenient characterization of such spaces appears in [6].

THEOREM 4 *(Edgar): Let X be a Banach space. Then $X < l_\infty$ if and only if whenever $x^{**} \in X^{**}$ is weak*-continuous on all bounded weak*-separable subsets of X^*, then $x^{**} \in X$.*

Suppose X is a separable Banach space. Then the unit ball of X^{**} is a bounded weak*-separable subset. Every functional x^{***} in X^{***} that is weak*-continuous on the unit ball of X^{**} actually belongs to X^*, so that Theorem 4 and Corollary 3 immediately give

COROLLARY 5: *If X is separable, then X^* has the perfect-PIP.*

A Banach space X has *property (C)* if and only if any collection of closed convex subsets of X with the countable intersection property has nonvoid intersection [2]. The class of spaces with property (C) include the K-analytic, weakly compactly generated, reflexive, and separable spaces, as well as various function spaces. For more details see [11]. By combining several results in [16], one can show that property (C) implies the PIP. We can get the following weaker result very easily, however, using Edgar's ordering.

THEOREM 6: *If X has property (C), then $X < l_\infty$ and thus has the perfect-PIP.*

Proof. Let x^{**} be an element of X^{**} that is weak*-continuous on all bounded weak*-separable subsets of X^*. We need to show that x^{**} is weak*-continuous on the unit ball of X^*. Suppose not. Then there exists a net (x^*_α), $\alpha \in \Gamma$, in the ball of X^* that converges weak* to 0, but for which $x^{**}(x^*_\alpha)$ does not converge to 0. By taking appropriate subnets, if necessary, and possibly replacing x^{**} by $-x^{**}$, we may assume without loss of generality that $x^{**}(x^*_\alpha) > \varepsilon$ for all α, for some $\varepsilon > 0$.

Let $A = \{x^*_\alpha : \alpha \in \Gamma\}$. Then 0 belongs to $\overline{A}^{w^*}$. For each x^* in A define A_{x^*} to be the set $\{ x : x^*(x) \geq 1 \}$ in X. Then each of these sets is non-empty, convex and closed with

$$\bigcap_{x^* \in A} A_{x^*} = \phi.$$

Thus there is a countable set $C \subset A$ with

$$\bigcap_{x^* \in C} A_{x^*} = \phi.$$

An easy application of the Hahn-Banach theorem now shows that 0 belongs to $\overline{\text{conv}(C)}^{w^*}$. Take a net (y^*_β) in the convex hull of C that converges weak* to 0. Each y^*_β has the form

$$y^*_\beta = \sum_{i=1}^{n} t_i(\beta) x^*_i(\beta)$$

where each $x^*_i(\beta)$ belongs to C and $\sum_{i=1}^{n} t_i(\beta) = 1$. But then

$$x^{**}(y^*_\beta) = \sum_{i=1}^{n} t_i(\beta) x^{**}(x^*_i(\beta)) \geq \sum_{i=1}^{n} t_i(\beta)\varepsilon = \varepsilon$$

which contradicts the fact that x^{**} is weak*-continuous on the bounded weak*-separable set $\overline{\text{conv}(C)}^{w^*}$.

Therefore x^{**} must be weak*-continuous on the ball of X^*, and thus belongs to X. By Theorem 4, we have $X < l_\infty$. ■

If x^{***} is an element of X^{***}, then it can be written in the form $x^{***} = x^* + y^{***}$ where x^* is an element of X^* and y^{***} is an element of X^{***} that satisfies $y^{***}(X) = 0$. Now x^*, as an element of X^{***}, is weak*-continuous on all subsets of X^{**}, so that if x^{***} is weak*-continuous on bounded weak*-separable subsets of X^{**}, then so, too, is y^{***}. Thus another way of viewing Edgar's theorem for dual Banach spaces is the following.

THEOREM 4* *(revisited): Let X be a Banach space. Then $X^* < l_\infty$ if and only if whenever $x^{***} \in X^{***}$ is weak*-continuous on all bounded weak*-separable subsets of X^{**} and $x^{***}(X) = 0$, then $x^{***} = 0$.*

We use this observation for proving the next sufficient condition for the perfect-PIP. This condition comes from Andrews' study [1] of the UPIP using the idea of a weak*-core. It may be viewed as a natural weakening of Mazur's condition on X^*, which is known to imply that X^* has the PIP [4]. Included in the hypotheses is the assumption that for the weak*-topology of X^*, each compact separable set is metrizable. Dual spaces whose preduals are weakly K-analytic exhibit this behavior.

THEOREM 7: *If every weak*-sequentially continuous functional on X^{**} is in the weak*-closure of a weak*-separable subset of X^*, and if in the weak*-topology of X^* every compact separable subset is metrizable, then $X^* < l_\infty$, and consequently X^* has the perfect-PIP.*

Proof. Let x^{***} be an element of X^{***} that is weak*-continuous on bounded weak*-separable subsets of X^{**} and that satisfies $x^{***}(x) = 0$ for all x in X. We want to show that $x^{***} = 0$. First note that x^{***} is weak*-sequentially continuous. Thus we may choose a weak*-separable subset S of X^* such x^{***} belongs to $\overline{S}^{w^*}$. Let Y be the weak*-closed linear span of S in X^*. Then the unit ball of Y is a weak*-compact, weak*-separable subset of X^*, and thus is weak*-metrizable in X^* by hypothesis.

Let (y_n^*) be a weak*-dense sequence in the unit ball of Y. For each integer m select a sequence $(z_{m,n})_{n=1}^{\infty}$ in X such that $||z_{m,n}|| = 1$ and

$$|y_m^*(z_{m,n})| \geq (1 - \frac{1}{n}) ||y_m^*||.$$

Let Z be the closed linear span of the set $\{z_{m,n} : m, n \in \mathrm{N}\}$. Then Z is a norming subspace for Y, that is,

$$||y^*|| = \sup |y^*(z)|$$

where the supremum is taken over all z of norm 1. If we now define $R : X^* \to Z^*$ to be the restriction operator, that is, $Rx^* = x^*|_Z$, then R is an isometry from Y to R(Y).

The functional $R^{**}x^{***}$ is weak*-continuous on all bounded weak*-separable subsets of Z^{**}, and thus belongs to Z^* since Z is separable. For each z in Z we have

$$R^{**}x^{***}(z) = x^{***}(R^*z) = x^{***}(z) = 0.$$

Therefore $R^{**}x^{***} = 0$.

We are now ready to show that $x^{***} = 0$. Towards this end, let x^{**} be in the unit ball of X^{**} and select a net (x_α) in the ball of X that converges weak* to x^{**}. For each α, define the linear function l_α on R(Y) by $l_\alpha(Ry) = y(x_\alpha)$. Note that

$$|l_\alpha(Ry)| = |y(x_\alpha)| \leq ||y||_{X^*} ||x_\alpha|| = ||Ry||_{Z^*} ||x_\alpha||$$

and thus each l_α is continuous on R(Y). Invoke the Hahn-Banach theorem to extend each l_α to a z_α^{**} in the ball of Z^{**}. Now let z^{**} be a weak*-cluster point of the net (z_α^{**}). There exists a subnet, which we continue to call (z_α^{**}), that converges weak* to z^{**}. Observe that for each y in Y we have

$$\begin{aligned} x^{**}(y) &= \lim_\alpha y(x_\alpha) = \lim_\alpha l_\alpha(Ry) \\ &= \lim_\alpha z_\alpha^{**}(Ry) = z^{**}(Ry) = R^*z^{**}(y). \end{aligned}$$

Thus $x^{**}(y) = R^*z^{**}(y)$ for each y in Y, and hence also for x^{***} since x^{***} is in the weak*-closure of a subset of Y. Therefore

$$x^{***}(x^{**}) = x^{***}(Rz^{**}) = R^{**}x^{***}(z^{**}) = 0.$$

Since this is true for each x^{**} in the ball of X^{**}, we must have $x^{***} = 0$, as desired. Therefore $X^* < l_\infty$. ■

3. Measurable cardinals, $l_1(\Gamma)$ and the perfect-PIP

We next turn our attention to $l_1(\Gamma)$, the space of real-valued summable functions on an abstract set Γ. Its behavior with respect to Edgar's ordering and the perfect-PIP depends on whether or not the cardinality of Γ is a measurable cardinal. A cardinal κ is said to be a real-valued (resp. 2-valued) measurable cardinal if there exists a set Γ having cardinality κ and a real-valued (resp. $\{0,1\}$-valued) measure defined on all subsets of Γ that vanishes on singletons. It is consistent with the standard axiom systems for set theory that neither real-measurable nor 2-valued measurable cardinals exist. There are models of ZFC, for example the "constructible sets", in which there is no measurable cardinal. The existence of a measurable cardinal, however, can never be shown to be consistent, even assuming the consistency of ZFC, by Godel's Theorem; that is, to show consistency of the existence of measurable cardinals within ZFC would be tantamount to showing the consistence of ZFC within itself, an impossibility. It is known that if a real-measurable cardinal exists, then Lebesgue measure has an extension to all subsets of the real line (see, for example, [16]). Thus the continuum would be real-measurable. Two-valued measurable cardinals, if they exist, must be enormous for the class of all non two-valued measurable cardinals is a closed class containing $_0$ (that is, it is strongly inaccessible) [10]. This means that the class is closed under all the standard operations for forming cardinals from given ones: addition, multiplication, the formation of suprema, exponentiation, and the passage from a given cardinal to its immediate successor or to any smaller cardinal.

Recall that the dual of $l_1(\Gamma)$ is $l_\infty(\Gamma)$ and that the dual of $l_\infty(\Gamma)$ may be identified with $ba(\Gamma, P(\Gamma))$, the space of bounded finitely additive set functions defined on all subsets of Γ. The following result is apparently part of folklore. As we cannot find a reference, however, we include its proof here.

LEMMA 8: $l_1(\Gamma) = ca(\Gamma, P(\Gamma))$ *if and only if card(Γ) is not a real-valued measurable cardinal.*

Proof. Let $Q : l_1(\Gamma) \to ba(\Gamma, P(\Gamma))$ denote the natural embedding of $l_1(\Gamma)$ into $l_1(\Gamma)^{**}$ given by

$$Qx(A) = \sum_{\alpha \in A} x_\alpha .$$

It is always the case that $Q(l_1(\Gamma))$ is contained in $ca(\Gamma, P(\Gamma))$. Now suppose that card(Γ) is a real-valued measurable cardinal. Then there exists a μ in $ca(\Gamma, P(\Gamma))$ such that $\mu \neq 0$ and $\mu(\{\alpha\}) = 0$ for each α in Γ. But then it is impossible to have $\mu = Qx$ for some nonzero x in $l_1(\Gamma)$, and so μ is not an element of $Q(l_1(\Gamma))$. Thus $l_1(\Gamma) \neq ca(\Gamma, P(\Gamma))$.

Conversely, suppose card(Γ) is not a real-valued measurable cardinal. Define $T : ca(\Gamma, P(\Gamma)) \to l_1(\Gamma)$ by $T\mu(\alpha) = \mu(\{\alpha\})$. Why is $T\mu$ an element of $l_1(\Gamma)$? Recall that

$$\sum_{\alpha \in \Gamma} |\mu(\alpha)| = \lim \sum_{\alpha \in F} |\mu(\alpha)|$$

where the limit is taken with respect to the net of finite subsets of Γ directed by inclusion. If this sum diverges, then there exist finite sets $F_1 \subset F_2 \subset \cdots$ such that

$$\sum_{\alpha \in F_n} |\mu(\alpha)| > n.$$

Since $|\mu|$ is countable additive, we would now obtain the contradiction

$$|\mu|(\bigcup_n F_n) = \lim_n |\mu|(F_n) = \lim_n \sum_{\alpha \in F_n} |\mu|(\alpha) \geq \lim_n \sum_{\alpha \in F_n} |\mu(\alpha)| = \infty.$$

Therefore we must have $T\mu \in l_1(\Gamma)$. Moreover, using the fact that for only countably many of the α's do we have $\mu(\alpha) \neq 0$, it is not difficult to see that T is a bounded operator. The operator T is also 1-1 since card(Γ) is not measurable, and onto. Indeed, we actually have that $T = Q^{-1}$. Thus $l_1(\Gamma) = ca(\Gamma, P(\Gamma))$. ■

When does $l_1(\Gamma)$ precede l_∞? The next result shows that this occurs if Γ is not "too large".

THEOREM 9: *If card(Γ) is not measurable, then* $l_1(\Gamma) < l_\infty$.

Proof. Suppose $\mu \in l_1(\Gamma)^{**}$ is weak*-continuous on every bounded separable subset of $l_1(\Gamma)^*$. We consider μ as an element of $ba(\Gamma, P(\Gamma))$. To show that μ belongs to $l_1(\Gamma)$, it suffices by Lemma 8 to show that μ is countably additive. Take a sequence of sets (A_n) in Γ decreasing to the empty set. Then χ_{A_n} converges weak* to 0 in $l_\infty(\Gamma)$. Now $\{\chi_{A_n} : n \in N\} \cup \{0\}$ is a bounded weak*-separable subset of $l_\infty(\Gamma)$ on which μ is continuous by hypothesis. This shows that $\mu(A_n)$ converges to 0. By Theorem 4, we conclude that $l_1(\Gamma) < l_\infty$. ■

This theorem shows that if measurable cardinals do not exist, then $l_1(\Gamma)$ has the perfect-PIP for all sets Γ. Let m_r denote the smallest real-valued measurable cardinal. We let c denote the cardinality of the continuum. Andrews [1] has shown that if $m_r \geq c$ then $l_1(\Gamma)$ has the UPIP for all sets Γ. Since his proof only involves a single measure, which one can take without any changes to be perfect, he actually shows that $l_1(\Gamma)$ has the perfect-PIP for all sets Γ. The proof is a modification of Edgar's proof showing that if $m_r \geq c$ then for any set Γ, the space $l_1(\Gamma)$ has Lebesgue-PIP [4]. Talagrand has remarked that the hypothesis $m_r \geq c$ is not needed in Edgar's proof since Kunen has shown that if $m_r \leq c$, then Axiom K holds and consequently *every* space has Lebesgue-PIP [8]. An interesting question is whether one can likewise remove the hypothesis $m_r \geq c$ from Andrews' theorem.

We do not know if the converse to Theorem 9 holds, but can offer the following result, which was indicated to us by Jerry Edgar.

THEOREM 10: *If $l_1(\Gamma) < l_\infty$, then card(Γ) is not two-valued measurable.*

Proof. Take an operator $T : l_1(\Gamma) \to l_\infty$. Since card($l_\infty$) = c there is a set Γ_1 with card(Γ_1) $\leq c$ and a map $\Theta : \Gamma \to \Gamma_1$ so that T factors as $\tilde{T}V$, where $\tilde{T} : l_1(\Gamma_1) : \to l_\infty$ and $V : l_1(\Gamma) \to l_1(\Gamma_1)$ are bounded operators with $V(e_\alpha) = e_{\Theta(\alpha)}$ on the unit vector basis of $l_1(\Gamma)$. To see this, take

$$\Gamma_1 = \{ T^{-1}(x) : x \in T(B_{l_1(\Gamma)})\}.$$

Now define $\Theta : \Gamma \to \Gamma_1$ by

$$\Theta(\alpha) = T^{-1}(x) \quad \text{if } T(e_\alpha) = x$$

and $\tilde{T} : l_1(\Gamma_1) \to l_\infty$ on the unit vector basis by

$$\tilde{T}(e_{T^{-1}(x)}) = x.$$

Suppose card(Γ) is two-valued measurable and let $\mu \in ca(\Gamma, P(\Gamma))$ be a non-zero 2-valued measure vanishing on the singletons. We claim that $V^{**}(\mu)$ is an element of $ca(\Gamma_1, P(\Gamma_1))$. To see why, take sets A_n in Γ_1 decreasing to the empty set. Then χ_{A_n} converges to 0 in the weak*-topology in $l_\infty(\Gamma_1)$. Therefore $V^*(\chi_{A_n})$ converges weak* to 0 in $l_\infty(\Gamma)$, and thus pointwise, yielding

$$\lim_n V^{**}(\mu)(\chi_{A_n}) = \lim_n \int V^*(\chi_{A_n}) d\mu = 0$$

by the Dominated Convergence Theorem. This demonstrates the countable additivity of $V^{**}(\mu)$.

Now $V^{**}(\mu)$ is also 2-valued. Since card(Γ_1) $\leq c$, card(Γ_1) is not a 2-valued measurable cardinal. Therefore, either $V^{**}(\mu) = 0$ or there exists a $\hat{\beta}$ in Γ_1 such that $V^{**}(\alpha)$ is concentrated at $\hat{\beta}$, that is, $V^{**}(\alpha) = e_{\hat{\beta}}$. In either case, $V^{**}(\mu)$ belongs to

$l_1(\Gamma_1)$, and so

$$T^{**}(\mu) = \tilde{T}^{**}V^{**}(\mu) \in \tilde{T}^{**}(l_1(\Gamma_1)) \subset l_\infty .$$

Since this holds for every operator T from $l_1(\Gamma_1)$ to l_∞, we see that

$$\mu \in \bigcap T^{**-1}(l_\infty).$$

On the other hand, the measure μ does not belong to $l_1(\Gamma)$ since it vanishes on singletons. Thus $l_1(\Gamma)$ does not precede l_∞ in Edgar's ordering. ■

4. Spaces not preceding l_∞

What are some other examples of sets that do not precede l_∞? Let ω_1 denote the first uncountable ordinal and let $[0,\omega_1]$ have the order topology. The space $C([0,\omega_1])$ consists of all continuous functions on this order interval with the maximum norm. A related space is $J(\omega_1)$ [5], the James-type Banach space that consists of all functions x on $[0,\omega_1]$ that satisfy

(i) $x(0) = 0$;

(ii) x is continuous on $[0,\omega_1]$;

(iii) $||x||_J = \sup(\sum_{i=1}^{n} |x(\gamma_i) - x(\gamma_{i-1})|^2)^{1/2} < \infty$;

where the supremum is taken over all finite sequences $\gamma_0 < \gamma_1 < \cdots < \gamma_n$ in $[0,\omega_1]$. Edgar and Zhao [7] have shown that

$$J(\omega_1) < C([0,\omega_1]) < l_\infty(\Gamma)$$

for all uncountable sets Γ. Moreover, they also showed that $J(\omega_1)$ does not precede l_∞, and therefore neither do the other two spaces (for uncountable Γ in the latter case).

Do any of these spaces have the perfect-PIP? By Theorem 1 it suffices to show that $J(\omega_1)$ fails the perfect-PIP in order to conclude that $C([O,\omega_1])$ and $l_\infty(\Gamma)$ for uncountable Γ also fail the perfect-PIP. We proceed to do so (assuming the Continuum Hypothesis).

THEOREM 11: (CH) $J(\omega_1)$ *fails the perfect-PIP.*

Proof. Define the function $\phi : [0,\omega_1) \to J(\omega_1)$ by $\phi(\alpha) = h_\alpha = \chi_{(\alpha,\omega_1]}$. The elements h_α form a basis for $J(\omega_1)$. Take a bijection $\Theta : [0,1] \to [0,\omega_1)$. Then $\phi \circ \Theta$ defines a function from the unit interval $[0,1]$ into $J(\omega_1)$.

A basis for $J(\omega_1)^*$ consists of the elements e_β given by $e_\beta(x) = x(\beta)$. Observe that

$$e_\beta \circ \phi(\alpha) = \langle e_\beta, \chi_{(\alpha,\omega_1]} \rangle = \chi_{(\alpha,\omega_1]}(\beta)$$

for each $\alpha < \omega_1$. Thus $e_\beta \circ \phi = \chi_{[0,\beta)}$ for $\beta < \omega_1$ while $e_{\omega_1} \circ \phi = 1$. Now for $\beta < \omega_1$

$$e_\beta \circ \phi \circ \Theta(t) = \chi_{[0,\beta)} \circ \Theta(t) = \begin{cases} 1 & \text{if } \Theta(t) \in [0,\beta) \\ 0 & \text{otherwise} \end{cases}$$
$$= \begin{cases} 1 & \text{if } t \in \Theta^{-1}[0,\beta) \\ 0 & \text{otherwise} \end{cases}$$

Therefore $e_\beta \circ \phi \circ \Theta$ equals 0 off a countable set for $\beta < \omega_1$, and equals 1 if $\beta = \omega_1$. This shows that $e_\beta \circ \phi \circ \Theta$ is measurable with respect to all perfect measures on $[0,1]$ for every β. Therefore $\phi \circ \Theta$ is scalarly measurable for all perfect measures. However, if we let λ denote Lebesgue measure on $[0,1]$, then

$$\int e_\beta \circ \phi \circ \Theta \, d\lambda = \begin{cases} 0 & \text{if } \beta < \omega_1 \\ 1 & \text{if } \beta = \omega_1 \end{cases}$$

Since e_β converges weak* to e_{ω_1}, this shows that $\phi \circ \Theta$ cannot be Lebesgue Pettis integrable and thus that $J(\omega_1)$ fails the perfect-PIP. (We have actually shown that $J(\omega_1)$ fails the UPIP.) ■

5. Conclusion

We have seen how Edgar's ordering may be used to verify that a space has the perfect-PIP if it can be shown that that space precedes l_∞. Conversely, we have looked at three spaces known not to precede l_∞ and have seen that they fail the perfect-PIP. An intriguing question is whether preceding l_∞ in Edgar's ordering characterizes the perfect-PIP. Unfortunately, the answer for arbitrary Banach spaces is no. Talagrand has constructed a space in [15] that has the perfect-PIP (assuming the Continuum Hypothesis) but does not precede l_∞. He actually constructed this example to illustrate that if the word "bounded" is removed from the hypothesis in Theorem 4, then the resulting condition, which characterizes Banach spaces whose weak topology is real compact [4], is not equivalent to the condition *with* the boundedness assumption. His space is not a dual space, however. If measurable cardinals exist, then the answer can still be no, even for dual spaces. We have seen that $l_1(\Gamma)$ does not precede l_∞ if card(Γ) is two-valued measurable but that if $m_r \geq c$ then $l_1(\Gamma)$ has the perfect-PIP for all sets Γ. Of course, if measurable cardinals do not exist, then it *is* the case that $l_1(\Gamma) < l_\infty$ for all sets Γ and $l_1(\Gamma)$ has the perfect-PIP for all Γ.

References

1. K. T. Andrews, *Universal Pettis integrability,* Can. J. Math. **37** (1985), 141-159.

2. H. H. Corson, *The weak topology of a Banach space,* Trans. Amer. Math. Soc. **101** (1961), 1-15.

3. G. A. Edgar, *Measurability in a Banach space, I,* Indiana Math. J. **26** (1977), 663-677.

4. G. A. Edgar, *Measurability in a Banach space, II,* Indiana Math. J. **28** (1979), 559-580.

5. G. A. Edgar, *A long James space,* Measure Theory, Oberwolfach, 1979, edited by D.Kolzow, Lecture Notes in Mathematics 794, Springer-Verlage, 1980.

6. G. A. Edgar, *An ordering for the Banach spaces,* Pacific J. Math. **108** (1983), 83-98.

7. G. A. Edgar and J. Zhao, *The ordering structure on Banach spaces,* Pacific J. Math. **116** (1985), 255-263.

8. D. Fremlin and M. Talagrand, *A decomposition theorem for additive set functions with applications to Pettis integrals and ergodic means,* Math. Z. **168** (1979), 117-142.

9. R. F. Geitz, *Pettis integration,* Proc. Amer. Math. Soc. **82** (1981), 81-86.

10. L. Gillman and M. Jerison, *Rings of Continuous Functions,* D. Van Nostrand Co., Inc., 1960.

11. R. Pol, *On a question of H. H. Corson and some related problems,* Fund. Math. **109** (1980), 143-154.

12. L. H. Riddle and E. Saab, *On functions that are universally Pettis integrable,* Illinois J. Math. **29** (1985), 509-531.

13. L. H. Riddle, E. Saab and J. J. Uhl, Jr., *Sets with the weak Radon-Nikodym property in dual Banach spaces,* Indiana Math. J. **32** (1983), 527-541.

14. V. V. Sazonov, *On perfect measures,* Amer. Math. Soc. Transl. (2) **48**

(1965), 229-254.

15. M. Talagrand, *Certaines formes lineaires pathologiques sur un espace de Banach dual,* Israeli J. Math. **35** (1980), 171-176.

16. M. Talagrand, Pettis Integral and Measure Theory, Memoirs Amer. Math. Soc. 51 (1984).

Martingale Proofs of a General Integral Representation Theorem

HASKELL ROSENTHAL*

Department of Mathematics
The University of Texas at Austin
Austin, Texas 78712

Abstract

Let K be a line-closed measure-convex bounded subset of a Banach space so that every relatively closed separable subset of K is analytic. By constructing certain martingales it is proved that if K has the Radon-Nikodým property, then for every k_0 in K there is a separable relatively closed convex subset K_0 of K and a Borel probability measure μ supported on the extreme points of L, for every relatively closed separable convex subset L of K with $L \supset K_0$, so that k_0 is the barycenter of μ; μ is uniquely determined by k_0 if and only if K is a simplex.

*This work was partially supported by NSF Grant DMS-8601752.

Introduction.

We give here a self-contained exposition of various known integral representation results, via the general theorem stated in the abstract. Both the existence and uniqueness parts of the theorem are obtained through the construction of certain martingales. Our main arguments and formulations are thus probabilistic in nature.

For the sake of orientation, we first recall the following representation result, due to E.G.F. Thomas [**26**].

Theorem. *Let K be a closed bounded measure-convex Souslin subset of a locally convex space and assume that K has the Radon-Nikodým property (the RNP). Then for every x in K, there is a Borel probability measure μ supported on the extreme points of K so that x is the barycenter of μ; this μ is uniquely determined (for every given x) if and only if K is a simplex.*

We show at the end of this section that this result follows from the theorem stated in the abstract. Of course these theorems are generalizations of important special cases discovered earlier. The seminal discovery of the compact metrizable case is due to Choquet [**5**], [**6**]; the existence assertion of Thomas' result in the setting of separable closed bounded convex subsets of a Banach space is due to Edgar [**8**], while the uniqueness assertion in this setting is due to Bourgin-Edgar [**3**] and Saint-Raymond [**23**]. For closed non-separable sets, the existence assertion of the result stated in the abstract is due to Mankiewicz [**18**]; while the uniqueness assertion is due to Bourgin-Edgar [**3**]. In the setting of possibly non-closed sets, existence and uniqueness results for convex subsets of locally convex spaces are discussed by Edgar in [**9**]. The existence assertion of the result stated in the abstract, for the case of certain bounded convex G_δ subsets of a separable Hilbert space, is due to Ghoussoub-Maurey [**15**]. (Also see Bourgin [**2**] for further references and background material.)

Let us now restrict ourselves to the Banach space case. Let $\mathbf{B}$ be a Banach space and $K \subset \mathbf{B}$ be as in the result stated in the abstract. Let $\mathcal{B}(K)$ denote the family of Borel subsets of K and $\mathcal{P}_t(K)$ the family of "tight" probabilities on $\mathcal{B}(K)$. That is, $\mu \in \mathcal{P}_t(K)$ provided μ is a probability measure on $\mathcal{B}(K)$ and there is a *separable* Borel subset B of $\mathbf{B}$ with $B \subset K$ and $\mu(B) = 1$. (As is well-known, it follows that in fact B may then be chosen to be σ-compact.) Now if $\mu \in P_t(K)$ and $x \in B$, x is called the *barycenter* of μ if $x = \int_K k\,d\mu(k)$, the integral interpreted in the Bochner sense. (The other terms in the statement of our representation theorem are defined below.) If X is the random variable defined on the probability space $(K, \mathcal{B}(K), \mu)$ by $X(k) = k$ for all $k \in K$, then we may say equivalently that x is the expected value of X; $x = \mathbb{E}X$. Conversely if X is a random variable (defined on some probability space) so that X is valued in Ext K (the set of extreme points of K) and $x \in \mathbb{E}X$, we obtain that x is the barycenter of μ where μ is the distribution of X. *By a martingale proof of the existence assertion of the representation theorem, we mean the construction (for a fixed $x \in K$) of a finite martingale $(X_n)_{n=0}^{\infty}$ valued in K, with $X_0 = x$, converging almost everywhere to a random variable X valued (almost surely) in* Ext K.

The first martingale proof of the existence part of Choquet's theorem (that is, the compact metrizable case in the result of Thomas formulated above) seems to have been given by Loomis [**17**], although his results are not formulated in this language. (Choquet's existence theorem is easily seen to be equivalent to the Banach space case; in fact, to the Hilbert space case. For an elementary non-martingale proof in this setting, see [**21**].) Edgar proved his representation theorem in [**8**] using a *transfinite* martingale. A martingale proof of Edgar's theorem in the (more general) setting of bounded measure-convex line-closed H_δ subsets of a separable Hilbert space, is given by Ghoussoub and Maurey in [**15**] (Proposition I.15).

We obtain our proof of the existence part of our theorem by constructing a dyadic martingale which has the desired properties. This martingale is rather different from the one given in [**15**]. However our proof that it works is inspired by the discussion given there. The construction itself is very easy to describe, and we do this now. Fix K a bounded convex non-empty subset of a Banach space: We introduce the function $\delta = \delta_K : K \to \mathbb{R}^+$ defined as follows: For $x \in K$,

$$\delta(x) = \sup\left\{ \left\|\frac{y-z}{2}\right\| : y, z \in K \text{ and } x = \frac{y+z}{2} \right\}. \tag{1}$$

Let $(\varepsilon_j)_{j=0}^{\infty}$ be a sequence of positive numbers with $\varepsilon_j \to 0$, and fix $x_0 \in K$. Choose $x_1, x_2 \in K$ with $x_0 = \frac{x_1 + x_2}{2}$ and $\|\frac{x_1 - x_2}{2}\| \geq \delta(x_0) - \varepsilon_0$. Next choose for each

$i = 1, 2$, x_{ij} in K with $x_i = \dfrac{x_{i1} + x_{i2}}{2}$ and $\|\dfrac{x_{i1} - x_{i2}}{2}\| \geq \delta(x_i) - \varepsilon_1$. Continuing in this fashion, we obtain a dyadic martingale $(X_n)_{n=0}^{\infty}$ starting at x_0 (*i.e.*, $X_0 \equiv x_0$); which converges almost everywhere to a K-valued random variable X, if K has the RNP and is line-closed. Assume in addition that K is *measure-convex* and every relatively-closed separable subset of K is *analytic* (*i.e.*, the continuous image of a Borel subset of some separable complete metric space); let K_0 denote the relatively-closed convex hull of the range of the above dyadic martingale (in K). We shall prove that X is valued almost surely in $\mathrm{Ext}\, K'$ for all separable relatively closed convex subsets K' of K with $K' \supset K$. We thus obtain the existence part of the theorem stated in the abstract, which we reformulate more precisely as follows:

The General Representation Theorem. *Let K be a bounded measure-convex line-closed subset of a Banach space B so that every separable relatively closed subset of K is analytic. Suppose K has the* RNP *and let $x_0 \in K$. There exist a separable relatively closed convex subset K_0 of K containing x_0 and a probability measure μ on the Borel subsets of B so that letting $\overline{\mu}$ be the completion of μ, then*

(a) *x_0 is the barycenter of μ*

(b) *$\overline{\mu}(\mathrm{Ext}\, L) = 1$ for any relatively closed separable convex subset L of K with $L \supset K_0$*

(c) *$L \supset K_0$ for all separable relatively closed convex subsets L of K with $\overline{\mu}(L) = 1$.*

Moreover K is a simplex if and only if for each $x_0 \in K$, the Borel probability measure μ satisfying (a) *and* (b) *(for some K_0 as above) is unique.*

Our proof that X (the limit random variable constructed above) is indeed $\mathrm{Ext}\, K$-valued (a.s.) is given in section 2. The demonstration involves a variation of the argument of Ghoussoub and Maurey sketched in [**15**, p.31]. The proof is completed following Lemma 2.4, and requires the basic result given (in the Borel case) by J.A. Johnson [**16**]: *if L is an analytic convex subset of a Banach space, a random variable is* $\mathrm{Ext}\, L$*-valued if and only if it is an extreme point of the set of L-valued random variables.* We state this in Lemma 2.3; it is a consequence of the von Neumann selection theorem [**19**] (as the proof in [**16**] shows). For the sake of completeness, in section 5 we give the deduction of Lemma 2.3 as well as a self-contained proof of the von Neumann theorem. In the last part of section 2, we construct some other martingales which work, including the one originally constructed for subsets of Hilbert space in [**15**]. The section concludes with a result (Proposition 2.6) which proves the universal measurability of the "δ-function" defined above and the "ρ-function" defined in [**15**].

We prove the uniqueness theorem in section 3, by obtaining a random-variable analogue of the directedness, in Choquet-order, of probability measures on a Choquet simplex. Let K be as in the first sentence of the General Representation Theorem, and let X and Y be K-valued random variables (defined on the same probability space). Let us say that $X \preceq Y$ if $\mathbb{E}(Y \mid X) = X$. (This notation won't be used in the remaining sections. As usual: $\mathbb{E}(Y \mid X)$ denotes the conditional expectation of Y given X. That is, let $\sigma(X)$ denote the σ-field of measurable sets generated by X. $\mathbb{E}(Y \mid X)$ is that $\sigma(X)$-measurable random-variable Z (determined a.s.) so that $\mathbb{E}(YI_A) = \mathbb{E}(ZI_A)$ for all $A \in \sigma(X)$. Thus $X \preceq Y$ means precisely that $\mathbb{E}(YI_A) = \mathbb{E}(XI_A)$ for all $A \in \sigma(X)$.) We prove in Theorem 3.4 that *if K is a simplex with the* RNP *and $\mu_1, \mu_2 \in \mathcal{P}_t(K)$ have the same barycenter, then there are K-valued random variables X_1, X_2, and X_3 (on $[0,1]$) with* $\operatorname{dist} X_i = \mu_i$ *and $X_i \preceq X_3$ for $i = 1,2$.* (This result appears to be below the surface in every special case of the theorem, including the one where K is a non-degenerate closed bounded interval of real numbers.) We go on to show that if X and Y are K-valued random variables with $X \in \operatorname{Ext} K$ a.s. and $X \preceq Y$, then $X = Y$ a.s.. The "non-trivial" uniqueness part of our representation theorem then follows immediately. Indeed, let K be a simplex with the RNP and suppose μ_1 and $\mu_2 \in \mathcal{P}_t(K)$ represent the same point of K. Choosing the random variables X_1, X_2 and X_3 as above, we obtain that $X_3 = X_i$ a.s. for $i = 1,2$. Hence $X_1 = X_2$ a.s., so $\mu_1 = \operatorname{dist} X_1 = \operatorname{dist} X_2 = \mu_2$.

Recall the Choquet-order $\preceq$ on $\mathcal{P}_t(K)$: for $\mu_1, \mu_2 \in \mathcal{P}_t(K)$, $\mu_1 \preceq \mu_2$ provided $\int \varphi \, d\mu_1 \leq \int \varphi \, d\mu_2$ for all bounded continuous convex functions φ on K. It follows easily that if X_1 and X_2 are K-valued random variables with $\mu_i = \operatorname{dist} X_i$ for $i = 1,2$, then if $X_1 \preceq X_2$, $\mu_1 \preceq \mu_2$. Indeed, if φ is a bounded continuous convex function on K, we have that $\int \varphi \, d\mu_1 = \mathbb{E}\varphi(X_1) = \mathbb{E}(\varphi(\mathbb{E}(X_2 \mid X_1)))$. But

$\varphi(\mathbb{E}(X_2 \mid X_1)) \le \mathbb{E}(\varphi(X_2) \mid X_1))$ a.s. by Jensen's inequality. Hence $\mathbb{E}\varphi(X_1) \le \mathbb{E}(\mathbb{E}(\varphi(X_2) \mid X_1)) = \mathbb{E}\varphi(X_2) = \int \varphi \, d\mu_2$. (This observation is due to Edgar [**9**]. We also note the nice converse proved by him in [**9**]: if $\mu_1 \preceq \mu_2$ with $\mu_i \in \mathcal{P}_t(K)$ for $i = 1, 2$, then there exist random variables X_1 and X_2 with $\mu_i = \operatorname{dist} X_i$ for $i = 1, 2$ and $X_1 \preceq X_2$.)

Thus Theorem 3.4 yields that $\mathcal{P}_t(K)$ is directed in the Choquet-order provided K is a simplex with the RNP. This result is due to Bourgain-Edgar for K closed [**3**] (and independently to Saint-Raymond for K separable and closed [**23**]). In Theorem 3.7, we reformulate 3.4 "intrinsically", in the closed separable setting (for the sake of simplicity). Assuming K has the RNP, this formulation asserts the existence of a probability measure λ on $K \times K$ with prescribed marginals μ_1 and μ_2, and the existence of a Borel-measurable function $h : K \times K \to K$ so that the vector-measure $h d\lambda$ has marginals $k \, d\mu_1$ and $k \, d\mu_2$, provided μ_1 and μ_2 have the same barycenter.

We prove Theorem 3.4 by constructing certain martingales in Lemma 3.2. The construction is valid in any simplex and may be of use in studying the structure of general simplexes in Banach spaces. The construction is based on an elementary combinatorial result, Lemma 3.1, whose formulation and proof do not involve the probabilistic and set-theoretic concepts central to the other results presented here.

Section 4 gives permanence properties and examples related to the class $\mathcal{R}$ (for "reasonable") of bounded line-closed measure convex analytic subsets of some Banach space. For example, Proposition 4.1 yields that if X, Y are separable Banach spaces and $T : X \to Y$ is a bounded one-one linear operator, then for K a bounded subset of X, $K \in \mathcal{R}$ if and only if $TK \in \mathcal{R}$; if $K \in \mathcal{R}$, then K has the RNP if and only if TK has the RNP. Of course $K \in \mathcal{R}$ provided K is a closed bounded convex subset of X. Then $TK \in \mathcal{R}$ and TK is a Borel set, but TK may evidently fail to be closed (in fact TK may fail to be Polish, even if T is a compact operator). The arguments and formulations in [**10**], [**14**] and [**15**] yield immediately that if $K \in \mathcal{R}$ and K has the RNP, then L has the RNP provided L is a convex H_δ-subset of K (for completeness, we give the proof in Proposition 4.2).

It is immediate from Proposition 4.1 that the class of RNP $K \in \mathcal{R}$ with K a relatively compact subset of Hilbert space, "codes" all RNP sets belonging to $\mathcal{R}$. Indeed, if X is a separable infinite-dimensional Banach space, simply let $T : X \to \ell^2$ be a one-one compact linear operator. Then for $K \in \mathcal{R}$ with $K \subset X$, K has the RNP if and only if TK has the RNP. Thus the study of the various open problems such as whether KMP and RNP are equivalent, in this generality, can be reduced to the Hilbert-space setting. We also note the remarkable result of Ghoussoub-Maurey which motivates our formulations [14]: *if K has the* RNP *and is a closed bounded subset of X, there is a one-one compact linear operator $T : X \to \ell^2$ so that TK is an H_δ-subset of ℓ^2.*

In Proposition 4.4, we give an example of a line-closed convex H_δ-subset L of a compact convex subset of Hilbert space so that L is not measure-convex. (Thus L has the RNP but does not belong to $\mathcal{R}$.) This answers a question of Ghoussoub and Maurey [15] in the negative. (Of course if $K \in \mathcal{R}$ and L is a line-closed measure-convex G_δ-subset of K, then L is analytic so also then $L \in \mathcal{R}$.) Finally, in Proposition 4.5, we given an example of a line-closed measure-convex G_δ-subset L of a compact convex subset of Hilbert space so that L has no extreme points. Thus L fails the RNP by our General Representation Theorem; this shows the necessity of the H_δ-hypothesis in Proposition 4.2. (It was recently brought to our attention that this same example is given in [5] (page 73, Remark 5.6)).

In section five, we summarize some properties of analytic sets, without proofs, and also give a fairly elementary proof of Lusin's theorem that analytic sets are universally measurable (via Theorem 5.2), and (as mentioned above), of von Neumann's theorem that a continuous map from one analytic set onto another admits a universally measurable right inverse (Theorem 5.7). Theorem 5.2 also gives immediately the standard result that probability measures on the Borel subsets of a Polish space have σ-compact supports. Our objective in this section is to give an accessible orientation to properties of analytic sets for those readers who may not have studied the deeper aspects of this subject. For an excellent concise treatment, see Appendices I and II of [7].

In the remainder of this introductory section, we give the definitions of the various concepts considered here, and deduce the theorem of Thomas stated at the beginning from our General Representation Theorem.

Although we feel the Representation Theorem is best *understood* in the Banach-space setting, it is nevertheless important to *formulate* things in the locally convex setting when possible. Let K be a Hausdorff space. By a K-valued random variable X we mean a function $X : [0,1] \to K$ which is a limit almost-everywhere of a sequence of K-valued simple Borel-measurable functions defined on $[0,1]$. ("Almost-everywhere" is with respect to Lebesgue-measure m; the unit-interval space is just for "concrete convenience"; definitions for general probability spaces are given in section 2.) If X is a K-valued random variable and $K' \subset K$, we say that X is K'-valued almost-surely if $(X \notin K')$ is of Lebesgue-measure zero. (Throughout we follow the usual probabilistic notation: if $X : A \to B$ is a given function and $S \subset B$, then $(X \in S)$ denotes the set of $\omega \in A$ with $X(\omega) \in B$. Also, if $(A, \mathcal{S}, \mu)$ is a probability space and $(X \in S)$ is measurable with respect to its completion $\overline{\mu}$, $P(X \in S)$ denotes $\overline{\mu}((X \in S))$.) Let L^1 denote the usual Banach space of equivalence classes of real-valued Lebesgue integrable function on [0,1]; let $\mathcal{P} = \{f \in L^1 : \mathbb{E}f = 1 \text{ and } f \geq 0 \text{ a.e.}\}$. We recall now the following definitions: *Let K be a convex subset of a locally convex space Y.* K is said to be

i) *line-closed* if $L \cap K$ is a closed subset of L for every line L in Y. (Equivalently, if $x, y \in Y$ satisfy $\lambda x + (1 - \lambda)y \in K$ for all $0 < \lambda < 1$, then $x, y \in K$).

ii) *measure-convex* if $\overline{co}\, W$ is a *compact* subset of K for every compact subset W of K. (As usual, $co\, S$ denotes the convex hull of a set S and $\overline{co}\, S$ denotes the closure of $co\, S$.)

K is said to have the

iii) *Radon-Nikodým Property* (the RNP) if for every bounded *affine* map $T : \mathcal{P} \to K$, there is a K-valued random variable X so that $Tf = \mathbb{E}(fX)$ for all $f \in \mathcal{P}$.

iv) *Martingale-RNP* if every bounded finite K-valued martingale converges almost-everywhere to a K-valued random variable.

(Here we take the martingale as defined on $[0,1]$; "finite" means that each term of the martingale is finite-valued.)

v) *Integral Representation Property* (the IRP) if for each relatively closed bounded convex subset K' of K and $x \in K'$, there is a K'-valued random variable X

so that $x = \mathbb{E}X$ and X is valued in $\operatorname{Ext} K''$ for all separable relatively closed convex subsets K'' of K with $P[X \in K''] = 1$.

vi) *Krein-Milman Property* (the KMP) if $K' = (\overline{\text{co}}\operatorname{Ext} K') \cap K'$ for all bounded relatively-closed convex subsets K' of K.

To motivate the various definitions, we recall some standard results. Let K be as above with K bounded.

Fact 1. (Fremlin and Pryce [**12**].) *K is measure-convex if and only if for every σ-compactly supported regular probability measure μ on the Borel subsets of K, $x = \int_K k\,d\mu(k)$ exists and belongs to K.*

(Thus in particular, bounded measure-convex sets K are σ-convex; that is, if $\lambda_1, \lambda_2, \ldots$ are non-negative scalars with $\sum \lambda_j = 1$ and $k_1, k_2, \ldots$ are elements of K, then $\sum \lambda_j k_j$ converges to an element of K.) Here, the integral is interpreted in the Pettis sense; that is, $f(x) = \int_K f(k)\,d\mu(k)$ for all $f \in X^*$. It follows that if X is a K-valued random variable, then $\mathbb{E}X$ exists and belongs to K. (And moreover if Y is a Banach space, this is equivalent to the measure-convexity of K.)

Fact 2. (cf. [**9**], [**20**].) *Let K be bounded and measure-convex. Then the following are equivalent.*

1) *K is line-closed and has the* RNP.
2) *K has the martingale* RNP.
3) *Every K-valued martingale converges almost everywhere to a K-valued random variable.*

(It is a classical fact that a closed bounded interval K satisfies 1)–3); *e.g.*, such a K satisfies 3) by the Doob martingale convergence theorem.)

Condition 3) was studied by Edgar in [**9**] and termed by him the *martingale convergence property*. The martingale-RNP was formulated in [**20**]; it is referred to in [**15**] as the property of being "martingale compact". We also note that as we have defined things, the open unit interval (0,1) has the RNP; an elementary probabilistic argument shows however that (0,1) fails the martingale RNP and in fact a convex subset of $\mathbb{R}$ satisfies the martingale RNP if and only if it is closed.

Now let K be as in the first sentence of our General Representation Theorem. We easily obtain from the Theorem that if K has the RNP, K has the IRP, hence if K is separable in addition, K also has the KMP. As in the "closed" setting, we do not know in this more general setting either if any of these implications can be reversed. Thus, if K has the IRP, does K have the RNP? If K has the KMP, does K have either the IRP or the KMP? Finally, we do not know if K has the KMP provided K is non-separable and has the RNP (which of course is true if K is *closed* by standard results). As noted above, it follows from the permanence properties discussed in section 4 that if K has the RNP whenever K has the KMP for K a line-closed measure-convex relatively compact Borel subset of Hilbert space, then the same is true for arbitrary bounded line-closed measure-convex separable Borel subsets K of a Banach space. In particular, the validity of this assertion would yield a positive solution to the famous open problem of whether the KMP implies the RNP for closed bounded convex subsets of a Banach space.

Evidently we obtain immediately that if K' is a convex bounded subset of a locally convex space such that K' is *affinely homeomorphic* to a K as in the first sentence of the Representation Theorem, then also K has the IRP if K' has the RNP. For example, it is easily seen that if K' is a bounded convex subset of a locally convex Frechet space, then K' is (uniformly) affinely homeomorphic to a bounded convex subset of a Banach space, so the General Representation Theorem applies in this setting.

We now give the deduction of the Thomas representation theorem, which we generalize somewhat as follows:

Corollary. *Let K be a bounded measure-convex line-closed Souslin subset of a locally convex space. If K has the* RNP, *K has the* IRP. *Moreover, if K has the* RNP, *then K is a simplex if and only if each element of K is the barycenter of a unique probability measure supported on* Ext K.

To prove the Corollary, we require the following standard result.

Lemma. *Let K be a bounded convex subset of a locally convex space so that K is the continuous image of a separable metric space. Then there exist a bounded convex subset K' of a separable Banach space and a continuous affine bijection*

$T : K \to K'$.

Proof of the Corollary. Let K be as in the statement of the Corollary; "K is Souslin" means that K is a continuous image of a Polish space (*i.e.*, of a complete separable metrizable space). Choose K' and K as in the Lemma. It then follows immediately that K' is analytic and line-closed. *Moreover K' is measure-convex.* To see this, it suffices to show by Fact 1 and the remarks following that if X' is a K'-valued random variable, then $\mathbb{E}X' \in K'$. Now we may assume that X' is Borel measurable. It follows since K is Souslin that also $X = T^{-1}X'$ is a K-valued random variable. Hence by Fact 1, $\mathbb{E}X \in K$ so of course $\mathbb{E}X' = T\mathbb{E}X \in K'$. Now assuming K has the RNP, we obtain immediately that K' also has the RNP (using either the definition and the above fact or directly observing that K' has the martingale-RNP since K has this property). Thus K' has the IRP, by our General Theorem. To see finally that K has the IRP, it suffices of course to show that if $x \in K$ itself, then there is a random variable X valued in $\operatorname{Ext} K$ with $x = \mathbb{E}X$. But simply choose X' a (Borel measurable) random variable valued in $\operatorname{Ext} K'$ with $Tx = \mathbb{E}X'$. Then evidently $X = T^{-1}X'$ is the desired random variable.

Suppose K has the RNP. By the usual standard reasoning, the set of Borel probability measures supported on $\operatorname{Ext} K$ is a simplex, so if each point of K is represented by a unique such measure, K is a simplex. Conversely if K is a simplex, so is K'. Hence K' has the uniqueness property by our representation theorem. Suppose then $x \in K$ and μ_1, μ_2 are probability measures supported on $\operatorname{Ext} K$ with x the barycenter of μ_i for $i = 1, 2$. But then letting $\nu_i(B) = T^{-1}(\mu_i(B))$ for $B \in \mathcal{B}(K')$, ν_1 and ν_2 are probability measures supported on $\operatorname{Ext} K$ with x the barycenter of ν_i for $i = 1, 2$.

(Standard reasoning yields that in fact the ν_i's have σ-compact support (see the remarks following the proof of Lemma 5.7 below), so in fact the ν_i's are supported on Borel subsets of our separable Banach space.) Hence by the General Representation Theorem, $\nu_1 = \nu_2$. Then if W is a σ-compact subset of K; TW is a Borel subset of K', and so $\mu_1(W) = \nu_1(TW) = \nu_2(TW) = \mu_2(W)$ since T is one-one. This implies $\mu_1 = \mu_2$. ∎

For the sake of completeness, we include a sketch of the proof of the Lemma.

Let K satisfy its hypotheses with $K \subset X$, X locally convex. We may assume K is symmetric. Indeed $K - K$ is also a bounded convex subset of X and $K - K$ is again the continuous image of a separable metric space. Once the result is obtained for $K - K$, we immediately get the conclusion for $K - k_0$, $k_0 \in K$ fixed; but then a trivial modification of the map T produces one defined on K itself, with the same range.

So we assume K is symmetric. Thus $0 \in K$. It follows that $\{0\}$ is a G_δ in K. (Indeed, K is a regular Hausdorff space; but if Ω is a regular Hausdorff space which is the continuous image of a separable metric space, then every closed subset of Ω is a G_δ subset.) Since X is locally convex, we may choose closed symmetric convex neighborhoods $U_1 \supset U_2 \supset \cdots$ of $\{0\}$ (in X) with $\{0\} = \bigcap_{n=1}^{\infty} U_n \cap K$. For each n, let $\|\cdot\|_n$ be the semi-norm on X with unit-ball equal to U_n. Since K is bounded, $\sup_{k \in K} \|k\|_n \overset{\mathrm{df}}{=} m_n < \infty$. Let $Y = \operatorname{span} K$ and define $\|\cdot\|$ on Y by $\|y\| = \sum_{n=1}^{\infty}(\|y\|_n / m_n 2^n)$ for all $y \in Y$. It follows that $\|\cdot\|$ is a *norm* on Y and of course $K \subset Ba(Y)$. We now let $\mathbf{B}$ equal the completion of $(Y, \|\cdot\|)$, K' equal K endowed with the norm-topology, and let $T : K \to K'$ be the identity map. It follows that T is continuous and thus has the desired properties. ∎

§2. Martingales converging to extreme-valued random variables.

Throughout this section, let $\mathbf{B}$ be a fixed Banach space and K a fixed non-empty bounded measure-convex line-closed subset of $\mathbf{B}$. For $(\Omega, \mathcal{S}, P)$ a probability space, let $\mathcal{S}(P)$ denote the family of sets measurable with respect to the completion of P (also denoted by P). The members of $\mathcal{S}(P)$ are called *events*; in case Ω is a Polish space and $\mathcal{S} = \mathcal{B}(\Omega)$, $(\Omega, \mathcal{S}(P), P)$ is called a Polish probability space and the elements of $\mathcal{B}(\Omega)$ itself are called Borel events; we shall find it convenient to assume that $(\Omega, \mathcal{S}, P)$ is *atomless*; that is, given E an event with $P(E) > 0$, there is an event $F \subset E$ with $0 < P(F) < P(E)$. (If $(\Omega, \mathcal{S}, P)$ is a Polish probability space, then it is atomless if and only if $P(\{x\}) = 0$ for all $x \in \Omega$.)

For B an event of positive probability, we let P_B denote the probability on $\mathcal{S}$ defined by

$$P_B(S) = P(S \cap B)/P(B) \quad \text{for all} \quad S \in \mathcal{S} .$$

(We sometimes denote $P_B(S)$ by $P(S \mid B)$.) Evidently then $(\Omega, \mathcal{S}, P_B)$ is atomless if $(\Omega, \mathcal{S}, P)$ is. If Ω is Polish and P is a probability on $\mathcal{B}(\Omega)$, then also P_B is a probability on $\mathcal{B}(\Omega)$ (although B may not be a Borel event), and so also $(\Omega, \mathcal{S}(P_B), P_B)$ is a Polish probability space. Now fix $(\Omega, \mathcal{S}, P)$ a probability space. By a $\mathbf{B}$-valued random variable X we mean a function $X : \Omega \to \mathbf{B}$ which is a limit almost-everywhere of a sequence of simple measurable functions. ($X : \Omega \to \mathbf{B}$ is called measurable if $X^{-1}(U) \in \mathcal{S}$ for all open $U \subset \mathbf{B}$ (equivalently, $X^{-1}(U) \in \mathcal{S}$ for all $U \in \mathcal{B}(\mathbf{B})$); for a Polish probability space, a measurable function $X : \Omega \to \mathbf{B}$ with separable range will be called a *Borel* random variable.) Standard results yield that $X : \Omega \to \mathbf{B}$ is a random variable if and only if X is measurable with respect to the completion of P and X has essentially separable range; *i.e.*, there is a separable Borel set $B \subset \mathbf{B}$ with $P(X \in B) = 1$.

Given a $\mathbf{B}$-valued random variable X, we define the *distribution of* X, denoted $\operatorname{dist} X$, to be the probability measure μ on $\mathcal{B}(\mathbf{B})$ given by $\mu(E) = P(X \in E)$ for all $E \in \mathcal{B}(\mathbf{B})$. We note that there is then a closed separable subset S of $\mathbf{B}$ with $\mu(S) = 1$ and thus $(S, \mathcal{S}(\nu), \nu)$ is a Polish probability space, where $\nu = \mu \mid \mathcal{B}(S)$.

If X is an integrable $\mathbf{B}$-valued random variable (for example, if X is uniformly bounded), we let

$$\mathbb{E}(X) = \mathbb{E}(X; P) = \int_\Omega X\, dP .$$

For B an event of positive probability, we denote $\mathbb{E}(X; P_B)$ also by the various notations $\mathbb{E}_B(X; P) = \mathbb{E}_B(X)$; of course by definition $\mathbb{E}(X; P_B) = \mathbb{E}(X I_B)/P(B)$. We note that since K is measure-convex, if B is an event with $P(B) > 0$ and X is a K-valued random-variable, then $\mathbb{E}_B X \in K$.

If $\mathcal{Y}$ is a σ-algebra of events and X is an integrable **B**-valued random variable, we denote the conditional expectation of X with respect to $\mathcal{Y}$, as usual, by $\mathbb{E}(X \mid \mathcal{Y})$. Standard results give that $\mathbb{E}(X \mid \mathcal{Y})$ is a well-defined **B**-valued random variable on $(\Omega, \mathcal{Y}, P \mid \mathcal{Y})$. If E is an event, $\mathbb{E}(I_B \mid \mathcal{Y})$ is denoted by $P(B \mid \mathcal{Y})$ and called the conditional probability of B given $\mathcal{Y}$. (For $B \subset W$, W an arbitrary set, I_B is the *indicator* of B; $I_B(w) = 1$ if $w \in B$, $I_B(w) = 0$ if $w \notin B$.)

In case $\mathcal{Y}$ is a *finite* algebra, it's worth noting that if $A_1, \ldots, A_n$ are distinct atoms of $\mathcal{Y}$ with $P(\bigcup_{i=1}^n A_i) = 1$, then

$$\mathbb{E}(X \mid \mathcal{Y}) = \sum_{i=1}^{n} (\mathbb{E}_{A_i} X) I_{A_i} \quad \text{a.s.} . \tag{2}$$

Suppose now m is Lebesgue measure on the Lebesgue-measurable subsets of the unit interval and P is the product Lebesgue measure $m \times m$ on the Lebesgue-measurable subsets of the square; *i.e.*, $(\Omega, \mathcal{S}(P), P) = ([0,1] \times [0,1], \mathcal{S}(m \times m), m \times m)$. We refer to this probability space as the "standard unit-square space". Similarly, $([0,1], \mathcal{S}(m), m)$ is the "standard unit-interval space". We note in passing that in fact if $(\Omega, \mathcal{S}(P), P)$ is a Polish probability space, there is an Ω-valued random variable X on the standard interval space with $P = \operatorname{dist} X$.

We denote the σ-algebra of vertical sets on the square by $\mathcal{A}$. That is, $\mathcal{A} = \{A \times [0,1] : A \in \mathcal{S}(m)\}$. It is worth pointing out that the $\mathcal{A}$-measurable random-variables can simply be identified with the random-variables defined on [0,1] as in the Introduction, and we shall make this identification. We also note that if X is an integrable B-valued random variable on the square-space, then regarding $\mathbb{E}(X \mid \mathcal{A})$ as defined on [0,1], we have for m-almost all s that

$$\mathbb{E}(X \mid \mathcal{A})(s) = \int X(s,t)\, dm(t) . \tag{3}$$

Finally, for each $n = 1, 2, \ldots$ we let $\mathcal{D}_n$ denote the nth dyadic field in [0,1], regarded as being a field in $[0,1] \times [0,1]$ if we are working on the square. Precisely, for each j, $1 \le j \le n$, (resp. $j = 2^n$) let

$$D_j^n = \left[\frac{j-1}{2^n}, \frac{j}{2^n}\right) \times [0,1] \quad \text{(resp. } \left[\frac{j-1}{2^n}, \frac{j}{2^n}\right] \times [0,1] \text{).}$$

then $\mathcal{D}_n$ denotes the algebra generated by the $\{D_j^n : 1 \le j \le 2^n\}$ (and of course the D_j^n's are then distinct atoms of $\mathcal{D}_n$ with $P(\bigcup_{j=1}^n D_j^n) = 1$). A sequence (X_n) of B-valued random variables defined on $[0,1]$ is called a standard dyadic martingale if for all n, X_n is $\mathcal{D}_n$-measurable and $\mathbb{E}(X_{n+1} \mid \mathcal{D}_n) = X_n$.

From now on, we shall assume that all random-variables are defined on a fixed atomless probability space. We begin with a simple lemma giving alternate ways of computing the function $\delta = \delta_K$ defined in (1) of the Introduction.

Lemma 2.1. *Let $x \in K$. Let $\mathcal{V}_x$ denote the family of all* B*-valued random variables U for which $x + U$ is K-valued a.e. and $\mathbb{E}U = 0$. Then*

$$(4) \quad \begin{cases} \delta(x) = \sup\{\|y - x\| : y \in K \text{ and there is a } z \in K \text{ with } x = \frac{y+z}{2}\ \} \\ \qquad = \sup\{\|\mathbb{E}_B U\| : U \in \mathcal{V}_x \text{ and } B \text{ is an event with } P(B) = \frac{1}{2}\ \} \\ \qquad = \sup\{\|\mathbb{E}_B U\| : U \in \mathcal{V}_x \text{ and } B \text{ is an event with } P(B) \ge \frac{1}{2}\ \}. \end{cases}$$

Remark. The family $\mathcal{V}_x$ was introduced in [**15**].

Proof. The first equality is trivial since if $x = \frac{y+z}{2}$, $y - x = \frac{y-z}{2}$. To prove the remaining equalities, it suffices to show that

$$(5) \quad \begin{cases} \text{given } y, z \in K \text{ with } x = \frac{y+z}{2}, \text{ there are a } U \in \mathcal{V}_x \\ \text{and an event } B \text{ with } P(B) = \frac{1}{2} \text{ so that } \mathbb{E}_B U = y - x \end{cases}$$

and

$$(6) \quad \begin{cases} \text{given } U \in \mathcal{V}_x \text{ and } B \text{ an event with } P(B) \ge \frac{1}{2}, \text{ there are } y, z \in K \\ \text{with } x = \frac{y+z}{2} \text{ and again } y - x = \mathbb{E}_B U. \end{cases}$$

To see (5), let $V = \frac{y-z}{2}$; *i.e.*, V is a constant random variable. Now let B be an event with $P(B) = \frac{1}{2}$, set $r = I_B - I_{\sim B}$ and then let $U = rV$. Since $x \pm V$ is

K-valued (for either choice of sign) and evidently $\mathbb{E}U = 0$, $U \in \mathcal{V}_x$, and of course $\mathbb{E}_B U = \frac{1}{2} \cdot 2 \cdot \frac{y-z}{2} = y - x$.

To see (6), if $P(B) = 1$, we simply take $y = x = z$, so exclude this trivial case and let $y = \mathbb{E}_B(x+U)$ and $w = \mathbb{E}_{\sim B}(x+U)$. Both y and w are in K since K is measure-convex. But trivially $y - x = \mathbb{E}_B U$; setting $z = 2x - y$, we have that $x = \frac{y+z}{2}$. We need to show that $z \in K$. We have, setting $\lambda = P(B)$, that

$$x = \mathbb{E}((x+U) \cdot I_B) + \mathbb{E}((x+U) \cdot I_{\sim B}) = \lambda y + (1-\lambda) w \ .$$

Hence $z = 2\lambda y + 2(1-\lambda)w - y = (2\lambda - 1)y + 2(1-\lambda)w \in K$ since $2\lambda - 1 + 2(1-\lambda) = 1$ and $0 \leq 2\lambda - 1 \leq 1$. ∎

Evidently, the least-trivial part of the Lemma consists in its consequence that

(7) $\|E_B U\| \leq \delta(x)$ for any $U \in \mathcal{V}_x$ and any event B with $P(B) \geq \frac{1}{2}$.

We shall require the following *conditional* version of (7).

Lemma 2.2. *Let Y be a finite K-valued random variable, $\mathcal{Y}$ a finite field of events so that Y is $\mathcal{Y}$-measurable, V a B-valued random variable so that $Y+V$ is K-valued with $\mathbb{E}(V \mid \mathcal{Y}) = 0$, and B an event independent of $\mathcal{Y}$ with $P(B) \geq \frac{1}{2}$. Then*

$$\|\mathbb{E}(VI_B \mid \mathcal{Y})\| \leq P(B)\delta(Y) \text{ a.e. .} \tag{8}$$

Remark. Since $Y+V$ is K-valued and Y is trivially integrable, V is also integrable so $\mathbb{E}(V \mid \mathcal{Y})$ makes sense. To say that B is *independent of* $\mathcal{Y}$ means that $P(B \mid \mathcal{Y}) = P(B)$ a.e.

Proof. Let $A_1, \ldots, A_n$ be distinct atoms of $\mathcal{Y}$ with $P(\bigcup_{i=1}^n A_i) = 1$. Since Y is $\mathcal{Y}$-measurable, for each i, $Y \mid A_i$ is constant, so choose y_i so that $Y \mid A_i \equiv y_i$. Now fixing i and setting $A = A_i$, we have that $y_i + V$ is K-valued a.s. with respect to the atomless probability P_A and moreover $\mathbb{E}(V; P_A) = \mathbb{E}(I_A V)/P(A) = 0$ since

$$\begin{aligned} \mathbb{E}(I_A V) = \mathbb{E}(\mathbb{E}(I_A V \mid \mathcal{Y})) &= \mathbb{E}(I_A \cdot (\mathbb{E}V \mid \mathcal{Y})) \\ &= \mathbb{E}(I_A \cdot 0) = 0 \ . \end{aligned}$$

Thus $V \in \mathcal{V}_{y_i}$ (with respect to P_A), so by (7), since $P_A(B) = P(B) \geq \frac{1}{2}$, $\|\mathbb{E}_B(V; P_A) \leq \delta(y_i)$. Since $\mathbb{E}_B(V; P_A) = \mathbb{E}_A(I_B V)/P(B)$, we thus have

$$\|\mathbb{E}_{A_i}(I_B V)\| \leq P(B)\delta(y_i) \ . \tag{9}$$

(8) follows immediately from (9) upon applying (2).

The next result is fundamental; it is formulated and proved (in the Borel case) in [16] (generalizing earlier work in [24]). We give a self-contained proof in section 5 below. The result says that for appropriate L, a random-variable is Ext L-valued if and only if it is an extreme point of the set of L-valued random variables.

Lemma 2.3. *Let L be a convex analytic subset of B and X an L-valued random variable. The following are equivalent.*

a) X is valued in Ext L a.e.

b) If Y and Z are L-valued random variables with $X = \dfrac{Y+Z}{2}$ a.e., *then $Y = Z = X$* a.e.

The following result is obtained readily from 2.3; its formulation is a variation of an argument sketched in [15].

Lemma 2.4. *Let L be as in 2.3 with L bounded, and let X be an L-valued random variable defined on the standard unit interval space. Assume that X is not valued in* Ext L a.e.

(a) There exist L-valued random variables Y and Z (on $[0,1]$) with $X = \frac{Y+Z}{2}$ and *$\mathbb{E}Y \neq \mathbb{E}Z$.*

(b) Regarding X as defined on the unit square, there exist a $\mathbf{B}$-valued random variable U and an event B (with respect to the unit-square space) so that

(i) $X + U$ is L-valued

(ii) $\mathbb{E}(U \mid \mathcal{A}) = 0$

(iii) $\mathbb{E}(I_B U) \neq 0$ and *$P(B \mid \mathcal{A}) = \frac{1}{2}$* a.e.

Proof. (a). By 2.3 we may choose L-valued random variables Y' and Z' (on $[0,1]$) with $X = \frac{Y'+Z'}{2}$ and $Y' \neq Z'$ (a.e.). Hence setting $W = Y' - Z'$, W is a bounded random variable with $W \neq 0$ a.e. Thus we need only prove the (standard) result that there exists an event A with $\mathbb{E}(I_A W) \neq 0$. Then we simply set

$$Y = I_A Y' + I_{\sim A} X$$

and

$$Z = I_A Z' + I_{\sim A} X \, .$$

The following elegant argument for obtaining such an event A was suggested by D. Alspach. Since W is (a.e.) separably valued, we may choose a countable family of closed balls $B_1, B_2, \ldots$ in $\mathbf{B}$ with $0 \notin \cup B_i$ so that

$$(10) \qquad (W \neq 0) = \Big(W \in \bigcup_i B_i\Big) \text{ (up to an event of probability zero)} .$$

Since $P(W \neq 0) > 0$, it follows that for some i, $P(W \in B_i) > 0$. Then setting $A = (W \in B_i)$, we have $\mathbb{E}_A W \in B_i$ since B_i is closed convex, hence $\mathbb{E}(I_A W) = P(A)\mathbb{E}_A W \neq 0$.

(b). Let r be a real-valued random variable independent of $\mathcal{A}$ with $P(r = 1) = P(r = -1) = \frac{1}{2}$. (A natural explicit choice is $r(s,t) = I_{[0,\frac{1}{2})}(t) - I_{[\frac{1}{2},1]}(t)$ for all $(s,t) \in [0,1] \times [0,1]$.) Let $V = \frac{Y-Z}{2}$, with Y, Z as in part (a). Thus V is $\mathcal{A}$-measurable. Note that $X \pm V$ is an L-valued random variable (for either choice of sign); it follows that defining $U = rV$, then U satisfies (i). Since V is $\mathcal{A}$-measurable, $\mathbb{E}(U \mid \mathcal{A}) = V \cdot \mathbb{E}(r \mid \mathcal{A}) = V \cdot 0 = 0$ a.e., so (ii) holds. Finally, let $B = [r = 1]$; then of course $P(B \mid \mathcal{A}) = \frac{1}{2}$ a.e. and since $I_B U = I_B V$,

$$\begin{aligned}\mathbb{E}(I_B U) = \mathbb{E}(I_B V) &= \mathbb{E}\Big(\mathbb{E}\big[(I_B V) \mid \mathcal{A}\big]\Big) \\ &= \mathbb{E}\big(V\mathbb{E}(I_B \mid \mathcal{A})\big) = \tfrac{1}{2}\mathbb{E}V \neq 0 \text{ by (a), completing the proof of 2.4.}\end{aligned}$$ ∎

Proof of the existence part of the General Representation Theorem. Assume that K satisfies the hypotheses of the Theorem. Let $x_0 \in K$, $\delta = \delta_K$, $(\varepsilon_j)_{j=0}^\infty$ be a sequence of positive numbers with $\varepsilon_j \to 0$, and (X_n) be the standard dyadic martingale constructed in the Introduction. Precisely, we have $X_0 = x_0$. If $n \geq 0$ and X_n a K-valued $\mathcal{D}_n$-measurable random variable has been constructed, say $X_n = \sum_{j=1}^{2^n} x_j^n I_{D_j^n}$, then for each j, setting $x_j = x_j^n$, choose y_j and z_j in K with $x_j = \dfrac{y_j + z_j}{2}$ and $\|y_j - x_j\| \geq \delta(x_j) - \varepsilon_{n+1}$. Then define $X_{n+1} = \sum_{j=1}^{2^n}(y_j I_{D_{2j-1}^{n+1}} + z_j I_{D_{2j}^{n+1}})$. Thus of course $\mathbb{E}(X_{n+1} \mid \mathcal{D}_n) = X_n$, so $(X_n)_{n=0}^\infty$ is a martingale with respect to $(\mathcal{D}_n)_{n=0}^\infty$. We note the basic defining property of (X_n): for all n,

$$(11) \qquad \|X_{n+1} - X_n\| \geq \delta(X_n) - \varepsilon_n \text{ a.e.}$$

Now let

$$K_0 = K \cap \overline{co}\Big(\bigcup_{n=0}^\infty X_n(\Omega)\Big) = K \cap \overline{co}\{x_j^n : 1 \leq j \leq 2^n , \ n = 0, 1, 2, \ldots\} .$$

Since K has the RNP, (X_n) converges almost surely to a K_0-valued random variable X; let μ be its distribution. Then we have that $\int_K k\,d\mu = \mathbb{E}X = x_0$; *i.e.*, x_0 is the barycenter of μ. Next let L be a relatively closed separable convex subset of K with $L \supset K_0$. To prove (b) of the Theorem, we must prove that X is valued in Ext L almost surely. Suppose this is false. Choose then an event B and a $\mathbf{B}$-valued random variable U satisfying (i)–(iii) of Lemma 2.4.

Since $X_n \to X$ a.e., $\mathbb{E}\|X_n - X\| \to 0$ by the bounded convergence theorem, so of course $\mathbb{E}\|X_{n+1} - X_n\| \to 0$; this implies in virtue of (11) that

$$\mathbb{E}\delta(X_n) \to 0\,. \tag{12}$$

We next have the following crucial claim. Fix n; then

$$\|\mathbb{E}[(X - X_n + U)I_B \mid \mathcal{D}_n]\| \le \tfrac{1}{2}\delta(X_n) \quad \text{a.e.} \tag{13}$$

To see this, let $Y = X_n$, $\mathcal{Y} = \mathcal{D}_n$, and $V = X - X_n + U$. We need only verify that the hypotheses of Lemma 2.2 are satisfied. $Y + V = X + U$ is L-valued by (i) of Lemma 2.4 and of course B is independent of $\mathcal{Y}$ since B is independent of $\mathcal{A}$. Now

$$\begin{aligned}
\mathbb{E}(V \mid \mathcal{Y}) &= \mathbb{E}[(X - X_n + U) \mid \mathcal{D}_n] = \mathbb{E}((X - X_n) \mid \mathcal{D}_n) + \mathbb{E}(U \mid \mathcal{D}_n) \\
&= 0 + \mathbb{E}((\mathbb{E}U \mid \mathcal{A}) \mid \mathcal{D}_n) \\
&= 0 + 0 = 0\,.
\end{aligned}$$

Thus (13) follows. Now we have that

$$\begin{aligned}
\|\mathbb{E}(X - X_n + U)I_B\| &= \|\mathbb{E}(\mathbb{E}(X - X_n + U)I_B \mid \mathcal{D}_n)\| \\
&\le \|\mathbb{E}\Big(\|(\mathbb{E}(X - X_n + U)I_B \mid \mathcal{D}_n)\|\Big)\| \quad \text{(by Jensen's inequality)} \\
&\le \tfrac{1}{2}\mathbb{E}\delta(X_n) \qquad \text{(by (13))}.
\end{aligned}$$

Hence we deduce from (12) that

$$\|\mathbb{E}(X - X_n + U)I_B\| \to 0\,. \tag{14}$$

But as already pointed out, $\|\mathbb{E}(X - X_n)\| \le \mathbb{E}\|X - X_n\| \to 0$. Hence using the triangle inequality, we obtain that $\mathbb{E}UI_B = 0$, contradicting (iii) of 2.4.

To obtain the final claim of the existence part of the Theorem, let L be a separable relatively closed convex subset of K with $\mu(L) = 1$. Then of course X is valued in L almost surely, so it follows that for each fixed n, since L is measure-convex, $\mathbb{E}(X \mid \mathcal{D}_n) = X_n \in L$ a.e. Thus $x_j^n \in L$ for all j and n with $1 \leq j \leq 2^n$, so $L \supset K_0$. ∎

We next consider other martingales which also yield a proof of the existence assertion in the General Representation Theorem. The first one is a slight variation of our dyadic martingale. Rather than working with the negligible quantities (ε_n), we could instead fix $0 < \tau < 1$ and construct the dyadic martingale (X_n) so that

$$\|X_{n+1} - X_n\| \geq \tau\delta(X_n) \quad \text{a.e. for all } n . \tag{15}$$

It then follows that (12) holds, and we again obtain that the limit random variable X has the desired properties. The Loomis construction [17] is carried out in this manner with $\tau = \frac{1}{2}$ (for K compact). It seems worth pointing out that when K is compact, one can in fact choose the dyadic martingale (X_n) satisfying (15) with $\tau = 1$; *i.e.*, no "tolerance" is necessary in this case.

We next consider a weighted dyadic martingale which converges a bit more rapidly than the one constructed above. Again, let L satisfy the hypotheses of the Representation Theorem. If K is non-compact, let (ε_n) be as above; however, if K is compact, set $\varepsilon_n = 0$ for all n. Now staring from x_0 in K, choose at the first stage y_1 and y_2 with $x_0 = \dfrac{y_1 + y_2}{2}$ and $\|y_1 - x_0\| \geq \delta(x_0) - \varepsilon_0$. Then define $t_1, t_2 > 0$ by

$$t_i = \sup\{t > 0 : x_0 + t(y_i - x_0) \in K\} . \tag{16}$$

Since K is bounded and line-closed, we have that $x_i \in K$, where

$$x_i = x_0 + t_i(y_i - x_0) \quad \text{for} \quad i = 1, 2 . \tag{17}$$

Note that in fact $t_i \geq 1$ for $i = 1, 2$; by renumbering if necessary, let's assume that $t_1 \leq t_2$. Then, setting $x_1' = 2x_0 - x_1$, we have that x_1' lies on the line segment joining x_0 to x_2, so by convexity x_1' is in K and

$$\begin{aligned} \left\|\frac{x_1' - x_1}{2}\right\| = \|x_1 - x_0\| = t_1\|y_1 - x_0\| &\geq \|y_1 - x_0\| \\ &\geq \delta(x_0) - \varepsilon_0 . \end{aligned} \tag{18}$$

(Thus in the compact case, we must have that $y_1 = x_1$.) Now let $X_0 \equiv x_0$ and let X_1 be a random variable (on [0,1] say) so that $P(X_1 = x_1) = \lambda$ and $P(X_1 = x_2) = 1-\lambda$, where $\lambda = \frac{t_2}{t_1+t_2}$. Then of course $\mathbb{E}X_1 = \lambda x_1 + (1-\lambda)x_2 = x_0$ and

$$\mathbb{E}\|X_1 - X_0\| = \frac{2t_2t_1}{t_1+t_2}\|y_1 - x_0\| \geq t_1\|y_1 - x_0\| \geq \delta(x_0) - \varepsilon_0 \tag{19}$$

by (18), since $\frac{2t_2}{t_1+t_2} \geq 1$. Now iterating this procedure, we obtain a martingale $(X_n)_{n=0}^{\infty}$ (on the unit interval) with respect to a sequence of finite fields $(\mathcal{Y}_n)_{n=0}^{\infty}$ so that for each n, $\mathcal{Y}_n$ has 2^n distinct atoms of positive probability and

$$\mathbb{E}(\|X_{n+1} - X_n\| \mid \mathcal{Y}_n) \geq \delta(X_n) - \varepsilon_n \quad \text{a.s.} \tag{20}$$

It then follows immediately from our proof that (X_n) converges a.e. to a random variable X with the desired properties. Indeed, since, $\mathbb{E}\|X_{n+1} - X_n\| = \mathbb{E}(\mathbb{E}\|X_{n+1} - X_n\| \mid \mathcal{Y}_n)$ for all n, (12) holds, and the rest of the proof is identical. Let us refer to the above martingale as the "longest-segment martingale". Both martingales have interesting properties in the finite-dimensional case.

Proposition 2.5. *Let $k \geq 2$, K be a convex body in $\mathbb{R}^k$, and $x_0 \in K$. Let (X_n) (resp. Y_n) be the dyadic (resp. longest-segment weighted dyadic) K-valued martingale constructed above (with "ε_n" $= 0$ for all n). Let then X and Y be random variables with $X_n \to X$ and $Y_n \to Y$.*

(a) (X_n) stops almost-surely. Hence X is a random variable supported on countably many extreme points.

(b) (Y_n) stops after at most k steps. Hence Y is a random variable supported on at most 2^k extreme points.

Remarks. (X_n) "stops almost-surely" means $X_{n+1} - X_n = 0$ for all n sufficiently large, almost surely. By construction we have $X_n = X_{n+1}$ if and only if X_n is an extreme point, so if $X_n = X_{n+1}$, $X_n = X_{n+j}$ for all j. Thus let τ be the first n so that $X_n = X_{n+1}$. Then τ is a random-variable and $X = X_\tau$ a.s. The motivation for the conclusion of (a) is that since $x_0 = \mathbb{E}X$, we obtain x_0 as a "natural" σ-convex combination of extreme points, while in (b) we have that $Y_k = Y_{k+1} = Y$ a.e. and so since $x_0 = \mathbb{E}Y$, we obtain x_0 as a finite linear combination of extreme points. Of course case (b) is just a standard proof of the classical theorem that $K = co\ \mathrm{Ext}\, K$.

Proof. For L a non-empty convex subset of K, let $\dim L$ equal the dimension of the affine space spanned by L ($= \dim \mathrm{span}(L - \ell_0)$ for any $\ell_0 \in L$). For $x \in K$, let

$$L_x = \left\{ y \in K : \text{ there is a } z \in K \text{ with } x = \frac{y+z}{2} \right\} . \tag{21}$$

It is easily seen that L_x is a convex subset of K; moreover if $x \in \partial K$ (the boundary of K), we have $L_x \subset \partial K$ and hence $\dim L_x \leq k-1$. Also $\delta_K(x) = \delta_{\overline{L}_x}(x)$. Now it follows by construction that if x_1, x_2 are the two values assumed by X_1 (resp. Y_1), then at least one of x_1, x_2 lie in ∂K (resp. both lie there). It follows that then $\dim L_{X_1} \leq k-1$ with probability at least $\frac{1}{2}$, $\dim L_{X_2} \leq k-2$ with probability at least $\frac{1}{2^2}$, and so X_k stops (*i.e.*, $X_k = X_{k+1}$) with probability at least $\frac{1}{2^k}$. Thus for any j and event A in $\mathcal{D}_j$, we have that $P([X_{k+j} = X_j] \mid A) \geq \frac{1}{2^k}$, hence

$$P([X_{k+j} \neq X_j] \mid A) \leq 1 - \tfrac{1}{2^k} \stackrel{\text{df}}{=} \eta . \tag{22}$$

(Indeed, fixing A, let $G_\ell^r = (\dim L_{X_\ell} \leq r) \mid A$. Then $P(G_{\ell+1}^{r-1} \mid G_\ell^r) \geq \frac{1}{2}$, whence by induction $P(G_{\ell+i}^{r-i} \mid G_\ell^r) \geq \frac{1}{2^i}$, $1 \leq i \leq r$, so $P(G_k^0 \mid G_j^k) \leq \frac{1}{2^k}$.) Now let $X_j' = X_{k+j}$, $j = 1, 2, \dots$, and let A_j be the event $X_j' \neq X_{j+1}'$. It follows from (22) that $P(A_{j+1} \mid A) \leq \eta$, that is (since $A_{j+1} \subset A_j$), $P(A_{j+1}) \leq \eta P(A_j)$, so

$$P(A_j) \leq \eta^j \quad \text{for all } j . \tag{23}$$

Evidently then $P(\bigcap_{j=1}^\infty A_j) = 0$, so $P(\bigcup_{j=1}^\infty [X_j' = X_{j+1}'])$. Hence (X_j') and so (X_j) stops almost surely. As far as (Y_j) is concerned, we have immediately that $\dim L_{Y_j} \leq k - j$ *everywhere*, so in fact $\dim L_{Y_k} \equiv 0$, completing the proof. ∎

The last martingale we wish to discuss in some detail is a slight variation of the one constructed by Ghoussoub-Maurey in **[15]**. To motivate its definition, consider the function

$$\begin{aligned} \delta_K'(x) = \delta'(x) = \sup\{\ & \lambda \|y - x\| + (1-\lambda)\|z - x\| : \\ & y, z \in K\,,\ 0 < \lambda < 1\,,\ \text{ and } x = \lambda y + (1-\lambda) z \ \} . \end{aligned}$$

We have that $2\delta(x) \geq \delta'(x) \geq \delta(x)$ for all $x \in K$; the construction for the "longest segment" martingale can be easily modified to obtain a weighted dyadic martingale (X_n) (starting at a given x_0) so that letting the $\mathcal{Y}_n$'s be as in that discussion, $\mathbb{E}(\|X_{n+1}-X_n\| \mid \mathcal{Y}_n) \geq \delta'(X_n)-\varepsilon_n$ a.s. But in fact *any* weighted dyadic martingale (X_n) satisfying (20) will again converge almost surely to the desired random variable X, assuming K satisfies the hypotheses of the General Representation Theorem. The martingale construction in **[15]** is far from being dyadic, but the proof that it works is in fact simpler than the dyadic case.

Define the function $D = D_K : K \to \mathbb{R}^+$ as follows:

$$\text{(24)} \qquad \text{For } x \in K,\ D(x) = \sup \sum_{i=1}^{n} \lambda_i \|x_i - x\|$$

where the supremum is taken over all $n \geq 1$, scalars λ_i with $\lambda_i > 0$ for all i and $\sum_{i=1}^n \lambda_i = 1$, and points $x_1, \ldots, x_n$ in K with $x = \sum_{i=1}^n \lambda_i x_i$.

Again, $D(x) = 0$ if and only if $x \in \operatorname{Ext} K$. Since K is measure convex, one obtains via a simple approximation argument that

$$\text{(25)} \qquad \begin{aligned} \mathrm{D(x)} &= \sup\{\mathbb{E}\|X - x\| : X \text{ is a } K\text{-valued r.v. with } \mathbb{E}X = x\} \\ &= \sup\{\mathbb{E}\|U\| : U \text{ is a } \mathbf{B}\text{-valued r.v. with } x + U \in K \text{ and } \mathbb{E}U = 0\}\ . \end{aligned}$$

(The second equality follows trivially from the first.) Again let K be as in the hypotheses of the General Representation Theorem, $x_0 \in K$, and (ε_n) as above. Using the definition of D, construct a finite K-valued martingale $(X_n)_{n=0}^\infty$ with $X_0 \equiv x_0$ so that for all n,

$$\text{(26)} \qquad \mathbb{E}(\|X_{n+1} - X_n\| \mid \mathcal{Y}_n) \geq D(X_n) - \varepsilon_n \quad \text{a.e.}$$

where $\mathcal{Y}_n$ is the field generated by $X_1, \ldots, X_n$. Assume (as we may) that for each n, every value of X_n is taken with positive probability, and set $K_0 = K \cap \overline{co}(\bigcup_{n=1}^\infty X_n(\Omega))$. It follows again that since K has the RNP, (X_n) converges almost surely to a K_0-valued random variable X, and as before, if $\mu = \operatorname{dist} X$ and $\mu(L) = 1$ with L separable relatively closed and convex, then since $X_n = \mathbb{E}(X \mid \mathcal{Y}_n)$, we have $X_n \in L$ for all n and hence again $L \supset K_0$. Finally, suppose L is a relatively closed separable convex subset of K with $L \supset K_0$ and suppose X is not Ext L-valued (a.e.). (As above, we work on the standard-unit-interval probability space and regard the

X_n's and X as $\mathcal{A}$-measurable random variables defined on the unit square.) Choose then a random variable U satisfying i) and ii) of Lemma 2.4(b) with $U \neq 0$ (a.e.); hence $\mathbb{E}\|U\| > 0$. Now (25) yields the following analogue of Lemma 2.2: let Y, $\mathcal{Y}$, and V be as in its hypotheses; then

$$\mathbb{E}(\|V\| \mid \mathcal{Y}) \leq D(Y) \text{ a.e.} . \tag{27}$$

Now fixing n and applying this to $Y = X_n$, $\mathcal{Y} = \mathcal{Y}_n$ and $V = X - X_n + U$, we obtain

$$\mathbb{E}(\|X - X_n + U\| \mid \mathcal{Y}_n) \leq D(X_n) \text{ a.e.} . \tag{28}$$

Taking expectations and applying Jensen's inequality, we have

$$\mathbb{E}\|X - X_n + U\| \leq \mathbb{E}D(X_n) . \tag{29}$$

Of course it follows that $\mathbb{E}\|X - X_n\| \to 0$ and hence by (26), $\mathbb{E}D(X_n) \to 0$; but then by the triangle inequality and (29), $\mathbb{E}\|U\| = 0$, a contradiction. ∎

Remark. One could as well, for each $1 \leq p \leq \infty$, define the function $D_p : K \to \mathbb{R}$ by

$$D_p^p(x) = \sup\{\mathbb{E}\|X - x\|^p : X \text{ is a } K\text{-valued r.v. with } \mathbb{E}X = x\} .$$

Here too the appropriate finite martingale satisfying

$$\mathbb{E}(\|X_{n+1} - X_n\|^p \mid \mathcal{Y}_n) \geq D_p^p(X_n) - \varepsilon_n$$

will have the desired properties. We have taken $D_1 = D$ above; the function defined in [**15**] is D_2. (The argument in [**15**] is given for K a bounded measure-convex H_δ-subset of ℓ^2, and uses the special properties of the ℓ^2-norm.)

The final topic we discuss in this section involves the measurability properties of the functions δ_K and D_K. Assume now that $\mathbf{B}$ is *separable*. A subset A of $\mathbf{B}$ is called *universally measurable* if A is measurable with respect to the completion of μ for every Borel probability measure μ on $\mathbf{B}$. We denote the (σ-algebra) of all universally measurable subsets of $\mathbf{B}$ by $\mathcal{U}(\mathbf{B})$.

It is worth pointing out that $A \in \mathcal{U}(\mathbf{B})$ *if and only if* $(X \in A)$ is an *event* for any $\mathbf{B}$-valued random variable X (and here we can restrict ourselves to the standard unit-interval space). It thus follows that if $\mathbf{B}'$ is another separable Banach space, $\varphi : \mathbf{B} \to \mathbf{B}'$ is a universally measurable, and X is a $\mathbf{B}$-valued random variable, then $\varphi(X)$ is a $\mathbf{B}$-valued random variable. A classical theorem of Lusin, to be reviewed in section 5, yields easily that analytic sets are universally measurable. We denote the σ-algebra generated by the analytic subsets of $\mathbf{B}$ by $\mathcal{A}(\mathbf{B})$. We now drop our assumption that K is measure-convex.

Proposition 2.6. *Let K be a bounded analytic convex subset of $\mathbf{B}$. Then δ_K and D_K are universally measurable functions. Moreover δ_K is upper semi-continuous and hence Borel-measurable in case K is compact.*

The Proposition may be equivalently phrased: if X is a K-valued random variable, then $\delta_K(X)$ and $D_K(X)$ are (real-valued) random variables.

Remark. In fact we prove that δ_K and D_K are $\mathcal{A}(\mathbf{B})$ measurable; *i.e.*, $f^{-1}(U) \in \mathcal{A}(\mathbf{B})$ for all Borel subsets U of $\mathbf{B}$, for $f = \delta_K$ or D_K.

Proof. We first treat the function $\delta = \delta_K$. For $\eta > 0$, let $F_\eta = \{x \in K : \delta(x) \geq \eta\}$; also let $G_\eta = \{x \in K :$ there are $y, x \in K$ with $x = \frac{y+z}{2}$ and $\|\frac{y-z}{2}\| \geq \eta\}$. If K is compact, $F_\eta = G_\eta$ is closed, so δ is upper semi-continuous. For the general case, we claim that F_η is analytic, which implies the universal measurability of δ. Since $F_\eta = \bigcap_{n=2}^{\infty} G_{(1-\frac{1}{n})\eta}$, it suffices to prove that G_η is itself analytic. Let W_η be the subset of $K \times K \times K = K^3$ defined by

$$W_\eta = \{(x, y, z) \in K^3 : x = \frac{y+z}{2} \text{ and } \|\frac{y-z}{2}\| \geq \eta\} .$$

Then W_η is a relatively closed subset of K^3 and so W_η is also analytic. But G_η equals the projection of W_η onto the first coordinate, so G_η is analytic, being a continuous image of an analytic set.

For the function D, we again let $F_\eta = \{x \in K : D(x) \geq \eta\}$ and prove that F_η is analytic for all $\eta > 0$. This time, for $n \geq 2$, let $C_n = \{(\lambda_1 \ldots \lambda_n) \in \mathbb{R}^n : \sum \lambda_i = 1$

and $\lambda_i \geq 0$ for all $i\}$, and set

$$G_{\eta,n} = \Big\{ x \in K : \text{ There exist } (\lambda_1 \dots \lambda_n) \in C_n \text{ and }$$
$$(x_1 \dots x_n) \in K^n \text{ with } x = \sum \lambda_i x_i \text{ and } \sum \lambda_i \|x_i - x\| \geq \eta \Big\}$$

Then it follows that

$$F_\eta = \bigcap_{m=2}^{\infty} \bigcup_{n=2}^{\infty} G_{\eta(1-\frac{1}{m}),n} ,$$

hence we need only show for fixed $\eta > 0$ and $n \geq 2$ that $G_{\eta,n}$ is analytic. Now simply let W be the subset of $K \times K^n \times C_n$ defined by

$$W = \Big\{ (x, \vec{y}, \vec{\lambda}) : x = \sum_{i=1}^{n} \lambda_i y_i \text{ and } \sum_{i=1}^{n} \lambda_i \|x_i - x\| \geq \eta \Big\} .$$

(Here $\vec{y} = (y_1 \dots y_n) \in K^n$ and $\vec{\lambda} = (\lambda_1 \dots \lambda_n) \in C_n$.) Again, W is a closed subset of $K \times K^n \times C_n$, which is also analytic, and of course as before, $G_{\eta,n}$ is the projection of W on the first coordinate, so $G_{\eta,n}$ is analytic. ∎

§3. The uniqueness part of the Representation Theorem.

We begin with a purely combinatorial elementary lemma concerning certain arrays in a vector lattice. The lemma yields the basic ingredient for constructing the martingales which prove the uniqueness result.

Lemma 3.1. *Let $(X, \leq)$ be a vector lattice. For $1 \leq j \leq 2^n$, $n = 0, 1, 2, \ldots$, let (u_j^n) and (v_j^n) be families of non-negative elements of X with $u_1^0 = v_1^0$ so that*

$$u_j^{n-1} = u_{2j-1}^n + u_{2j}^n \quad \text{and} \quad v_j^{n-1} = v_{2j-1}^n + v_{2j}^n \tag{30}$$

for all $1 \leq j \leq 2^n$, $n = 1, 2, \ldots$. Then for $1 \leq i, j \leq 2^n$, $n = 0, 1, 2, \ldots$, there exists a family $(w_{i,j}^n)$ of non-negative elements of X so that $w_{1,1}^0 = u_1^0$ and for all $n \geq 1$,

$$\sum_{k,k'=0}^{1} w_{2i-k,2j-k'}^n = w_{i,j}^{n-1} \quad \text{for all } 1 \leq i, j \leq 2^{n-1} \tag{31}$$

and

$$\sum_{i=1}^{2^n} w_{i,j}^n = u_j^n \quad \text{for all } 1 \leq j \leq 2^n \tag{32a}$$

$$\sum_{j=1}^{2^n} w_{i,j}^n = v_i^n \quad \text{for all } 1 \leq i \leq 2^n\ . \tag{32b}$$

Before dealing with the proof of 3.1, we first recall two standard facts about vector lattices.

Fact 1. *Let $n \geq 2$ and $x, u_1, \ldots, u_n$ be non-negative elements of X with $x \leq u_1 + \cdots + u_n$. There exist $v_1, \ldots, v_n$ in x with $0 \leq v_i \leq u_i$ for all i and $x = v_1 + \cdots + v_n$.*

The condition formulated in Fact 1 is known as the Riesz Decomposition Property. It has the following consequence.

Fact 2. *Let $k, \ell \geq 1$ and $x_1, \ldots, x_k$, $y_1, \ldots, y_\ell$ be non-negative elements of X with $\sum_{i=1}^k x_i = \sum_{j=1}^\ell y_j$. There exist (z_{ij}), $1 \leq i \leq k$, $1 \leq j \leq \ell$ non-negative elements of X with $\sum_{j=1}^\ell z_{ij} = x_i$ for all $1 \leq i \leq k$ and $\sum_{i=1}^k z_{ij} = y_j$ for all $1 \leq j \leq \ell$.*

Remark. Both of these Facts are due to F. Riesz. We refer the reader to **[24]** for a nice exposition of them as well as of other standard results about vector lattices.

Proof of Lemma 3.1. We construct the $w_{i,j}^n$'s by induction on n. Set $w_{11}^0 = u_1^0$. Since $u_1' + u_2' = u_1^0 = v_1^0 = v_1' + v_2'$, the existence of the w_{ij}^n's satisfying (31) and

(32) for $n = 1$, follows immediately from Fact 2. Now suppose $m > 1$ and the w_{ij}^{m}'s have been constructed satisfying (31) and (32) for $n = m - 1$.

First fix i with $1 \leq i \leq 2^{m-1}$. We claim that we may choose non-negative elements $x_{\ell,j}$ for $\ell = 2i - 1$ or $2i$ and $1 \leq j \leq 2^{m-1}$, so that

$$\sum_{j=1}^{2^{m-1}} x_{\ell,j} = v_{\ell}^{m} \text{ for } \ell = 2i-1 \text{ or } 2i \text{ and} \tag{33a}$$

$$\sum_{\ell=2i-1}^{2i} x_{\ell,j} = w_{i,j}^{m-1} \text{ for } 1 \leq j \leq 2^{m-1}\,. \tag{33b}$$

Indeed, this follows from the hypotheses of 3.1, the induction hypothesis and Fact 2, for we have that

$$\sum_{\ell=2i-1}^{2i} v_{\ell}^{m} = v_{i}^{m-1} = \sum_{j=1}^{2^{m-1}} w_{i,j}^{m-1}\,.$$

Now fix j, $1 \leq j \leq 2^{m-1}$. We next claim that for $1 \leq i \leq 2^{m}$ we may choose $w_{i,2j-1}^{m}$ with

$$0 \leq w_{i,2j-1}^{m} \leq x_{i,j} \text{ for all } 1 \leq i \leq 2^{m} \text{ and} \tag{34}$$

$$\sum_{i=1}^{2^{m}} w_{i,2j-1}^{m} = u_{2j-1}^{m}\,. \tag{35}$$

This follows from Fact 1, once we show that

$$u_{2j-1}^{m} \leq \sum_{i=1}^{2^{m}} x_{i,j} \tag{36}$$

But we have that

$$\begin{aligned}
\sum_{i=1}^{2^{m}} x_{i,j} &= \sum_{i=1}^{2^{m-1}} (x_{2i-1,j} + x_{2i,j}) \\
&= \sum_{i=1}^{2^{m-1}} w_{i,j}^{m-1} && \text{by (33)} \\
&= u_{j}^{m-1} && \text{by the induction hypothesis} \\
&= u_{2j-1}^{m} + u_{2j}^{m} \geq u_{2j-1}^{m} && \text{by (30), proving (36).}
\end{aligned}$$

Finally, define $w^m_{i,2j}$ by

$$w^m_{i,2j} = x_{i,j} - w^m_{i,2j-1} \quad \text{for all } 1 \le i \le 2^m . \tag{37}$$

It follows from (34) that $w^m_{i,2j} \ge 0$ for all i. Now for $1 \le i \le 2^{m-1}$ we have that

$$\sum_{k=0}^{1} \sum_{k'=0}^{1} w^m_{2i-k,2j-k'} = \sum_{k=0}^{1} x_{2i-k,j} = w^{m-1}_{i,j} \quad \text{by (37) and (33b).}$$

Hence (31) holds for $n = m$. For $1 \le i \le 2^m$, we have that

$$\begin{aligned} \sum_{j=1}^{2^m} w^m_{i,j} &= \sum_{j=1}^{2^{m-1}} x_{i,j} \quad \text{by (37)} \\ &= v^m_i \quad \text{by (33a)} . \end{aligned}$$

Hence (32) holds for $n = m$. Finally, (32a) holds for the odd j's, by (35). Suppose then $1 \le j \le 2^{m-1}$. We have that

$$\begin{aligned} \sum_{i=1}^{2^m} w^m_{i,2j} &= \sum_{i=1}^{2^m} x_{i,j} - \sum_{i=1}^{2^m} w^m_{i,2j-1} \quad \text{by (37)} \\ &= \sum_{i=1}^{2^{m-1}} w^{m-1}_{i,j} - u^m_{2j-1} \quad \text{by (33b) and (35)} \\ &= u^{m-1}_j - u^m_{2j-1} \quad \text{by the induction hypothesis} \end{aligned}$$

and (30). Hence (32a) holds for both odd and even j's, for $n = m$, completing the proof of 3.1. ∎

Now let K be a simplex in a linear space X. That is, K is a convex subset of X so that $K \times \{1\}$ is the base of a lattice cone in $X \times \mathbb{R}$. We use 3.1 to construct certain finite martingales valued in K. (Of course we are interested in applying this in the setting of certain subsets of Banach spaces. However, since we only deal with finite martingales, there is no reason to insist on a topology being present.) We say that K is in algebraically general position in X if there is a linear functional $f : X \to \mathbb{R}$ with $f \mid K \equiv 1$. We recall that if K is a simplex in algebraically general position in X with $X =$ linear span K, then X is a vector lattice under the order induced by the cone generated by K (cf. [21] for a recent exposition of these classical ideas).

Lemma 3.2. *Let K be a simplex, $k_0 \in K$, and let $(\underline{X}_n)$, $(\underline{Y}_n)$ be standard dyadic K-valued martingales with $\underline{X}_0 = k_0 = \underline{Y}_0$. There exist increasing sequences of finite fields of Borel subsets of $[0,1]$, $(\mathcal{X}_n)$, $(\mathcal{Y}_n)$ and $(\mathcal{Z}_n)$, and K-valued martingales (X_n), (Y_n) and (Z_n) with respect to these respective fields, with the following properties:*

1. $\operatorname{dist}(\mathcal{X}_n) = \operatorname{dist}(\mathcal{Y}_n) = \operatorname{dist}(\mathcal{D}_n)$ *and* $\operatorname{dist}(X_n) = \operatorname{dist}(\underline{X}_n)$, $\operatorname{dist}(Y_n) = \operatorname{dist}(\underline{Y}_n)$.
2. $\mathcal{X}_n \cup \mathcal{Y}_n \subset \mathcal{Z}_n$ *for all* n.
3. $\mathbb{E}(Z_n \mid \mathcal{X}_n) = X_n$ *and* $\mathbb{E}(Z_n \mid \mathcal{Y}_n) = Y_n$ *for all* n.

Proof. We may assume without loss of generality that K is in algebraically general position in a linear space X with $X = \operatorname{span} K$. Let $\le$ be the order relation induced by K and let f be a linear functional on X with $f \mid K \equiv 1$. Now for $1 \le i \le 2^n$, let x_i^n and y_i^n be the unique elements of X so that for all n,

$$\underline{X}_n = \sum_{i=1}^{2^n} x_i^n I_{D_i^n} \quad \text{and} \quad \underline{Y}_n = \sum_{i=1}^{2^n} y_i^n I_{D_i^n} \,. \tag{38}$$

Set $\mathcal{X}_1^0 = \mathcal{Y}_1^0 = \mathcal{Z}_{1,1}^0 = [0,1]$. Also let $X_0 \equiv k_0 \equiv Y_0 \equiv Z_0$. We shall choose by induction on n, Borel subsets of [0,1], $(\mathcal{Z}_{i,j}^n)$, $1 \le i,j \le 2^n$, $(\mathcal{Y}_i^n)$, $1 \le i \le 2^n$ and $(\mathcal{X}_j^n)$, $1 \le j \le 2^n$ and elements $(z_{i,j}^n)$, $1 \le i,j \le 2^n$ of K with the following properties for all $n = 1, 2, \ldots$.

$$\begin{gathered}(\mathcal{Z}_{i,j}^n) \text{ is a partition of [0,1]; that is, } \bigcup_{i,j=1}^{2^n} \mathcal{Z}_{i,j}^n = [0,1] \\ \text{and } \mathcal{Z}_{i,j}^n \cap \mathcal{Z}_{i,j}^n = \emptyset \text{ if } (i,j) \ne (i',j') \,.\end{gathered} \tag{39}$$

$$\mathcal{X}_j^n = \bigcup_{i=1}^{2^n} \mathcal{Z}_{i,j}^n \text{ for all } 1 \le j \le 2^n \text{ and } \mathcal{Y}_i^n = \bigcup_{j=1}^{2^n} \mathcal{Z}_{i,j}^n \text{ for all } 1 \le i \le 2^n \,. \tag{40}$$

$$P(\mathcal{X}_j^n) = P(\mathcal{Y}_j^n) = \frac{1}{2^n} \text{ for all } 1 \le j \le 2^n \,. \tag{41}$$

(42) Condition 3 of 3.2 holds, where

$$X_n = \sum_{i=1}^{2^n} x_i^n I_{\mathcal{X}_i^n}$$

$$Y_n = \sum_{i=1}^{2^n} y_i^n I_{\mathcal{Y}_i^n}$$

$$Z_n = \sum_{i,j=1}^{2^n} z_{i,j}^n I_{\mathcal{Z}_{i,j}^n}$$

and $\mathcal{X}_n$, $\mathcal{Y}_n$, and $\mathcal{Z}_n$ are the fields generated by $\{\mathcal{X}_j^n : 1 \le j \le 2^n\}$, $\{\mathcal{Y}_j^n : 1 \le j \le 2^n\}$ and $\{\mathcal{Z}_{i,j}^n : 1 \le i,j \le 2^n\}$, respectively.

Letting $\gamma_{i,j}^n = P(\mathcal{Z}_{i,j}^n)$, then for $1 \le i,j \le 2^{n-1}$,

$$\text{(43)} \qquad \begin{aligned} \gamma_{i,j}^{n-1} z_{i,j}^{n-1} &= \sum_{k,k'=0}^{1} \gamma_{2i-k,2j-k'}^n z_{2i-k,2j-k'}^n \quad \text{and} \\ \mathcal{Z}_{i,j}^{n-1} &= \bigcup_{k,k'=0,1} \mathcal{Z}_{2i-k,2j-k}^n \, . \end{aligned}$$

We note that (40) and the last equality of (43) easily yield

$$\text{(44)} \quad \mathcal{X}_j^{n-1} = \mathcal{X}_{2j-1}^n \cup \mathcal{X}_{2j}^n \text{ and } \mathcal{Y}_j^{n-1} = \mathcal{Y}_{2j-1}^n \cup \mathcal{Y}_{2j}^n \text{ for all } 1 \le j \le 2^{n-1} \, .$$

Hence it follows that 1. of 3.2 holds; of course (43) yields that (Z_n) is a martingale with respect to $(\mathcal{Z}_n)$, and hence (X_n) and (Y_n) are martingales with respect to $(\mathcal{X}_n)$ and $(\mathcal{Y}_n)$, respectively.

Set $u_i^n = 2^{-n} x_i^n$ and $v_i^n = 2^{-n} y_i^n$ for $1 \le i \le 2^n$, $n = 0,1,2,\dots$. Since $(\underline{X}_n)$ and $(\underline{Y}_n)$ are dyadic martingales it follows immediately that (u_i^n) and (v_i^n) satisfy the hypotheses of Lemma 3.1. Hence we may choose for $1 \le i,j \le 2^n$; $n = 0,1,2,\dots$ a family $(w_{i,j}^n)$ of non-negative elements of X satisfying the conclusion of Lemma 3.1. Now let $n \ge 1$ and $1 \le i,j \le 2^n$. Set $\gamma_{i,j}^n = f(w_{i,j}^n)$. If $\gamma_{i,j}^n \ne 0$, *define* $z_{i,j}^n$ by

$$\text{(45)} \qquad \gamma_{i,j}^n z_{i,j}^n = w_{i,j}^n \, .$$

If $\gamma_{i,j}^n = 0$, let $z_{i,j}^n = k_0$. It follows immediately that $z_{i,j}^n$ is thus a well defined element of K (and of course $\gamma_{i,j}^n \ge 0$).

We now construct the various sets $\mathcal{X}_j^n$, $\mathcal{Y}_j^n$ and $\mathcal{Z}_{k,j}^n$ by induction so that

$$\gamma_{i,j}^n = P(\mathcal{Z}_{i,j}^n) \text{ for all } n \text{ and } 1 \le i,j \le 2^n . \tag{46}$$

Suppose then $n \ge 1$, and the Borel sets $\mathcal{Z}_{i,j}^{n-1}$ have been constructed with $(\mathcal{Z}_{i,j}^{n-1})$, $1 \le i,j \le 2^{n-1}$ a partition of $[0,1]$. Now fix $1 \le i,j \le 2^{n-1}$. If $\gamma_{i,j}^{n-1} = 0$, let $\mathcal{Z}_{2i-k,2j-k'}^n = \mathcal{Z}_{i,j}^{n-1}$ for $k,k' = 0$ or 1. Suppose $\gamma_{i,j}^{n-1} \ne 0$. Then it follows from (31), (46) and the definition of the $\gamma_{i,j}^n$'s that

$$P(\mathcal{Z}_{i,j}^{n-1}) = \gamma_{i,j}^{n-1} = \sum_{k,k'=0}^{1} \gamma_{2i-k,2j-k'}^n . \tag{47}$$

We may thus choose a partition $\{\mathcal{Z}_{2i-k,2j-k'}^n : k,k' = 0,1\}$ of Borel subsets of $\mathcal{Z}_{i,j}^{n-1}$ with $P(\mathcal{Z}_{2i-k,2j-k'}^n) = \gamma_{2i-k,2j-k'}^n$ for all $k,k' = 0,1$. We now have by (31) and the definition of the $z_{i,j}^n$'s that (43) holds. Of course we now *define* $\mathcal{X}_j^n$ and $\mathcal{Y}_i^n$ by (40), for $1 \le i,j \le 2^n$.

Now (32a) and (32b) yield immediately that

$$\sum_{i=1}^{2^n} \gamma_{i,j}^n z_{i,j}^n = 2^{-n} x_j^n \text{ for all } 1 \le j \le 2^n \tag{48a}$$

and

$$\sum_{j=1}^{2^n} \gamma_{i,j}^n z_{i,j}^n = 2^{-n} y_i^n \text{ for all } 1 \le i \le 2^n . \tag{48b}$$

Letting $1 \le j \le 2^n$ and applying the functional f to (48a), we obtain that

$$P(\mathcal{X}_j^n) = \sum_{i=1}^{2^n} P(\mathcal{Z}_{i,j}^n) = \sum_{i=1}^{2^n} \gamma_{i,j}^n = 2^{-n} . \tag{49}$$

(48a) and (49) yield that $\mathbb{E}(Z_n \mid \mathcal{X}_n) = x_j^n$ on the event $\mathcal{X}_j^n$, which of course implies $\mathbb{E}(Z_n \mid \mathcal{X}_n) = X_n$. Similary, we obtain that $P(\mathcal{Y}_j^n) = \frac{1}{2^n}$ for all $1 \le j \le 2^n$ and $\mathbb{E}(Z_n \mid \mathcal{Y}_n) = Y_n$. Thus (39) – (43) hold, completing the inductive construction and proof of Lemma 3.2. ∎

Let now K be a fixed subset of a fixed Banach space $\mathbf{B}$. *We assume from now on that K is bounded, measure-convex, and line-closed.*

Lemma 3.3. *Let $\mu \in \mathcal{P}_t(K)$. There exists a standard dyadic martingale (X_n) valued in K so that*

$$\text{(50)} \qquad \lim_{n\to\infty} \mathbb{E}\varphi(X_n) = \int \varphi\, d\mu \ \text{ for all bounded continuous functions } \varphi \text{ on } \mathbf{B}.$$

Proof. By a standard probabilistic result, we may choose a Borel random variable $X : [0,1] \to K$ with $\mu = \operatorname{dist} X$. For $n = 0,1,2,\ldots$ define X_n by $X_n = \mathbb{E}(X \mid \mathcal{D}_n)$. (Thus $X_n = \sum_{i=1}^{2^n} x_i^n I_{D_i^n}$ where $x_i^n = 2^n \mathbb{E}(X I_{D_i^n})$ for all $1 \le i \le 2^n$.)

Since K is measure convex, X_n is valued in K for all n. Since K is bounded, $X_n \to X$ a.e., which implies (50). ∎

We are now prepared for the main result of this section.

Theorem 3.4. *Let K be a simplex with the* RNP, *and let $\mu, \nu \in \mathcal{P}_t(K)$ have the same barycenter. There exist K-valued random variables X, Y, and Z on the standard unit-interval space so that*

$$\text{(51)} \qquad \operatorname{dist} X = \mu \quad , \quad \operatorname{dist} Y = \nu$$

and

$$\text{(52)} \qquad \mathbb{E}Z \mid X = X \quad , \quad \mathbb{E}Z \mid Y = Y \, .$$

Remark. It is easily seen (as we show later) that if K satisfies the conclusion of 3.4, then K is a simplex provided K is as in the hypotheses of the General Representation Theorem.

Proof. By the preceeding lemma, we may choose standard dyadic martingales $(\underline{X}_n)$ and $(\underline{Y}_n)$ so that

$$\text{(53)} \qquad \lim_{n\to\infty} \mathbb{E}(\varphi(\underline{X}_n)) = \int \varphi\, d\mu \quad \text{and} \quad \lim_{n\to\infty} \mathbb{E}(\varphi(\underline{Y}_n)) = \int \varphi\, d\nu$$

for all bounded continuous functions φ on $\mathbf{B}$. Now choose $(\mathcal{X}_n)$, $(\mathcal{Y}_n)$, $(\mathcal{Z}_n)$ and (X_n), (Y_n), (Z_n) as in the statement of Lemma 3.2. Since K has the RNP, we obtain the existence of K-valued random variables X, Y and Z so that $X_n \to X$, $Y_n \to Y$ and $Z_n \to Z$ a.e. Since $\operatorname{dist}(X_n) = \operatorname{dist}(\underline{X}_n)$ and $\operatorname{dist}(Y_n) = \operatorname{dist}(\underline{Y}_n)$, it follows that for all bounded continuous functions φ on $\mathbf{B}$ and all n,

$$\text{(54)} \qquad \mathbb{E}\varphi(X_n) = \mathbb{E}\varphi(\underline{X}_n) \quad \text{and} \quad \mathbb{E}\varphi(Y_n) = \mathbb{E}\varphi(\underline{Y}_n) \, .$$

But also $\mathbb{E}\varphi(X_n) \to \mathbb{E}\varphi(X)$ and $\mathbb{E}\varphi(Y_n) \to \mathbb{E}\varphi(Y)$ as $n \to \infty$. Hence by (53) and (54), $\mathbb{E}\varphi(X) = \int \varphi \, d\mu$ and $\mathbb{E}\varphi(Y) = \int \varphi \, d\nu$, which proves (51). Now let $\mathcal{X} = \sigma(\bigcup_{n=1}^{\infty} \mathcal{X}_n)$ and $\mathcal{Y} = \sigma(\bigcup_{n=1}^{\infty} \mathcal{Y}_n)$. (For any family of subsets $\mathcal{F}$ a given set, $\sigma(\mathcal{F})$ denotes the σ-algebra generated by $\mathcal{F}$.) To prove (52), it suffices to show that $\mathbb{E}(Z \mid \mathcal{X}) = X$ and $\mathbb{E}(Z \mid \mathcal{Y}) = Y$. Now we have that

$$\mathbb{E}(Z \mid \mathcal{X}_n) \to \mathbb{E}(Z \mid \mathcal{X}) \text{ a.e. as } n \to \infty . \tag{55}$$

Fixing n, then

$$\begin{aligned} \mathbb{E}(Z \mid \mathcal{X}_n) &= \mathbb{E}\big((\mathbb{E}Z \mid \mathcal{Z}_n) \mid \mathcal{X}_n\big) \\ &= \mathbb{E}(Z_n \mid \mathcal{X}_n) \\ &= X_n \end{aligned}$$

by conditions 2 and 3 of Lemma 3.2 and the fact that $Z_n = \mathbb{E}(Z \mid \mathcal{Z}_n)$. It thus follows from (55) and the definition of X that $\mathbb{E}(Z \mid \mathcal{X}) = X$ a.e.; similary, $\mathbb{E}(Z \mid \mathcal{Y}) = Y$, completing the proof. ∎

Lemma 3.5. *Let K be analytic with the* RNP, *X and Z K-valued random variables (on the same probability space), A an event of positive probability, and $E = (0 < P(A \mid X))$. Then*

$$\frac{\mathbb{E}(I_A Z \mid X)}{P(A \mid X)} \in K \text{ on } E \text{ a.s.} .$$

Remark. Evidently it follows that $\mathbb{E}(Z \mid X) \in K$. This is rather easily seen directly, without the assumption that K is analytic. The Lemma also holds if K is just assumed to be closed (and convex, bounded of course). However the fact that we deal here with possibly non-closed sets seems to require the RNP. We also note that if $E_0 = [0 < P(A \mid X) < 1]$ is of positive probability, then 3.5 yields that

$$\frac{\mathbb{E}(I_A Z \mid X)}{P(A \mid X)} \in K \text{ on } E_0 .$$

It is this assertion that we need in the sequel. (We also need the following standard martingale convergence theorem: *if X is an integrable* **B**-*valued random variable and $\mathcal{X}, \mathcal{X}_1, \mathcal{X}_2, \ldots$ are σ-fields with $\mathcal{X}_n \subset \mathcal{X}_{n+1}$ for all n and $\mathcal{X} = \sigma(\bigcup_{n=1}^{\infty} \mathcal{X}_n)$, then* $\mathbb{E}(X \mid \mathcal{X}_n) \to \mathbb{E}(X \mid \mathcal{X})$ a.s.)

Proof. Let $\mathcal{X} = \sigma(X)$. Now we may choose $\mathcal{X}$-measurable K-valued simple (*i.e.*, finite) random variables (X_n) with $X_n \to X$ a.s. By changing X on a set of probability zero, we may assume that $X_n \to X$ everywhere and hence that $\mathcal{X} = \sigma(\bigcup_{n=1}^{\infty} \mathcal{X}_n)$ where $\mathcal{X}_n = \sigma(X_1, \ldots, X_n)$ for all n. Now define $Y_n = \mathbb{E}(I_A Z \mid \mathcal{X}_n)$ for all n.

It then follows that

$$Y_n \to Y \text{ a.s., where } Y = \mathbb{E}(I_A Z \mid X) . \tag{56}$$

Indeed, this follows from the martingale convergence theorem, since $Y_n = \mathbb{E}(\mathbb{E}(I_A Z \mid X) \mid \mathcal{X}_n) = \mathbb{E}(Y \mid \mathcal{X}_n)$ for all n.

Now define $\varphi_n = \mathbb{E}(I_A \mid \mathcal{X}_n)$ for all n. Then we have that $\varphi_n \to P(A \mid X)$ a.s. (again by the martingale convergence theorem). Hence we have that there is an event $E_1 \subset E$ with $P(E \sim E_1) = 0$ so that on E_1, $\varphi_n > 0$ for all n sufficiently large and

$$\frac{Y_n}{\varphi_n} \longrightarrow \frac{Y}{P(A \mid X)} . \tag{57}$$

Now let $U_n = \mathbb{E}_A(Z \mid \mathcal{X}_n)$ for all n. ($\mathbb{E}_A(Z \mid \mathcal{X}_n)$ denotes the conditional expectation of Z with respect to $\mathcal{X}_n$, on our probability space endowed with the probability P_A.) Then (U_n) is a finite K-valued martingale, so there is a K-valued random variable U so that $U_n \to U$ P_A-a.s. We claim that

$$U = \frac{Y}{P(A \mid X)} \quad P_A\text{-a.s.} \tag{58}$$

We prove this by explicitly exhibiting the martingales (U_n), (Y_n) and (φ_n). For each n, let $(X_i^n)_{i=1}^{m_n}$ be the distinct atoms of $\mathcal{X}_n$. By changing X on a set of probability zero, we may assume without loss of generality that $(X_i^n)_{i=1}^{m_n}$ is a partition of our probability space. Now let $G_n = \{i : P(A \cap X_i^n) > 0\}$. We have for all n that

$$U_n = \sum_{i \in G_n} \frac{\mathbb{E}(Z I_{A \cap X_i^n})}{P(A \cap X_i^n)} I_{A \cap X_i^n} \quad P_A\text{-a.s.} \tag{59}$$

Now if $w \in E_1$, then for n sufficiently large, $\varphi_n(w) > 0$; since

$$\varphi_n(w) = \sum_{i \in G_n} (P(X_i^n))^{-1} P(A \cap X_i^n) I_{X_i^n}(w) ,$$

it follows that $w \in \bigcup_{i\in G_n} X_i^n$ and hence $Y_n(w) = \sum_{i\in G_n} (P(X_i^n))^{-1}\mathbb{E}(ZI_A I_{X_i^n}) I_{X_i^n}(w)$. We obtain that on E_1,

$$\frac{Y_n}{\varphi_n} = \sum_{i\in G_n} \frac{\mathbb{E}(ZI_{A\cap X_i^n})}{P(A\cap X_i^n)} I_{X_i^n} \text{ for all } n \text{ sufficiently large.} \tag{60}$$

Thus on $E_1 \cap A$, we have by (59) and (60) that $Y_n/\varphi_n = U_n$ for n sufficiently large, whence $U = Y/P(A \mid X)$ (almost surely) on $E_1 \cap A$, by (57) and the fact that $U_n \to U$ P_A-a.s. But $P(E_1 \cap A) = P(A)$, proving (58).

We haven't needed the assumption that K is analytic up to this point, but it seems we need it now. Since K is analytic, K is universally measurable, and hence the event $F \stackrel{\text{df}}{=} ((Y/P(A \mid X)) \in K) \cap E$ is measurable with respect to the completion of $P \mid \mathcal{X}$. It follows that we may choose $E' \in \mathcal{X}$ with $E' \subset E$ and $F\Delta E'$ of probability zero. We obtain from (58) that $P(A \sim E') = 0$. But then $P(A \mid X) = P(A\cap E' \mid X) = P(A \mid X) I_{E'}$ a.s. This implies by the definition of E that $P(E \sim E') = 0$; hence $Y/P(A \mid X) \in K$ on E a.s. ∎

We need one last preliminary result, before giving the proof of the uniqueness assertion of the Representation Theorem.

Proposition 3.6. *Let K be analytic with the* RNP. *Assume that X and Z are K-valued random variables (on the same probability space) so that $X \in \operatorname{Ext} K$ a.s. and $\mathbb{E}(Z \mid X) = X$ a.s. Then $X = Z$ a.s.*

Remark. We obtain the same conclusion if K is assumed instead to be closed bounded convex.

Proof of 3.6. Suppose the conclusion of 3.6 is false. As noted in our proof of Lemma 2.4, we may then choose A an event of positive probability so that

$$\mathbb{E}I_A X \neq \mathbb{E}I_A Z \ . \tag{61}$$

Now define B by

$$B = [0 < P(A \mid X) < 1] \ . \tag{62}$$

Then setting $A_1 = B \cap A$, it follows that also $B = [0 < P(A \mid X) < 1]$. We claim that

$$\mathbb{E}(I_{A_1} X) \neq \mathbb{E}(I_{A_1} Z) \ . \tag{63}$$

Indeed, let $A_2 = A \cap \sim B$. Then

$$P(A_2 \mid X) = P(A \mid X) \cdot I_{\sim B} = P(A \mid X) I_{[P(A|X)=1]} = I_{[P(A|X)=1]} \text{ a.e.}$$

This shows that A_2 is X-measurable; hence

$$\mathbb{E}(I_{A_2} Z \mid X) = I_{A_2} \mathbb{E}(Z \mid X) = I_{A_2} X ,$$

so $\mathbb{E} I_{A_2} Z = \mathbb{E} I_{A_2} X$. Thus were (63) false, we would obtain, since $I_{A_1} + I_{A_2} = I_A$, that (61) is false. This proves (63). We may thus replace A by A_1 and so assume that $A \subset B$. It follows of course that $P(B \sim A) \neq 0$ and moreover

$$\begin{aligned} B &= [0 < P((B \sim A) \mid X) < 1] \\ &= [0 < P((\sim A) \mid X) < 1]. \end{aligned}$$

Now define random variables X_1 and X_2 by

$$(64) \quad \begin{cases} X_1 = \dfrac{\mathbb{E}(I_A Z \mid X)}{P(A \mid X)} I_B \\ \text{and} \\ X_2 = \dfrac{\mathbb{E}(I_{\sim A} Z \mid X)}{P(\sim A) \mid X)} I_B . \end{cases}$$

We have by the preceding lemma that X_i are well-defined random variables valued in K, on the event B (almost surely), for $i = 1, 2$. Since $\mathbb{E}(Z \mid X) = X$, we thus obtain that

$$(65) \qquad X I_B = P(A \mid X) X_1 + (1 - P(A \mid X)) X_2 \text{ a.s.}$$

Finally we claim that

$$(66) \qquad P(E) > 0 \text{ where } E = (X I_B \neq X_1)$$

Were this false, we would have that

$$(67) \qquad P(A \mid X) X I_B = \mathbb{E}(I_A Z \mid X) \text{ a.s.}$$

But since $A \subset B$, $P(A \mid X) X = P(A \mid X) X I_B$ a.s. and of course $\mathbb{E}(I_A X \mid X) = P(A \mid X) X$, hence by (67),

$$\mathbb{E}(I_A X) = \mathbb{E}(\mathbb{E}(I_A X \mid X)) = \mathbb{E}(P(A \mid X) X) = \mathbb{E}(\mathbb{E}(I_A Z \mid X)) = \mathbb{E}(I_A Z) ,$$

contradicting (61).

Thus (66) holds. But then (62) and (65) yield that X is *not* extreme valued, contradicting the basic assumption of the proposition. ∎

We are now prepared for the proof of the uniqueness assertion of the General Representation Theorem. Let K be as in its hypotheses and suppose first that K is a simplex. Let $k_0 \in K$ and let μ_1 and μ_2 be probability measures on $\mathcal{B}(\mathbf{B})$ so that for $i = 1, 2$, $k_0 = \int_K k \, d\mu_i(k)$ and there exists K_i a relatively closed separable convex subset of K with

(68)

$\overline{\mu}_i(\operatorname{Ext} L) = 1$ for any relatively closed separable convex subset L of K with $L \supset K_i$

(where $\overline{\mu}_i$ denotes the completion of μ_i). Now let $K_0 = \overline{co}(K_1 \cup K_2) \cap K$ and let $\nu_i = \mu_i \mid (\mathcal{B}(\mathbf{B}) \cap K_0)$. It obviously suffices to prove that $\nu_1 = \nu_2$. Now $K_0 \supset K_i$ for $i = 1, 2$ and K_0 is separable; hence by (68), $\overline{\nu}_i(\operatorname{Ext} K_0) = 1$ for $i = 1, 2$. By Theorem 3.4, we may choose K_0-valued random variables X_1, X_2 and Z (on the standard unit-interval space) with $\operatorname{dist} X_i = \nu_i$ and $\mathbb{E} Z \mid X_i = X_i$ for $i = 1, 2$. It follows for $i = 1, 2$ that $X_i \in \operatorname{Ext} K_0$ a.s. Since K_0 is analytic by the assumptions of the General Theorem, we obtain by Proposition 3.6 that $X_1 = Z = X_2$ a.s. Hence $\nu_1 = \operatorname{dist} X_1 = \operatorname{dist} X_2 = \nu_2$.

Suppose conversely that each $x_0 \in K$ is the barycenter of a unique $\mu \in \mathcal{P}_t(K)$ satisfying (b) of the Representation Theorem (for some K_0 as in its statement). The usual standard arguments show that in fact if G denotes the family of all such $\mu \in \mathcal{P}_t(K)$, then G is a simplex (when regarded as a subset of the linear space of all signed measures on $\mathcal{B}(\mathbf{B})$ with separable support). The map $A : G \to K$ defined by: $A(\mu) = \int k \, d\mu$ for $\mu \in G$, is thus an affine bijection between G and K, so K is also a simplex. This completes the proof of the General Representation Theorem. ∎

Remark. Suppose K is as in the hypotheses of the General Representation Theorem. Then if K satisfies the conclusion of Theorem 3.4, the subsequent discussion shows that each element of K is the barycenter of a unique element of G, as above, hence K is a simplex. Suppose that K is a Riesz Decomposition Set; that is, if X denotes the linear span of $K \times \{1\}$ in $\mathbf{B} \times \mathbb{R}$, with $\leq$ the order relation induced by the cone generated by $K \times \{1\}$, then $(X, \leq)$ satisfies the Riesz Decomposition Property (as given in Fact 1 at the beginning of this section). The proofs of 3.1 and 3.2 only used this property, hence we obtain that K satisfies the conclusion of Theorem 3.4, so K is a simplex. In particular, a closed bounded convex subset of **B** is a simplex provided it has the RNP and is a Riesz Decomposition Set.

We conclude this section with an "intrinsic" reformulation of Theorem 3.4 (which for simplicity we phrase only in the setting of closed separable sets).

Theorem 3.7. *Let K be a closed bounded separable simplex with the* RNP, *and let $\mu, \nu \in \mathcal{P}(K)$ have the same barycenter. There exist a $\lambda \in \mathcal{P}(K \times K)$ and a Borel measurable function $h : K \times K \to K$ so that for all Borel sets $B \subset K$,*

$$\lambda(B \times K) = \mu(B) \text{ and } \lambda(K \times B) = \nu(B) \tag{69}$$

and

$$\int_{B \times K} h \, d\lambda = \int_B k \, d\mu(k) \text{ and } \int_{K \times B} h \, d\lambda = \int_B k \, d\nu(k) \,. \tag{70}$$

Theorem 3.7 has non-trivial content if for example K is a closed bounded interval of real numbers. 3.7 then asserts that if μ and $\nu \in \mathcal{P}(K)$ have the same first moment, there exist $\lambda \in \mathcal{P}(K \times K)$ and $h : K \times K \to K$ Borel measurable so that the "x" and "y" marginals of λ are μ and ν respectively, while the "x" and "y" marginals of $h \cdot \lambda$ are $i \cdot \mu$ and $i \cdot \nu$ respectively, where $i : K \to K$ is the identity function.

We note that for K as in its statement, 3.7 implies Theorem 3.4. For $(K \times K, \mathcal{B}(K \times K), \lambda)$ is a Polish probability space. Define K-valued random variables X, Y and Z on $K \times K$ by $Z = h$ and $X(k_1, k_2) = k_1$, $Y(k_1, k_2) = k_2$ for all $(k_1, k_2) \in K \times K$. Then (69) asserts that $\operatorname{dist} X = \mu$ and $\operatorname{dist} Y = \nu$, while (70) asserts that $\mathbb{E} Z \mid X = X$ and $\mathbb{E} Z \mid Y = Y$.

It thus follows that conversely if K has the RNP, is closed bounded convex separable, and satisfies the conclusion of 3.7, then K is a simplex. (In particular, if $n \geq 2$ and K is a closed bounded convex subset of $\mathbb{R}^n$ which is not a simplex, then K fails 3.7.)

Proof of 3.7. Choose K-valued Borel random variables X, Y and Z on the standard unit-interval space, satisfying the conclusion of Theorem 3.4. Now we may assume that Z is $\sigma(X,Y)$-measurable. Indeed, let $\widetilde{Z} = \mathbb{E}(Z \mid \sigma(X,Y))$. Then since K is closed, $\widetilde{Z}$ is K-valued and $\mathbb{E}(\widetilde{Z} \mid X) = \mathbb{E}(Z \mid X) = X$ by (52), and similarly $\mathbb{E}(\widetilde{Z} \mid Y) = Y$, so we just replace Z by $\widetilde{Z}$ if necessary. It then follows by a standard result in probability theory that there exists a Borel measurable function $h : K \times K \to K$ with $Z = h(X,Y)$ a.s. (For completeness, we sketch a proof of this immediately following this discussion.)

Now let $\lambda = \operatorname{dist}(X,Y)$. That is, $\lambda \in \mathcal{P}(K \times K)$ is defined by: $\lambda(B) = P((X,Y) \in B)$ for any $B \in \mathcal{B}(K \times K)$. Then for any Borel set $B \subset K$, $\lambda(B \times K) = P[(X,Y) \in B \times K] = P(X \in B) = \mu(B)$ and

$$
\begin{aligned}
\int_{B\times K} h\, d\lambda &= \mathbb{E}\big(h(X,Y)I_{[(X,Y)\in B\times K]}\big) \\
&= \mathbb{E}(Z I_{[X\in B]}) \\
&= \mathbb{E}\big(\mathbb{E}(Z I_{[X\in B]} \mid X)\big) \\
&= \mathbb{E}\big((\mathbb{E}Z \mid X) I_{[X\in B]}\big) \\
&= \mathbb{E}(X I_{[X\in B]}) \quad \text{by (52)} \\
&= \int_B k\, d\mu(k) \quad \text{since } \mu = \operatorname{dist} X\, .
\end{aligned}
$$

The same argument shows the other equalities in (69) and (70), proving Theorem 3.7. ∎

We finally give the standard result used in the above discussion.

Lemma 3.8. *Let M be a Polish space, and X, Y, Z be M-valued Borel random variables (defined on a fixed Polish probability space) with Z $\sigma(X,Y)$-measurable. There exists a Borel measurable function $h : M \times M \to M$ so that $Z = h(X,Y)$ a.s.*

Proof. Let $U = (X,Y)$. U is thus an $M \times M$-valued Borel random variable. It follows that there is a Borel subset W of the range of U with $U \in W$ almost surely. (To see this, we can either use the fact that the range of U, being analytic, is measurable with respect to the distribution of U, or apply Lusin's theorem to find compact subsets $K_1 \subset K_2 \subset \cdots$ of our probability space with $P(\cup K_j) = 1$ and

$U \mid K_j$ continuous for all j. Then $W = \bigcup_{j=1}^{\infty} U(K_j)$ works.) Let $\mathcal{Z} = \{h(X,Y) : h : M \times M \to M$ is Borel-measurable$\}$. Now any simple $\sigma(X,Y)$-measurable M-valued variable V belongs to $\mathcal{Z}$. Indeed, by definition, $\sigma(X,Y)$ is the family of all events E of the form $E = [(X,Y) \in B]$, where $B \in \mathcal{B}(M \times M)$. Thus we may choose n, disjoint Borel subsets $B_1, \ldots, B_n$ of $M \times M$, and $\alpha_1, \ldots, \alpha_n$ in M with $\alpha_1, \ldots, \alpha_n$ the distinct values of V so that $V = \alpha_i$ on the event $E_i \overset{\text{df}}{=} [(X,Y) \in B_i]$ and $\sum_{i=1}^{n} P(E_i) = 1$. Define h by $h \mid B_i = \alpha_i$, $1 \leq i \leq n$, and $h \mid\sim \bigcup_{i=1}^{n} B_i = \alpha_1$. Then $V = h(X,Y)$.

Now choose a sequence (Z_n) of $\sigma(X,Y)$-measurable M-valued simple random variables with $Z_n \to Z$ pointwise (everywhere). For each n, choose $h_n : M \times M \to M$ Borel measurable with $Z_n = h_n(X,Y)$, It follows that (h_n) converges pointwise on W. Indeed, for each $w \in W$, choose W with $w = (X(w), Y(w))$; then $h_n(w) = h_n(X(w), Y(w)) = Z_n(w)$ and so $\lim_{n\to\infty} h_n(w) = \lim Z_n(w) = Z(w)$. Let $m_0 \in M$ and define h by $h(v) = \lim_{n\to\infty} h_n(v)$ if $v \in W$; $h(v) = m_0$ if $v \notin W$. Then h is a Borel measurable function and $Z = h(X,Y)$ on the event $E \overset{\text{df}}{=} [U \in W]$; since $P(E) = 1$, $Z = h(X,Y)$ a.s. ∎

§4. Some permanence properties and examples.

Let us say (just for the sake of an abbreviated terminology) that a subset K of a Banach space is *reasonable* provided K is *bounded, line-closed, measure-convex, and analytic*. Evidently every relatively-closed convex subset of a reasonable set is also reasonable. Our first permanence result gives in particular that the class of reasonable sets is closed under one-one affine continuous images. More generally, a line-closed affine continuous image of a reasonable set is also reasonable. Of course if K is a separable closed bounded convex subset of a Banach space X, then K is reasonable. Thus if Y is a Banach space and $T : X \to Y$ is a bounded linear operator, TK is reasonable and a Borel set provided $T \mid K$ is one-one. However T could be compact and still TK could fail to be Polish even though T were one-one on all of X. Also T could be compact and TK line-closed and hence reasonable, yet TK could fail to be a Borel set. That is, the family of reasonable Borel sets forms *the* natural class of sets containing the closed bounded convex subsets of a separable Banach space and closed under one-one affine continuous images. The more general class of reasonable sets arises when we consider line-closed affine continuous images of closed bounded convex sets.

Proposition 4.1. *Let K and L be bounded convex subsets of Banach spaces X and Y respectively, with K separable, and let $T : K \to L$ be an affine continuous surjection.*

(a) *Suppose K is reasonable. Then L is reasonable provided T is one-one or L is line-closed.*

(b) *Suppose there is a one-one affine continuous map $\widetilde{T} : \overline{K} \to X$ with $\widetilde{T} \mid K = T$. Then K is reasonable provided L is reasonable.*

(c) *Suppose T is one-one and K is reasonable. Then K has the* RNP *if and only if L has the RNP.*

Remarks.

1. Suppose X is separable and $T : X \to Y$ is a one-one bounded linear operator. It follows that for bounded subsets K of X, K is reasonable if and only if TK is reasonable.

2. Supposse we just assume that L is a compact subset of Y, K is a reasonable subset of X, and $T : K \to L$ is an affine continuous surjection. Then L is analytic. Since K is measure-convex and bounded, K is σ-convex. It follows immediately that L is σ-convex and hence L is bounded. Our proof of (a) yields that in fact L is also measure-convex. However L need not be line-closed. For example, let X be a separable non-reflexive Banach space and $T : X \to \mathbb{R}$ a continuous non-norm-achieving linear functional on X. Then $T(BaX)$ is a non-line-closed subset of $\mathbb{R}$. We include the "line-closed" property in the formulation of the class of reasonable sets, to insure that a reasonable set has the RNP if and only if it has the martingale-RNP.

Proof of 4.1.

(a) L is analytic since it is the continuous image of an analytic set. We next show that L is measure-convex. We use the "barycentric" equivalent formulation of this concept. Thus, let $\mu \in \mathcal{P}(L)$ and let $y = \int_L \ell \, d\mu(\ell)$. y is a well-defined element of Y; we must show that y belongs to L. Now we have that given $\varepsilon > 0$,

(71) there is a compact convex subset K_ε of K with $\mu(TK_\varepsilon) > 1 - \varepsilon$.

Indeed, we may choose C_ε a compact subset of K with $\mu(TC_\varepsilon) > 1 - \varepsilon$. (We obtain this standard result in Theorem 5.2 (see specifically Lemma 5.6)). Since K is measure convex, $K_\varepsilon \overset{\text{df}}{=} \overline{co}\, C_\varepsilon \subset K$, which yields (71). We next observe that there exists a finite or countable sequence $L_1, L_2, \ldots$ of disjoint Borel subsets of L with $\mu(\cup L_j) = 1$ and

(72) $\lambda_j \overset{\text{df}}{=} \mu(L_j) > 0$ and $\overline{co}\, L_j$ a compact subset of L for all j.

Indeed, first let $Y_j = TK_{1/j}$ for all j, where K_ε is given by (71). Then $\mu(\bigcup_{j=1}^\infty Y_j) = 1$ and Y_j is a compact convex subset of L for all j. Next, let $\widetilde{L}_1 = Y_1$ and $\widetilde{L}_j = Y_j \sim \bigcup_{i=1}^{j-1} Y_i$ for all $j > 1$. Finally, let $k_1 < k_2 < \cdots$ be an enumeration of the set of j's with $\mu(\widetilde{L}_j) > 0$ and set $L_j = \widetilde{L}_{k_j}$ for all j. Then the sequence $L_1, L_2, \ldots$ has the desired properties.

Now let $y_j = \lambda_j^{-1} \int_{L_j} \ell \, d\mu(\ell)$ for all j. It follows that $y_j \in \overline{co}\, L_j$ and hence

$y_j \in L$ for all j. Evidently

$$y = \int_{\bigcup_j L_j} \ell \, d\mu(\ell) = \sum_j \int_{L_j} \ell \, d\mu(\ell) ,$$
$$= \sum_j \lambda_j y_j .$$

As noted in the remark preceding this proof, L is σ-convex; hence $y \in L$. Thus we obtain that L is reasonable provided it is line-closed. Now suppose T is one-one. To see that L is line-closed, let u, v in Y be such that $\lambda u + (1-\lambda)v \in L$ for all $0 \leq \lambda < 1$. It suffices to show that $u \in L$. Since T is one-one, letting $z = T^{-1}v$, there exists an $x \in X$ so that $T^{-1}(u, v] = (x, z]$ (where for a, b in a linear space W, $(a, b]$ denotes the set of all w of the form $w = \lambda a + (1-\lambda)b$, $0 \leq \lambda < 1$). Indeed, this is clear geometrically; perhaps the simplest analytic proof is to let $\widetilde{T} : \text{Aff}\, K \to Y$ be the unique affine extension of T to Aff K, the affine-span of K. $\widetilde{T}$ is one-one since T is, and we just let $x = \widetilde{T}^{-1}u$. Thus $\lambda x + (1-\lambda)z \in K$ for all $0 \leq \lambda < 1$, so $x \in K$ since K is line-closed. Since $T^{-1}(u, v] = (x, z]$, it follows that $u = Tx \in L$. This proves (a).

(b) The fact that T itself is one-one implies by the same argument as in (a), that K is line-closed since L has this property. Since $\overline{K}$ is a Polish space and L is analytic, $\widetilde{T}^{-1}(L)$ is analytic. Since $\widetilde{T}$ is one-one, $K = \widetilde{T}^{-1}(L)$ is thus analytic. To see that K is measure-convex, let C be a compact subset of K. Then TC is a compact subset of L, hence $\overline{co}\, TC \subset L$, so $W \overset{\text{df}}{=} \widetilde{T}^{-1}(\overline{co}\, TC) \subset K$ since $\widetilde{T}$ is one-one. But W is a closed convex subset of X, hence $\overline{co}\, C \subset W \subset K$, proving K is measure convex.

(c) Suppose first that K has the RNP and $V : \mathcal{P} \to L$ is an affine map. Then letting $U = T^{-1}V$, $U : \mathcal{P} \to K$ is an affine map and so there is a K-valued random variable F so that $UF = \mathbb{E}(fF)$ for all $f \in \mathcal{P}$. But then letting $G = TF$, G is an L-valued random variable, and for $f \in \mathcal{P}$,

$$Vf = TUf = T\mathbb{E}(fF) = \mathbb{E}(fTF) = \mathbb{E}(fG) ,$$

showing that G indeed "represents" V.

Next suppose L has the RNP. We prove that K has the RNP, using the same argument as the one given for the proof of Theorem 1.1 of [1]. Let $U : \mathcal{P} \to K$ be an affine map. Then letting $V = TU$, $V : \mathcal{P} \to L$ is an affine map and so we may choose an L-valued random-variable G representing V; that is,

$$Vf = \mathbb{E}(fG) \text{ for all } f \in \mathcal{P} . \tag{72}$$

It follows that setting $F = T^{-1}G$, then F is a K-valued random variable representing U. Indeed, we have (see the proof of Lusin's theorem in section 5) that there is a σ-compact subset W of K with $V \in TW$ a.s. We could thus assume without loss of generality that V is a Borel random variable valued in TW everywhere, hence F is valued in W everywhere. Then if B is a Borel subset of W, TB is a Borel subset of Y, and thus $F^{-1}(B) = G^{-1}(TB)$ is Borel; so in fact F is a Borel random variable. Finally, since K is measure-convex,

$$\mathbb{E}(fF) \in K \text{ for all } f \in \mathcal{P} . \tag{73}$$

Indeed, it follows easily that $\mathbb{E}(fF) \in K$ for all $\mathcal{D}$-measurable f; since K is σ-convex and line-closed, (73) follows by the main result of [**20**] (see also [**22**]). Then if $f \in \mathcal{P}$,

$$\begin{aligned} TUf &= \mathbb{E}(fG) && \text{by (72)} \\ &= \mathbb{E}(fTF) \\ &= T\mathbb{E}(fF) && \text{by the continuity of } T \text{ and (73).} \end{aligned}$$

Hence since T is one-one, $Uf = \mathbb{E}(fF)$, completing the proof that K has the RNP. ∎

We next give a permanence property for certain subsets of RNP-sets. We first recall the following concept, given in [**14**], [**15**].

Definition. Let K be a convex subset of a locally convex space. A subset L of K is called an H_δ-subset of K if there exists a sequence $K_1, K_2, \dots$ of subsets of K so that

(74) K_j is a relatively-closed convex subset of K for all j and $L = K \sim \bigcup_{j=1}^{\infty} K_j$.

It is evident that if K is a separable convex subset of a Banach space, then L is an H_δ-subset of K for every relatively closed subset L of K. The next result thus yields in particular that L has the RNP (and is reasonable) provided L is a relatively closed convex subset of a reasonable set with the RNP.

Proposition 4.2. *Let K be a reasonable subset of a Banach space, and suppose K has the* RNP. *Then L has the* RNP *provided L is a convex H_δ-subset of K.*

Proof. Let $T : \mathcal{P} \to L$ be an affine map. Since K has the RNP, there is a K-valued random-variable X, representing T. That is, $Tf = \mathbb{E}fX$ for all $f \in \mathcal{P}$. We claim that $X \in L$ a.s. Now choose subsets $K_1, K_2, \ldots$ of K satisfying (74). Since each K_j is relatively closed, it is an analytic set (in fact if A is analytic, B is also analytic for all $B \in \mathcal{B}(A)$). Hence $[X \in K_j]$ is an event and it suffices to prove that it is an event of probability zero, for all j. Suppose this were false; choose j with $A \stackrel{\text{df}}{=} (X \in K_j)$ of positive probability. Then letting $f = (P(A)^{-1})I_A$, we have that $Tf = \mathbb{E}fX = \mathbb{E}_A f \in K_j$, for K_j is measure-convex. Then $Tf \notin L$, a contradiction. ∎

Combining 4.1(c) and 4.2, we obtain immediately

Corollary 4.3. *Let K and L be reasonable subsets of certain Banach spaces, and let $T : K \to L$ be a one-one continuous affine map. If L has the* RNP*and TK is an H_δ-subset of L, then K has the* RNP.

Remarks.

1. The elegant argument for 4.2 (in different language) is due to Edgar and Wheeler ([**10**, Theorem 4.13]); where they prove that a Banach space X has the RNP provided X is a weak* H_δ-subset of X^{**}. H_δ-subsets of various classes of sets were introduced by Ghoussoub and Maurey in [**14**] and [**15**], and they used the above reasoning to obtain Corollary 4.3 in the case where K and L are closed. Of course it follows from 4.2 (as noted in [**14**]) that a convex H_δ-subset of a compact convex set (in a Banach space) has the RNP. The results of [**14**] in fact yield immediately the following remarkable result: *Let X be a separable Banach space, Y any infinite-dimensional Banach space, and K a closed bounded convex subset of X. Then the following are equivalent:*

(a) K has the RNP

(b) there exists a one-one compact linear operator $T : X \to Y$ so that TK is an H_δ-subset of $\overline{TK}$.

In particular, it is thus obtained in **[14]** that every closed bounded convex separable set with the RNP can be "coded" onto a reasonable H_δ- subset of a compact convex subset of Hilbert space.

If K is reasonable and L is an H_δ-subset of K, then L is of course analytic. It follows from a result of Fremlin-Talagrand **[13]** that L is σ-convex; however L need not be reasonable, even if K is compact and L is line-closed. (This answers a question of Ghoussoub and Maurey in the negative (see p.30 of **[15]**).

Proposition 4.4. *There exists a compact convex subset K of Hilbert space and a line-closed convex H_δ-subset L of K with L non-measure-convex.*

Proof. Let Δ denote the Cantor set; regard $\mathcal{P}(\Delta)$ as a subset of $C(\Delta)^*$, endowed with the weak*-topology. (Recall that $\mathcal{P}(\Delta)$ denotes the set of probability measures on the Borel subsets of Δ.) $\mathcal{P}(\Delta)$ is thus weak*-compact convex. We exhibit a line-closed weak*-H_δ subset W of $\mathcal{P}(\Delta)$ with W non-weak*-measure-convex. (W itself is thus a counter-example to the question in **[15]**.) Then let $T : \ell^2 \to C(\Delta)$ be a compact linear operator with dense range, let $K = T^*\mathcal{P}(\Delta)$ and $L = T^*W$. Setting $\varphi = T^* \mid \mathcal{P}(\Delta)$, then φ is an affine-homeomorphism between $\mathcal{P}(\Delta)$ in its weak*-topology and K in its norm topology. It follows that K and L have the desired properties.

Now fix $m \in \mathcal{P}(\Delta)$ with m continuous; that is, $m(\{w\}) = 0$ for all $w \in \Delta$. Define W by

$$W = \{\mu \in \mathcal{P}(\Delta) : \mu \perp m\} . \tag{75}$$

W is trivially convex. Since W is norm-closed in $\mathcal{P}(\Delta)$ (endowed with the usual norm-topology on all signed Borel measures), W is line-closed.

Next let $0 < \delta < 1$, and define U_δ by

(76)

$$U_\delta = \{\mu \in \mathcal{P}(\Delta) : \text{ there is a clopen subset } A \text{ of } \Delta \text{ with } m(A) < \delta \text{ and } \mu(A) > 1 - \delta\}$$

It is evident that $\mathcal{P}(\Delta) \sim U_\delta$ is a weak*-compact convex subset of $\mathcal{P}(\Delta)$, for all

$\delta < 1$. We thus obtain that W is a weak*-H_δ subset of $\mathcal{P}(\Delta)$ by establishing

$$W = \bigcap_{n=1}^{\infty} U_{1/n} \ . \tag{}$$

Suppose first $\mu \in W$, and let $\delta > 0$. We may choose by regularity a compact subset E of Δ with $\mu(E) > 1 - \delta$ and $m(E) = 0$. Then simply choose A clopen with $E \subset A$ and $m(A) < \delta$. Thus $\mu \in U_\delta$. Hence $\mu \in \bigcap_{n=1}^{\infty} U_{1/n}$. For the converse, suppose $\mu \in \bigcap_{n=1}^{\infty} U_{1/n}$, let $\varepsilon > 0$, and choose for each n a clopen subset $A_{n,\varepsilon}$ of Δ with $\mu(A_{n,\varepsilon}) > \dfrac{1-\varepsilon}{2^n}$ and $m(A_{n,\varepsilon}) < \frac{\varepsilon}{2^n}$. Then setting $A_\varepsilon = \bigcap_{n=1}^{\infty} A_{n,\varepsilon}$, it follows that $m(A_\varepsilon) = 0$ while $\mu(\sim A_\varepsilon) < \sum_{n=1}^{\infty} \frac{\varepsilon}{2^n} = \varepsilon$; that is, $\mu(A_\varepsilon) > 1 - \varepsilon$. Hence letting $A = \bigcup_{n=1}^{\infty} A_{1/n}$, then $m(A) = 0$ and $\mu(A) = 1$, showing $m \perp \mu$, so $\mu \in W$.

Finally, define $\tau : \Delta \to W$ by $\tau(w) = \delta_w$ for all $w \in \Delta$ (where δ_w is the point-mass measure at w). Then τ is a homeomorphism from Δ to $\tau(\Delta)$. Thus $\tau(\Delta)$ is weak*-compact. We have that $\tau(\Delta) = \operatorname{Ext} \mathcal{P}(\Delta)$, so $\overline{co}^*\tau(\Delta)$, the weak*-closed convex hull of $\tau(\Delta)$, equals $\mathcal{P}(\Delta)$, and hence since $m \in \overline{co}^*\tau(\Delta)$, W is not weak*-measure convex. (In fact, we also have $m = \int_\Delta \tau(x)\, dm(x)$, regarding this as a weak*-integral. That is, m can be regarded as a measure on W, but the w^*-barycenter of τ (with respect to m) does not belong to W). ∎

We finally give an example showing the necessity of the H_δ-assumption in 4.2. (We recently found out that the same example which we use is given in **[15]** (page 73, Remark 5.6).)

Proposition 4.5. *There exist a compact convex subset K of Hilbert space and a reasonable G_δ-subset L of K so that L has no extreme points. Thus L fails the* KMP *and hence L fails the* RNP.

Proof. Let Δ and $\mathcal{P}(\Delta)$ be as above (with $\mathcal{P}(\Delta)$ endowed with the weak*-topology). Let $\mathcal{P}_c(\Delta) = \{\mu \in \mathcal{P}(\Delta) : \mu \text{ is continuous, } i.e., \mu(\{x\}) = 0 \text{ for all } x \in \Delta\}$. It suffices to show that $\mathcal{P}_c(\Delta)$ is weak*-measure-convex and line-closed. Standard evident reasoning yields that $\mathcal{P}_c(\Delta)$ has no extreme points; letting T be the same map and K the same set as in the preceding proof, we then obtain that $L \overset{\text{df}}{=} T^*\mathcal{P}_c(\Delta)$ has the desired properties. Rather than proving this directly, we prefer to show the following:

$$
(78) \qquad \begin{cases} \text{there exist a closed bounded convex subset } G \text{ of a Banach space} \\ \text{and an affine homeomorphism } T : \mathcal{P}_c(\Delta) \to G \\ \text{from } \mathcal{P}_c(\Delta) \text{ in its } w^*\text{-topology onto } G \text{ in its norm-topology.} \end{cases}
$$

It follows immediately that since G is line-closed measure-convex, $\mathcal{P}_c(\Delta)$ is also line-closed and w^*-measure convex. Moreover, $\mathcal{P}_c(\Delta)$ is thus w^*-Polish and hence a w^*-G_δ-subset of $\mathcal{P}(\Delta)$ (in fact we show this first directly).

We identify Δ with $\{0,1\}^N$. Let D denote the family of all finite sequences of 0's and 1's (including the empty sequence). Let $U_\phi = D$ and for $k \geq 1$, $\alpha = (\alpha_1, \ldots, \alpha_k) \in D$, let $|\alpha| = k$ and $U_\alpha = \{x \in D : x(i) = \varepsilon_i \text{ for all } 1 \leq i \leq k\}$. Of course U_α is a clopen subset of D for all α; moreover we have easily that for (μ_n), $\mu \in \mathcal{P}(\Delta)$, $\mu_n \to \mu$ in $\mathcal{P}(\Delta)$ if and only if $\mu_n(U_\alpha) \to \mu(U_\alpha)$ for all $\alpha \in D$. For each $n \geq 1$ and $\varepsilon > 0$, let

$$
(79) \qquad U_{n,\varepsilon} = \{\mu \in \mathcal{P} : \mu(U_\alpha) < \varepsilon \ \text{ for all } \ \alpha \ \text{ with } \ |\alpha| = n\} .
$$

It is evident that $U_{n,\varepsilon}$ is a relatively open subset of $\mathcal{P}(\Delta)$ for all n and ε. The fact that $\mathcal{P}_c(\Delta)$ is a weak* G_δ now follows from the next assertion.

(80)

$$
\text{For } \ \mu \in \mathcal{P}(\Delta)\,,\ \mu \in \mathcal{P}_c(\Delta) \Longleftrightarrow \mu \in \bigcap_{m=1}^{\infty} \bigcup_{n=1}^{\infty} U_{n,1/m} \Longleftrightarrow \mu(U_\alpha) \to 0 \ \text{ as } \ |\alpha| \to \infty\,.
$$

To see this, suppose the final condition fails. Then choose $\delta > 0$ and $\alpha_1, \alpha_2, \ldots$ in D with $|\alpha_n| \to \infty$, yet $\mu(U_{\alpha_j}) > \delta$ for all j. Suppose

$$
(81) \qquad \text{there exists an } x \in \Delta \text{ with } x \text{ belonging to } U_{\alpha_j} \text{ for infinitely many } j.
$$

By passing to a subsequence, we can assume without loss of generality that $|\alpha_j| < |\alpha_{j+1}|$ and $x \in U_{\alpha_j}$ for all j. But this implies that $U_{\alpha_{j+1}} \subset U_{\alpha_j}$ for all j, and moreover that $\{x\} = \bigcap_{j=1}^{\infty} U_{\alpha_j}$. Hence $\mu(\{x\}) = \lim_j \mu(U_{\alpha_j}) \geq \delta > 0$. Thus $\mu \in \mathcal{P}_c(\Delta)$. Since the final condition of (80) trivially implies the second one, we need only show that the second one implies the first one. But if $\mu \notin \mathcal{P}_c(\Delta)$, choose $x \in \Delta$ with $\mu(\{x\}) > 0$. Choose m with $\frac{1}{m} < \mu(\{x\})$. Then for all n, there n α with $|\alpha| = n$ and $x \in U_\alpha$; hence $\mu(U_\alpha) \geq \frac{1}{m}$ and so $\mu \notin U_{n,\frac{1}{m}}$. Thus $\bigcup_{n=1}^{\infty} U_{n,\frac{1}{m}}$, proving (80).

Now define $T : \mathcal{P}_c(\Delta) \to c_0(D)$ by $(T\mu)(\alpha) = \mu(U_\alpha)$ for all $\mu \in \mathcal{P}_c(\Delta)$. ($c_0(D)$ notes the set of $f : D \to \mathbb{R}$ with $f(\alpha) \to 0$ as $|\alpha| \to \infty$, under the sup-norm. Of ourse $c_0(D)$ is isometric to c_0 since D is countable. The third condition of (80) ields that $T\mu$ indeed belongs to $c_0(D)$ for $\mu \in D$.)

It is immediate that T is an affine map; since the closed linear span of the indicators of the U_α's equals $C(\Delta)$, T is one-one. It is trivial that T^{-1} is continuous. To see that T is continuous, let (μ_n), μ in $\mathcal{P}_c(D)$ and suppose $\mu_n \to \mu$. Let $\varepsilon > 0$ and by (80), choose m so that

$$\mu(U_\alpha) < \tfrac{\varepsilon}{2} \text{ if } |\alpha| = m \, . \tag{82}$$

Since there are finitely many α with $|\alpha| \leq m$, we may choose an ℓ so that $n \geq \ell$ implies

$$\big|\mu_n(U_\alpha) - \mu(U_\alpha)\big| < \tfrac{\varepsilon}{2} \text{ for all } n \geq \ell \text{ and } \alpha \text{ with } |\alpha| \leq m \, . \tag{83}$$

Suppose then $|\alpha| > m$. Choose β with $|\beta| = m$ and $U_\alpha \subset U_\beta$. Then we have that

$$\begin{aligned} \mu_n(U_\alpha) \leq \mu_n(U_\beta) &< \tfrac{\varepsilon}{2} + \mu(U_\alpha) && \text{by (83)} \\ &< \varepsilon && \text{by (82)}\,. \\ \text{Moreover } \mu(U_\alpha) \leq \mu(U_\beta) &< \varepsilon && \text{again by (82).} \end{aligned}$$

It follows since $\mu_n(U_\alpha)$ and $\mu(U_\alpha)$ are non-negative that $|\mu_n(U_\alpha) - \mu(U_\alpha)| < \varepsilon$. Thus $T\mu_n \to T\mu$. It remains to show that $T\mathcal{P}_c(\Delta)$ is closed. Suppose then that (μ_n) is a sequence in $\mathcal{P}_c(\Delta)$ and $g \in c_0(D)$ is such that $T\mu_n \to g$. It follows in particular that $\lim_{n\to\infty} \mu_n(U_\alpha) = g(\alpha)$ for all $\alpha \in D$. This implies that there is a $\mu \in \mathcal{P}(\Delta)$ with $\mu_n \to \mu$ weak* and $\mu(U_\alpha) = g(\alpha)$ for all $\alpha \in D$. Hence $\mu(U_\alpha) \to 0$ as $|\alpha| \to \infty$, so $\mu \in \mathcal{P}_c(\Delta)$ by (80), and hence $T\mu = g$ by the continuity of T. Thus $g \in T\mathcal{P}_c(\Delta)$, completing the proof. ∎

Remark. Of course it thus follows that if L is the set we obtain which satisfies the conclusion of Proposition 4.5, then L is affinely homeomorphic to a closed bounded convex subset of c_0.

§5. A review of some properties of analytic sets.

In this section we review some classical results about analytic sets, without proofs. We also give a fairly elementary proof of Lusin's theorem that analytic sets are universally measurable, and a direct proof of von Neumann's selection theorem. Our objective is to provide an accessible and elementary summary of these results, without going into certain deeper aspects such as proofs involving Souslin schemes. For an excellent concise treatment of the results discussed here, we refer the reader to Appendices I and II of [7].

For a topological space X, $\mathcal{B}(X)$ denotes the family of Borel subsets of X, that is, the σ-algebra of sets generated by the open subsets of X. Recall that X is Polish if X is homeomorphic to a complete separable metric space.

Now let X be Polish and K be a subset of X. K is said to be

i) *analytic* if there is a Polish space Y and a continuous $\varphi : Y \to X$ with $\varphi(Y) = K$

ii) *coanalytic* if $X \sim K$ is analytic

iii) *universally measurable* if K is measurable with respect to the completion of μ for every $\mu \in \mathcal{P}(X)$.

(Here $\mathcal{P}(X)$ denotes the family of probability measures on $\mathcal{B}(X)$. We also use the terminology: $\mu \in \mathcal{P}(X)$ is *supported* on $K \subset X$ if there is a $B \in \mathcal{B}(X)$ with $B \subset K$ and $\mu(B) = 1$.) We shall say that a metrizable topological space Z is *absolutely analytic* if Z is a continuous image of some Polish space. It is trivial that Z is absolutely analytic if and only if for any (resp. some) Polish space X and any (resp. some) subset K of X homeomorphic to Z, K is an analytic subset of X. (In a similar way, one can define a space to be absolutely coanalytic, Borel, or universally measurable. The same result then holds, but then is not trivial.)

$\mathbb{N}$ denotes the set of natural numbers; $\mathbb{N}^{\mathbb{N}}$ denotes the set of functions from $\mathbb{N}$ to itself, endowed with the Tychonoff topology. ($\mathbb{N}^{\mathbb{N}}$ is often referred to as the Baire-null space, cf. [7].) It is a standard classical fact that $\mathbb{N}^{\mathbb{N}}$ is homeomorphic to the set of irrational numbers in [0,1]; of course $\mathbb{N}^{\mathbb{N}}$ is a Polish space.

The next theorem summarizes several useful results about Polish spaces and analytic sets. There is no attempt made at non-redundancy in the properties formulated.

orem 5.1. *Let X and Y be Polish spaces, $\varphi : X \to Y$ a Borel measurable p, and $K \subset X$.*

) *If K is analytic, K is a continuous image of $\mathbb{N}^{\mathbb{N}}$.*

) *If K is a Borel set, K is analytic. In fact, K is a one-one continuous image of a closed subset of $\mathbb{N}^{\mathbb{N}}$.*

c) *If K is both analytic and coanalytic, K is a Borel set.*

d) *If $L \subset X$ is analytic and $K \in \mathcal{B}(L)$, K is analytic.*

e) *If $K_1, K_2, \ldots$ are analytic subsets of X and $K = \bigcap_{i=1}^{\infty} K_i$ or $K = \bigcup_{i=1}^{\infty} K_i$, K is analytic.*

f) *If K is analytic, K is universally measurable.*

g) *If K is analytic, so is $\varphi(K)$.*

h) *If φ is one-one and K is a Borel set, so is $\varphi(K)$.*

i) *If $L \subset Y$ is analytic and $K = \varphi^{-1}(L)$, then K is analytic.*

j) *K is Polish if and only if K is a G_δ in X.*

The above list is of course far from exhaustive. For example, we have not mentioned the projective nature of analytic sets. Thus *e.g.*, if K is an analytic subset of a Polish space X and Y is any uncountable Polish space, there exists a G_δ-subset G of $Y \times X$ so that $K = \pi G$, where $\pi : Y \times X \to X$ is the second coordinate projection. Indeed, it can be shown that $\mathbb{N}^{\mathbb{N}}$ is homeomorphic to a subset L of Y. Let $\tau : L \to K$ be a continuous surjection. Then if G is the graph of τ, G is homeomorphic to $\mathbb{N}^{\mathbb{N}}$ and hence is a G_δ-subset of $Y \times X$ with the desired property. The above characterization of analytic sets is particularly useful when Y is compact; we use this implicitly in our proof of 5.1 (f) below (which is the only one of the results in 5.1 we prove here).

We have also deliberately left out formulations involving Souslin schemes; these give perhaps the deepest insight into the *proofs* of some of the assertions of 5.1, but are not needed for the applications of 5.1 that we use in this paper. Briefly, a Souslin scheme of subsets of a set Ω is a family $(S_\alpha)_{\alpha \in \mathcal{T}}$ of subsets of Ω, where $\mathcal{T}$ denotes the "infinitely-branching tree" consisting of all finite sequences of $\mathbb{N}$. The *kernel* S of the scheme $(S_\alpha)_{\alpha \in \mathcal{T}}$ is defined by $S = \bigcup_{f \in \mathbb{N}^{\mathbb{N}}} \bigcap_{i=1}^{\infty} S_{(f(1),\ldots,f(i))}$. Some results about Souslin schemes: if X is Polish and $K \subset X$ is analytic, K is the kernel of a Souslin scheme of closed subsets of X. On the other hand, if K is the kernel of a Souslin scheme of analytic subsets of X, K is also analytic (this generalizes Theorem 5.1 (e) and gives a transparent proof of 5.1(i)). If $(\Omega, \mathcal{S}, \mu)$ is a σ-finite complete measure space, then the members of $\mathcal{S}$ are closed under the Souslin operation, that is, the operation of taking kernels of Souslin schemes. (The latter result, due to Lusin, yields 5.1(f) as well as the consequence that the class of universally measurable subsets of a Polish space is closed under the Souslin operation.)

We establish 5.1(f) by giving an elementary proof of the following result.

Theorem 5.2. *Let M and L be Polish spaces, $\varphi : M \to L$ a continuous map, and $\mu \in \mathcal{P}(L)$. There exist a σ-compact subset F of M and a G_δ-subset G of L so that $\varphi(M) \subset G$ and $\mu(G \sim \varphi(F)) = 0$.*

(Note that since φ is continuous, $\varphi(F)$ is σ-compact; hence G and $\varphi(F) \in \mathcal{B}(L)$.)

Letting φ be the identity map, we obtain immediately

Corollary 5.3. *Let M be Polish and $\mu \in \mathcal{P}(M)$. Then μ has σ-compact support.*

Of course 5.1(f) also follows immediately. For let K be an analytic subset of the Polish space X and let $\mu \in \mathcal{P}(X)$. Choose M Polish and $\varphi : M \to X$ continuous with $\varphi(M) = K$. Choose F and G as in the theorem. Since then $\varphi(F) \subset K \subset G$ and $\mu(G \sim \varphi(F)) = 0$, K is measurable with respect to the completion of μ.

We also then obtain the following generalization.

Corollary 5.4. *Let K be absolutely analytic, Y Polish, $\varphi : K \to Y$ a continuous map, and $\mu \in \mathcal{P}(Y)$. There exists a σ-compact subset W of K so that $\mu(\varphi(W)) = \bar{\mu}(\varphi(K))$ where $\bar{\mu}$ denotes the completion of μ.*

Proof. Recall that by definition, K is homeomorphic to an analytic subset of some Polish space. Hence we may choose X Polish and $\psi : X \to K$ a continuous surjection. Thus $\varphi(K) = \varphi\psi(X)$ is analytic and hence $\bar{\mu}$-measurable. Using Theorem 5.2, choose F a σ-compact subset of X and G a G_δ-subset of Y with $G \supset \varphi(K)$ and $\mu(G \sim \varphi\psi(F)) = 0$. Then $W \overset{\text{df}}{=} \psi(F)$ has the desired properties. ∎

We now pass to the proof of Theorem 5.2. We first use the standard elementary result that every separable metric space admits a metrizable compactification. *Let now M, L and μ be as in the hypotheses of* 5.2. *Also, let us set $A = \varphi(M)$.*

nma 5.5. *Let Y be a metrizable compactification of L. There exist a compact rizable space X, a G_δ-subset Z of X, a homeomorphism $i : M \to Z$, and a ntinuous map $f : X \to Y$ so that $\varphi = fi$.*

ote that the conclusion of 5.5 gives that $A = f(Z)$. 5.5 of course allows us to pass the compact setting, which we use in obtaining the key Lemma 5.6 below.

Proof of 5.5. Let $\widetilde{M}$ be a metrizable-compactification of M, let $X = \widetilde{M} \times Y$, set Z =graph $\varphi = \{(u, \varphi(u)) : u \in M\}$, and define f by $f(u,v) = v$ for all $(u,v) \in X$. Now define $i : M \to Z$ by $i(u) = (u, \varphi(u))$ for all $u \in M$. Then i is a homeomorphism, hence Z is Polish and so is a G_δ-subset of X. Evidently f is continuous and $\varphi = fi$. ∎

Now L is a G_δ-subset of Y, so there is a unique $\tilde{\mu} \in \mathcal{P}(Y)$ with $\tilde{\mu} \mid \mathcal{B}(L) = \mu$. Of course $\tilde{\mu}(Y \sim L) = 0$. It now obviously suffices to prove that

$$(84) \qquad \begin{cases} \text{there is a } \sigma\text{-compact subset } F \text{ of } Z \text{ and a } G_\delta\text{-subset } G \text{ of } Y \\ \text{with } \tilde{\mu}(G \sim f(F)) = 0 \end{cases}$$

Indeed, once this is done; we set $F' = i^{-1}(F)$ and $G' = G \cap L$. Then F' is a σ-compact subset of M, G' is a G_δ-subset of L, $\varphi(F') = f(F)$, and $\mu(G' \sim f(F)) = \tilde{\mu}(G' \sim f(F)) = 0$, proving Theorem 5.2.

For ease in notation, we identify μ and $\tilde{\mu}$. Now let μ^* be the outer measure induced by μ. That is, μ^* is defined on the family of all subsets of Y by

$$\mu^*(S) = \inf\{\mu(B) : B \in \mathcal{B}(Y)\ ,\ S \subset B\} \text{ for all } S \subset Y\ .$$

By the regularity of μ, in fact we have that for $S \subset Y$, there is a G_δ-subset G of Y with $S \subset G$ and $\mu^*(S) = \mu(G)$. We next note the following standard fact:

$$(85) \qquad \text{Let } S_1 \subset S_2 \subset \cdots \text{ be subsets of } Y. \text{ Then } \mu^*\left(\bigcup_{j=1}^{\infty} S_j\right) = \lim_{j\to\infty} \mu^*(S_j).$$

To see this, set $S = \bigcup_{j=1}^{\infty} S_j$. Since μ^* is monotone, it is clear that the limit exists and is less than or equal to $\mu^*(S)$. For the other inequality, for each j, choose $B_j \in \mathcal{B}(Y)$ with $B_j \supset S_j$ and $\mu(B_j) = \mu^*(B_j)$. Now for each j, let $Y_j = \bigcap_{i=j}^{\infty} B_i$.

Then also $Y_j \in \mathcal{B}(Y)$, $S_j \subset Y_j \subset B_j$ so $\mu(Y_j) = \mu^*(S_j)$, *and* $Y_j \subset Y_{j+1}$ for all j. Hence since $\cup Y_j \supset S$,

$$\mu^*(S) \leq \mu(\cup Y_j) = \lim_{j\to\infty} \mu(Y_j) = \lim_{j\to\infty} \mu^*(S_j) .$$

We next give the rather delicate heart of this proof.

Lemma 5.6. *Let $\varepsilon > 0$. There exists a compact subset C_ε of Z with $\mu(f(C_\varepsilon)) \geq \mu^*(A) - \varepsilon$.*

(84) now follows easily. For we choose G a G_δ-subset of Y with $A \subset G$ and $\mu^*(A) = \mu(G)$. Then we set $F = \bigcup_{n=1}^\infty C_{1/n}$, with $C_{1/n}$ as in Lemma 5.6 for all n. Thus $\mu(f(F)) = \mu^*(A) = \mu(G)$. Hence $\mu(G \sim \varphi(F)) = 0$, proving (84).

Proof of Lemma 5.6. Choose $U_1 \supset U_2 \supset \cdots$ open subsets of X so that

$$Z = \bigcap_{j=1}^\infty U_j . \tag{86}$$

We shall construct compact subsets $K_1, K_2, \ldots$ of X with

$$K_j \subset U_j\ ,\ K_j \supset K_{j+1} \text{ and } \mu^*(f(K_j \cap Z)) > \mu^*(A) - \varepsilon \sum_{i=1}^j \frac{1}{2^i} \text{ for all } j . \tag{87}$$

Once this is done, we set $C_\varepsilon = \bigcap_{j=1}^\infty K_j$. Lemma 5.6 then follows by (86) and (87). Indeed, $C_\varepsilon \subset Z$ by (86) and (87). Since $K_n \supset K_{n+1}$ for all n, and $f(K_1)$ is compact, $f(C_\varepsilon) = \bigcap_{n=1}^\infty f(K_n)$ and thus

$$\begin{aligned}\mu(f(C_\varepsilon)) = \lim_{n\to\infty} (f(K_n)) &\geq \lim_{n\to\infty} \mu^*(f(K_n \cap Z)) \\ &\geq \mu^*(A) - \varepsilon \qquad \text{by (87).}\end{aligned}$$

We now construct the desired K_j's. Since X is compact metrizable, every open subset U of X equals a countable increasing union of compact subsets. Thus for each j, we may choose compact sets $K_{j1}, K_{j2}, \ldots$ with

$$K_{j\ell} \subset K_{j\ell+1} \text{ for all } \ell \text{ and } U_j = \bigcup_{\ell=1}^\infty K_{j\ell} . \tag{88}$$

it follows from (86) and (88) that $A = \bigcup_{\ell=1}^{\infty} f(K_{1,\ell} \cap Z)$ and $f(K_{1,\ell} \cap Z) \subset$ ${}_{1,\ell+1} \cap Z)$ for all ℓ; hence by (85) we may choose an ℓ so that setting $K_1 = K_{1,\ell}$, n

$$\mu^*(f(K_1 \cap Z)) > \mu(A) - \frac{\varepsilon}{2} .$$

ow suppose $K_j \subset U_j$ has been chosen compact with

$$\mu^*(f(K_j \cap Z)) > \mu^*(A) - \varepsilon \sum_{i=1}^{j} \frac{1}{2^i} . \tag{89}$$

Then it follows from (86) and (88) that

$$K_j \cap Z = K_j \cap U_{j+1} \cap Z = \bigcup_{\ell=1}^{\infty} K_j \cap K_{j+1,\ell} \cap Z$$

and thus

$$f(K_j \cap Z) = \bigcup_{\ell=1}^{\infty} f(K_j \cap K_{j+1,\ell} \cap Z) \text{ with } f(K_j \cap K_{j+1,\ell} \cap Z) \subset f(K_j \cap K_{j+1,\ell+1} \cap Z) \text{ for all } \ell .$$

Hence by (85), we may choose an ℓ so that setting $K_{j+1} = K_j \cap K_{j+1,\ell}$, then

$$\mu^*(f(K_{j+1} \cap Z)) > \mu^*(f(K_j) \cap Z) - \frac{\varepsilon}{2^{j+1}} . \tag{90}$$

Of course then K_{j+1} is a compact subset of U_{j+1} and

$$\mu^*(f(K_{j+1} \cap Z)) > \mu^*(A) - \varepsilon \sum_{i=1}^{j+1} \frac{1}{2^i} \text{ by (90).}$$

This completes the inductive construction of the K_j's satisfying (87), thus proving Lemma 5.6 and hence Theorem 5.2. ∎

Remark. This argument implicitly used ideas from Choquet's theory of capacities [4]; it yields the standard result that for every Souslin space A, every probability μ on the Borel subsets of A has a σ-compact support. In fact, suppose Z is Polish and $f : Z \to A$ is a continuous surjection. Now letting μ^* be the outer-measure induced by μ, a standard result about outer measures yields that if $S_1 \supset S_2 \supset \cdots$ are subsets of A with $\bigcap_j S_j$ measurable, then $\mu(\bigcap_j S_j) = \lim_{j\to\infty} \mu^*(S_j)$. Choose X a metric compactification of Z and let $U_1, U_2, \ldots$ be as in the proof of 5.6. The argument for

5.6 yields the existence of compact subsets $K_1, K_2, \dots$ of X satisfying (87) (where of course now $\mu^*(A) = \mu(A) = 1$). The above-mentioned property then yields that letting $C_\varepsilon = \bigcap_{j=1}^\infty K_j$, then since $f(C_\varepsilon) \subset \bigcap_{n=1}^\infty f(K_n \cap Z) \subset f(\bigcap_{n=1}^\infty K_n) = f(C_\varepsilon)$, $\mu(f(C_\varepsilon)) = \lim_{n\to\infty} \mu^*(f(K_n \cap Z)) \geq 1-\varepsilon$. (So it isn't really necessary to choose a compactification of A.) Thus in fact we obtain the existence of a σ-compact subset F of Z with $\mu(f(Z)) = 1$. It also follows by similar reasoning that if W is a Souslin subset of A, then W is $\bar{\mu}$-measurable where $\bar{\mu}$ is the completion of μ; hence again, W is universally measurable (see below for the general definition).

We finally treat the von Neumann selection theorem and its consequence, Lemma 2.3 of Section 2. The concept of universal measurability may be extended to general spaces as follows. Call a subset E of a Hausdorff space M, *universally measurable*, if E is measurable with with respect to the completion of μ for every probability measure μ on $\mathcal{B}(M)$ with σ-compact support. (Note that as we have seen, if M is Souslin, every $\mu \in \mathcal{P}(M)$ has σ-compact support). Let $\mathcal{U}(M)$ denote the class of universally measurable subsets of M. If M and L are separable metric spaces, a function $f : M \to L$ is called universally measurable if $f^{-1}(U) \in \mathcal{U}(M)$ for all open $U \subset L$. The von Neumann selection theorem yields in particular that a continuous map from one Polish space onto another, admits a universally measurable right inverse.

Theorem 5.7. (von Neumann [**19**]) *Let M and Ω be absolutely analytic spaces and $\varphi : M \to \Omega$ be a continuous surjection. There exists a universally measurable function $f : \Omega \to M$ so that $\varphi f(x) = x$ for all $x \in M$. Moreover if M is compact, f may be chosen to be Borel measurable.*

(The final statement is due to Federer and Morse [**11**].)

Before giving the proof, we first deduce Lemma 2.3. We first need a fairly simple result.

Lemma 5.8. *Let (Ω, S, P) be a probability space, M a Polish space, and $X : \Omega \to M$ a given function. Then X is a random variable if and only if $(X \in U)$ is an event for all $U \in \mathcal{U}(M)$.*

Proof. According to the definitions we gave earlier, X is an M-valued random able *means* that there exists a sequence (X_n) of simple (*i.e.*, finite-M-valued) measurable functions with $X_n \to X$ $\overline{P}$-a.e., where $\overline{P}$ denotes the completion of However, since M is separable this is actually equivalent to the statement that is itself $\overline{P}$-measurable, that is, $(X \in U)$ is an event (*i.e.*, $\overline{P}$-measurable) for all $\in \mathcal{B}(M)$. Since $\mathcal{B}(M) \subset \mathcal{U}(M)$, we thus obtain the "if-assertion" of 5.8. For the only-if part", let $\mu = \operatorname{dist} X$ and $U \in \mathcal{U}(M)$. Now choose $B_1, B_2 \in \mathcal{B}(M)$ with $B_1 \subset U \subset B_2$ and $\mu(B_2 \sim B_1) = 0$. Hence letting $E_i = (X \in B_i)$, E_i are events for $i = 1, 2$, and $E_1 \subset (X \in U) \subset E_2$ with $\overline{P}(E_2 \sim E_1) = 0$. Hence $(X \in U)$ is an event. ∎

Proof of Lemma 2.3. Let L, $\mathbf{B}$, and X be as in the statement of 2.3.

(a) $\Rightarrow$ (b). This is essentially trivial. Suppose it were false that $X = Y$ a.e. Let $E = \left(X \neq Y \text{ and } X = \frac{Y+Z}{2}\right)$. then $P(E) > 0$, yet $X(E) \cap \operatorname{Ext} L = \emptyset$. This contradicts the assumption that $X \in \operatorname{Ext} L$ a.e.

(b) $\Rightarrow$ (a). We may assume that X has separable range. Thus we may assume $\mathbf{B}$ is separable. Next we note that $L \sim \operatorname{Ext} L$ is analytic. To see this, let $Z \subset L \times L$ be defined by $Z = \{(\ell_1, \ell_2) : \ell_i \in L \text{ for } i = 1, 2 \text{ and } \ell_1 \neq \ell_2\}$. Z is an analytic subset of $\mathbf{B} \times \mathbf{B}$, being a relatively open subset of $L \times L$. Define $\psi : Z \to L$ by $\psi(\ell_1, \ell_2) = \frac{\ell_1 + \ell_2}{2}$ for all $(\ell_1, \ell_2) \in Z$. It follows that $L \sim \operatorname{Ext} L = \psi(Z)$, and hence since ψ is obviously continuous, $L \sim \operatorname{Ext} L$ is analytic.

We thus have that $L \sim \operatorname{Ext} L \in \mathcal{U}(\mathbf{B})$, so also of course $\operatorname{Ext} L \in \mathcal{U}(\mathbf{B})$ and so $(X \in \operatorname{Ext} L)$ is an event. Now suppose $X \in \operatorname{Ext} L$ a.e., is false. Then we maychoose $W \in \mathcal{B}(\mathbf{B})$ with

$$W \subset L \sim \operatorname{Ext} L \quad \text{and} \quad P(X \in W) > 0\,. \tag{91}$$

Let $M = \psi^{-1}(W)$ and $\varphi = \psi \mid M$. Then M is analytic and $\varphi : M \to W$ is a continuous surjection, so by Theorem 5.2 we may choose $f : W \to M$ a universally measurable right inverse to φ. Let $\pi_i : \mathbf{B} \times \mathbf{B} \to \mathbf{B}$ be the coordinate projections; $\pi_i(b_1, b_2) = b_i$ for $(b_1, b_2) \in \mathbf{B} \times \mathbf{B}$, $i = 1, 2$. Then define for $i = 1, 2$, functions X_i on our probability space by

$$X_i = \pi_i \circ f \circ X I_{(X \in W)} + X I_{(X \notin W)}\,. \tag{92}$$

We then have that the X_i's are random variables satisfying

$$X = \frac{X_1 + X_2}{2} \quad \text{and} \quad P((X_1 \neq X)) > 0\,. \tag{93}$$

Indeed, if U is an open subset of $\mathbf{B}$, $X_i^{-1}(U) \cap (X \in W) = X^{-1}(f^{-1}(\pi_i^{-1}(U))) \cap (X \in W)$. Letting $V = f^{-1}(\pi_i^{-1}(U))$, V is a universally measurable subset of W; since $W \in \mathcal{B}(\mathbf{B})$, $V \in \mathcal{U}(\mathbf{B})$, and hence $X^{-1}(V) \cap (X \in W)$ is an event by the preceding lemma. Of course $X_i^{-1}(U) \cap (X \notin W)$ is also an event, so $X_i^{-1}(U)$ is, for $i = 1,2$. Thus the X_i's are random variables by the preceding lemma. Now it follows by the definition of f and the π_i's that for $w \in W$, $\dfrac{\pi_1 f(w) + \pi_2 f(w)}{2} = w$. This proves the equality in (93). Finally, $(X_1 \neq X) = W$, proving the inequality in (93) by (91). Thus (b) of Lemma 2.3 is contradicted. ∎

Remark. The above proof is a reformulation, in our setting, of an argument given by G. Edgar in [8].

Proof of Theorem 5.7. Let us first assume that $M = \mathbb{N}^{\mathbb{N}}$. (We shall see later that the result easily reduces to this.) For $k \geq 1$, $x_1, \ldots, x_k \in \mathbb{N}$, let $U_{x_1,\ldots,x_k} = (x_1, \ldots, x_k) \times \mathbb{N}^{\mathbb{N} \sim \{1,\ldots,k\}} = \{x \in \mathbb{N}^{\mathbb{N}} : x(i) = x_i,\ 1 \leq i \leq k\}$. We define f as follows: fix $\omega \in \Omega$. Let $x_1 = \min\{y(1) : y \in \mathbb{N}^{\mathbb{N}} \text{ and } \varphi(y) = \omega\}$. Then evidently $\omega \in \varphi(U_{x_1})$. Suppose $k \geq 1$, and $x_1, \ldots, x_k$ have been defined so that $\omega \in \varphi(U_{x_1,\ldots,x_k})$. Let $x_{k+1} = \min\{y(k+1) : y \in U_{x_1,\ldots,x_k} \text{ and } \varphi(y) = \omega\}$. Then evidently $\omega \in \varphi(U_{x_1,\ldots,x_{k+1}})$. This completes the inductive definition of the x_j's. Define $x \in \mathbb{N}^{\mathbb{N}}$ by $x(j) = x_j$ for all j. Now for each k, choose $y^k \in U_{x_1,\ldots,x_k}$ with $\varphi(y^k) = \omega$. Then $y^k \to x$ as $k \to \infty$. Hence $\varphi(y^k) \to \varphi(x)$, so we deduce that $\varphi(x) = \omega$. We *define* $f(\omega) = x$. Evidently f is an inverse to φ. We shall in fact show that f is $\mathcal{A}(\Omega)$-measurable, where $\mathcal{A}(\Omega)$ denotes the σ-algebra generated by the analytic subsets of Ω. In turn, (just for the precision of the description), let $\mathcal{A}_0$ denote the *algebra* of sets generated by the analytic subsets of Ω. We prove the following result by induction on k:

(94) For all $x_1, \ldots, x_k$ in $\mathbb{N}$, $f^{-1}(U_{x_1,\ldots,x_k})$ belongs to $\mathcal{A}_0$.

Since the $U_{x_1,\ldots,x_k}$'s are a base for the topology of $\mathbb{N}^{\mathbb{N}}$, (94) yields that f is $\mathcal{A}(\Omega)$-measurable.

The case $k = 1$: Let $x_1 \in \mathbb{N}$. We claim:

$$\text{(95)} \qquad f^{-1}(U_{x_1}) = \varphi(U_{x_1}) \sim \bigcup_{1 \le j < x_1} \varphi(U_j) .$$

(If $x_1 = 1$, we let $\bigcup_{1 \le j < x_1} \varphi(U_j) = \emptyset$.)

Now $\varphi(U_j)$ is an analytic subset of Ω for any integer j, hence (95) yields that $f^{-1}(U_{x_1}) \in \mathcal{A}_0$. Suppose first $\omega \in f^{-1}(U_{x_1})$. Then we have that there is a $y \in \mathbb{N}^{\mathbb{N}}$ with $y(1) = x_1$ and $\varphi(y) = \omega$. Hence $\omega \in \varphi(U_{x_1})$. Moreover if $z \in \mathbb{N}^{\mathbb{N}}$ with $\varphi(z) = \omega$, then $z(1) \ge x_1$; hence $\omega \notin \varphi(U_j)$ if $j < x_1$. Thus ω belongs to the set on the right side of the equality in (95). Conversely, if ω belongs to the right-hand side of (95), then $x_1 = \min\{y(1) : y \in \mathbb{N}^{\mathbb{N}} \text{ and } \varphi(y) = \omega\}$, hence $\omega \in f^{-1}(U_{x_1})$. Thus (95) is established.

Now let $k \ge 1$ and suppose (94) has been proved. Let $x_1, \ldots, x_{k+1} \in \mathbb{N}$. Now the definition of f and an analysis identical to the one given for the "$k = 1$"-case yield that

(96)

$$f^{-1}(U_{x_1,\ldots,x_{k+1}}) = f^{-1}(U_{x_1,\ldots,x_k}) \bigcap \Big[\varphi(U_{x_1,\ldots,x_{k+1}}) \sim \bigcup_{1 \le j < x_{k+1}} \varphi(U_{x_1,\ldots,x_k,j})\Big]$$

Again since $\varphi(U_{y_1,\ldots,y_{k+1}})$ is analytic for any choice of integers $y_1, \ldots, y_{k+1}$, (96) and our induction hypothesis yield that $f^{-1}(U_{x_1,\ldots,x_{k+1}}) \in \mathcal{A}_0$. Hence we have proved by induction that (94) holds for all k, proving the theorem when $M = \mathbb{N}^{\mathbb{N}}$. In general, however, we simply choose $g : \mathbb{N}^{\mathbb{N}} \to M$ a continuous surjection. Then let $h : \Omega \to \mathbb{N}^{\mathbb{N}}$ be an $\mathcal{A}(\Omega)$-measurable right inverse to $\varphi \circ g$. It follows that $f \overset{\text{df}}{=} g \circ h$ is an $\mathcal{A}(\Omega)$-measurable right inverse to φ.

If M is compact, suppose first $M = \{0,1\}^{\mathbb{N}}$. We then define f in exactly the same way as above. In this case, however, the sets $U_{x_1,\ldots,x_k}$ are compact subsets of $\{0,1\}^{\mathbb{N}}$ for $x_1, \ldots, x_k \in \{0,1\}$ and hence we obtain that $f^{-1}(U_{x_1,\ldots,x_k})$ belongs to the algebra generated by the compact subsets of Ω, for all k and $x_1, \ldots, x_k \in \{0,1\}$. Thus f is Borel measurable. Again if M is arbitrary compact metrizable, we just choose $g : \{0,1\}^{\mathbb{N}} \to M$ a continuous surjection, let $h : \Omega \to \{0,1\}^{\mathbb{N}}$ be a Borel-measurable inverse to $\varphi \circ g$, and set $f = g \circ h$ as before. This completes the proof of the von Neumann selection theorem. ∎

Remark. Of course this proof makes no use of the assumption that Ω is metrizable. Thus instead we can just assume Ω is Hausdorff. Moreover the last part of the (general) case allows us to replace M by any Souslin space. Let then $\mathcal{A}(\Omega)$ denote the σ-algebra generated by the Souslin subsets of Ω. We obtain that *if $\varphi : M \to \Omega$ is a continuous surjection, with M Souslin, there exists an $\mathcal{A}(\Omega)$-measurable right inverse $f : \Omega \to X$ for φ.*

Exploiting the "graph-trick", *i.e.*, the projective nature of analytic sets, we note finally the following generalization of 5.7. (This fact was not used in any of our other results.)

Corollary 5.8. *Let M and Ω be absolutely analytic spaces and $\varphi : M \to \Omega$ be a Borel-measurable surjection. There exists a universally measurable right inverse $f : \Omega \to M$ for φ.*

Proof. We may assme without loss of generality that M is Polish, and also let $\Omega' \supset \Omega$, with Ω' Polish. Define $i : M \to M \times \Omega'$ by $i(m) = (m, \varphi(m))$ for all $m \in M$, and let $G = i(M)$. Then G is a Borel subset of $M \times \Omega'$ and hence is absolutely analytic. Let $\psi(m, \omega) = \omega$ for all $(m, \omega) \in M \times \Omega'$; then $\psi \mid G : G \to \Omega$ is a continuous surjection. Choose by Theorem 5.7 a universally measurable right inverse for $\psi \mid G$. Then set $f = i^{-1}g$. It's clear that f is a right inverse for φ. Let U be an open subset of M. Then $f^{-1}(U) = g^{-1}(i(U))$. i is a one-one Borel measurable map, hence $i(U)$ is a Borel set. But then $g^{-1}(i(U)) \in \mathcal{U}(M)$. ∎

References

1. J. Bourgain and H. Rosenthal, *Applications of the theory of semi-embeddings to Banach space theory*, J. Funct. Anal. **52** (1983), 149–188.
2. R.D. Bourgin, "Geometric Aspects of Convex Sets with the Radon-Nikodým Property", Lecture Notes in Mathematics **993**, Springer-Verlag, New York, Berlin, Heidelberg, 1983.
3. R.D. Bourgin and G.A. Edgar, *Noncompact simplexes in Banach spaces with the Radon-Nikodým property*, J. Funct. Anal. (2) **23** (1976), 162–176.
4. G. Choquet, *Theory of capacities*, Ann. Inst. Fourier Grenoble **5** (1955), 131–295.
5. G. Choquet, *Unicité des représentations intégrales au moyen des points extrémaux dans les cônes convexes réticulés*, C.R. Acad. Sci. Paris **243** (1956), 555–557.
6. G. Choquet, *Existence des représentations intégrales au moyen des points extrémaux dans les cônes convexes réticulés*, C.R. Acad. Sci. Paris **243** (1956), 699–702.
7. J.L. Doob, "Classical Potential Theory and Its Probababilistic Counterpart", Springer-Verlag, New York, Berlin, Heidelberg, 1984.
8. G.A. Edgar, *A noncompact Choquet theorem*, Proc. Amer. Math. Soc. **49** (1975), 354–358.
9. G.A. Edgar, *On the Radon-Nikodým property and martingale convergence*, Springer-Verlag Lecture Notes in Mathematics **645** (1978), 62–76.
10. G.A. Edgar and R. Wheeler, *Topological properties of Banach spaces*, Pac. J. Math. **115** (1984), 317–350.
11. H. Federer and A.P. Morse, *Some properties of measurable functions*, Bull. Amer. Math. Soc. **49** (1943), 270–277.
12. D.H. Fremlin and I. Pryce, *Semiextremal sets and measure representation*, Proc. London Math. Soc. (3) **29** (1974), 502–520.
13. D.H. Fremlin and M. Talagrand, *On CS-closed sets*, Mathematika **26** (1979), 30–32.
14. N. Ghoussoub and B. Maurey, *G_δ-embeddings in Hilbert space*, J. Funct. Anal. **61** (1985), 72–97.
15. N. Ghoussoub and B. Maurey, *H_δ-embeddings in Hilbert space and optimization on G_δ-sets*, Memoirs Amer. Math. Soc. **349** (1986).
16. J.A. Johnson, *Extreme measurable selections*, Proc. Amer. Math. Soc. **44** (1974), 107–112.
17. L.H. Loomis, *Dilations and extremal measures*, Advances in Math. **17** (1975), 1–13.
18. P. Mankiewicz, *A remark on Edgar's extremal integral representation theorem*, Studia Math. **63** (1978), 259–265.
19. J. von Neumann, *On rings of operators. Reduction theory*, Annals of Math. (2) **50** (1949), 401–485.

20. H. Rosenthal, *Geometric properties related to the Radon-Nikodým property*, Seminaire d'Initiation a l'Analyse, University of Paris VI, 20e Année (1980-81), No.22, 14 pp.

21. H. Rosenthal, *On the Choquet representation theorem*, Longhorn Notes (1986-87), The University of Texas at Austin (to appear).

22. H. Rosenthal, *L^1-convexity*, Longhorn Notes (1986-87), The University of Texas at Austin (to appear).

23. J. Saint Raymond, *Représentation intégrale dans certains convexes*, Sem. Choquet, 14e Année, University of Paris VI (1974-75), No.2, 11 pp.

24. Z. Semadeni, "Banach Spaces of Continuous Functions", Volume I, Polish Scientific Publishers, Warsaw, 1971.

25. K. Sundaresan, *Extreme points of the unit cell in Lebesgue-Bochner function spaces*, Colloq. Math. **22** (1970), 111–119.

26. E.G.F. Thomas, *A converse to Edgar's theorem*, Springer-Verlag Lecture Notes in Mathematics **794** (1979), 497–512.

Printed in the United States
By Bookmasters